STRUCTURE-FUNCTION ANALYSIS OF EDIBLE FATS

STRUCTURE-FUNCTION ANALYSIS OF EDIBLE FATS

SECOND EDITION

Edited by

ALEJANDRO G. MARANGONI
University of Guelph, Guelph, ON, Canada

Academic Press and AOCS Press
Academic Press is an imprint of Elsevier
125 London Wall, London EC2Y 5AS, United Kingdom
525 B Street, Suite 1650, San Diego, CA 92101, United States
50 Hampshire Street, 5th Floor, Cambridge, MA 02139, United States
The Boulevard, Langford Lane, Kidlington, Oxford OX5 1GB, United Kingdom

Published in cooperation with American Oil Chemists' Society www.aocs.org
Director, Content Development: Janet Brown

Notices

Knowledge and best practice in this field are constantly changing. As new research and experience broaden our understanding, changes in research methods, professional practices, or medical treatment may become necessary.

Practitioners and researchers must always rely on their own experience and knowledge in evaluating and using any information, methods, compounds, or experiments described herein. In using such information or methods they should be mindful of their own safety and the safety of others, including parties for whom they have a professional responsibility.

To the fullest extent of the law, neither the Publisher nor the authors, contributors, or editors, assume any liability for any injury and/or damage to persons or property as a matter of products liability, negligence or otherwise, or from any use or operation of any methods, products, instructions, or ideas contained in the material herein.

Library of Congress Cataloging-in-Publication Data
A catalog record for this book is available from the Library of Congress

British Library Cataloguing-in-Publication Data
A catalogue record for this book is available from the British Library

ISBN: 978-0-12-814041-3

For information on all Academic Press and AOCS Press publications
visit our website at https://www.elsevier.com/books-and-journals

Working together
to grow libraries in
developing countries

www.elsevier.com • www.bookaid.org

Publisher: Andre Gerhard Wolff
Acquisitions Editor: Nancy Maragioglio
Editorial Project Manager: Billie Jean Fernandez
Production Project Manager: Prem Kumar Kaliamoorthi
Cover Designer: Matthew Limbert

Typeset by TNQ Technologies

CONTENTS

9. Ultra-Small Angle X-ray Scattering: A Technique to Study Soft Materials 267

Fernanda Peyronel, David A. Pink

10. Fat Crystallization and Structure in Bakery, Meat, and Cheese Systems 287

Kristin D. Mattice, Alejandro G. Marangoni

CONTRIBUTORS

Nuria C. Acevedo
University of Guelph, Guelph, ON, Canada

Peyronel Fernanda
University of Guelph, Guelph, ON, Canada

Joamin Gonzalez-Gutierrez
University of Manitoba, Winnipeg, MB, Canada

Stefan H.J. Idziak
University of Waterloo, Waterloo, ON, Canada

Braulio A. Macias Rodriguez
University of Guelph, Guelph, ON, Canada

Farnaz Maleky
University of Guelph, Guelph, ON, Canada

Alejandro G. Marangoni
University of Guelph, Guelph, ON, Canada

Kristin D. Mattice
University of Guelph, Guelph, ON, Canada

Arun S. Moorthy
University of Guelph, Guelph, ON, Canada

Fernanda Peyronel
University of Guelph, Guelph, ON, Canada

David A. Pink
St. Francis Xavier University, Antigonish, NS, Canada

Keshra Sangwal
Lublin University of Technology, Lublin, Poland

Kiyotaka Sato
Hiroshima University, Higashi Hiroshima, Japan

Martin G. Scanlon
University of Manitoba, Winnipeg, MB, Canada

INTRODUCTION

Humans have used fats and oils in their diets since prehistoric times. Fats are important because they are a rich and concentrated source of calories—they are the most concentrated caloric source in nature (9 kcal/g); they provide essential fatty acids and carry many important water-insoluble micronutrients, such as vitamins A, E, K, phytosterols, beta-carotene, lutein, among many others. Fats also provide organoleptic characteristics to foods, which make those foods desirable by the consumer, including flavor, texture, lubricity, and satiety. Because of the usefulness of fats and oils as food ingredients, research is always oriented toward furthering our knowledge of their physical and chemical characteristics, both in the bulk and dispersed states.

Fats and oils are a subgroup of lipids. Lipids are organic compounds with the common characteristic of being soluble in nonpolar organic solvents, such as isobutanol and hexane, and generally insoluble in water. The terms "fat" and "oil" are used interchangeably. The use of each term is based on the physical state of the material at room temperature, which will be dependent on geographical latitude and altitude and the time of the year. In our view, a fat or oil should be defined at a TAG mixture that is solid or liquid at 25°C, respectively, to avoid ambiguities. Fats are not usually 100% solid, but rather mixtures of hard crystalline solids intimately associated with liquid oil, in the range 10%−90% (Bailey, 1950).

Chemically, fats are mixtures of usually more than 95% of triacylglycerol (TAG) molecules, and 1%−5% minor components, including phospholipids, glycolipids, free fatty acids, monoacylglycerols, diacylglycerols, tocopherols, phytosterols, to name a few. TAG molecules consist of a glycerol backbone with three fatty acids esterified to the three alcohol groups at specific locations referred to as sn-1, sn-2, and sn-3. These fatty acids can differ based on chain length, saturation, branching, and the presence of *trans* or *cis* double bonds. Methods to isolate TAGs and determine their fatty acid composition, fatty acid positional distribution within a TAG are well established (Christie and Han, 2010; The Lipid Library at http://lipidlibrary.aocs.org/).

TAGs are usually classified into three groups with respect to the fatty acids that are present. Monoacid TAGs have the same fatty acids present in positions sn-1, sn-2, and sn-3, such as tristearin and tripalmitin. TAG molecules with two or three different types of fatty acids are called diacid or triacid TAGs, both of which are referred to as mixed-acid TAGs. Diacid TAGs can be further placed into subgroups based on which fatty acid occupies position sn-2, making the TAG molecule thus symmetric or asymmetric. Fats, in turn, can be composed of only one type of TAG molecule (not commonly), or they can be a combination of many different TAG molecules. For example, milk fat contains a still unknown number of TAGs, definitely greater than 200 molecular species (Jensen, 1995; Gresti et al., 1993), while 80% of all TAGs in high-oleic sunflower oil are triolein molecules (Gunstone, 2004).

The most influential structural characteristics in a TAG molecule, which affect their physical properties (melting and crystallization behavior, solid fat content, polymorphism, nano- and microstructure, mechanical properties, oil-binding capacity) include the length of the fatty acid chains, the number, position and configuration of double bonds (saturated vs. unsaturated or *cis* vs. *trans*), and finally, the stereospecific position of the fatty acid on the glycerol backbone (Small, 1986). The most common TAGs contain fatty acids with chain lengths between 4 and 22 carbon atoms with zero to six double bonds (Gunstone, 2004). A characteristic of having a fat composed of a mixture of different TAG molecules is that the material does not display a unique melting point, but rather a melting range defined by chemical nature and molecular interactions of this complex mixture. This is due both to the great variety of molecular species present and because of the complex phase behavior between and among TAG species (Bailey, 1950; Rossel, 1967; Timms, 1984; Small, 1986). Moreover, TAGs can crystallize into different solid state structures, depending on crystallization conditions (temperature, shear, time), giving rise to polymorphism (Clarkson and Malkin, 1934; Chapman, 1962). Understanding the way that TAGs arrange and interact is critical if an understanding of the macroscopic functionality of fats is sought.

An interesting aspect of fats is that they behave like elastic solids until a deforming stress exceeds a certain value (yield value, yield force, or yield stress), at which point the product starts to flow like a viscous fluid (Haighton, 1959; deMan and deMan, 2002). This phenomenon, termed plasticity, arises from the fact that the crystallized material forms a fat crystal network that entraps liquid oil.

Many of the sensory attributes of fat and fat-structured food products, like spreadability, mouthfeel, texture, and flavor are strongly influenced by the physical characteristics of the fat crystal network. The mechanical and organoleptic properties, as well as the stability (shelf life) of materials such as chocolate, butter, margarine, and spreads are, to a great extent, determined on crystallization of the material. This is why we need to understand fat crystallization and its relationship to structure: to control and engineer material properties.

It is important to understand the structural organization present in a material and relate it to macroscopic properties because it can guide efforts in replacing ingredients, optimizing functionality, and improving health in a rational fashion. Ultimately, this will also result in a more focused, less frustrating, and less expensive endeavor.

The figure shown below summarizes our current knowledge of the structure of edible fats, from TAG molecules to progressively larger supramolecular assemblies until the macroscopic world is reached. This view summarizes the work of many researchers in the field, over the past 50 years, such as van den Tempel (1961), deMan and Beers

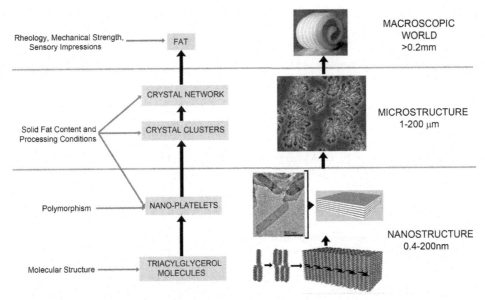

Figure 1 Structural hierarchy in a fat crystal network—from molecules to material.

(1987), Heertje (1993), Marangoni et al. (2012), and Ramel et al. (2016). The complete view of this structural hierarchy was not possible until the recent characterization of the nanoscale in fats (Acevedo and Marangoni, 2010a,b), Pink et al. (2013), Peyronel et al. (2013) (Fig. 1).

In this book, we have strived to bring together a group of experts and expert practitioners of the art and science of crystallization, structure, molecular interactions, phase behavior, and mechanics. They will offer a complete but practical perspective of how to analyze, characterize, and understand fat structure and properties. This purpose of this book is not to review and summarize exhaustively what has happened before but point toward where we should be heading to.

REFERENCES

Acevedo, N.C., Marangoni, A.G., 2010a. Characterization of the nanoscale in triacylglycerol crystal networks. Cryst. Growth Des. 10, 3327–3333.

Acevedo, N.C., Marangoni, A.G., 2010b. Towards nanoscale engineering of triacylglycerol crystal networks. Cryst. Growth Des. 10, 3334–3339.

Bailey, A.E., 1950. Bailey's Industrial Oil and Fat Products. Interscience Publishers, New York.

Christie, W.W., Han, X., 2010. Lipid Analysis: Isolation, Separation, Identification and Lipidomic Analysis. Woodhead Publishing Ltd., Cambridge.

Chapman, D., 1962. The polymorphism of glycerides. Chem. Rev. 62, 433–456.

Clarkson, C.E., Malkin, T., 1934. Alternation in long-chain compounds. Part II. An X-ray and thermal investigation of the triglycerides. J. Chem. Soc. 666–671.

deMan, J.M., Beers, A.M., 1987. Fat crystal networks: structure and rheological properties. J. Text. Stud. 18, 303–318.

deMan, J.M., deMan, L., 2002. Texture of fats. In: Marangoni, A.G., Narine, S.S. (Eds.), Physical Properties of Lipids. Marcel Dekker, New York, pp. 191–217.

Gresti, J., Bugaut, M., Maniongui, C., Bezard, J., 1993. Composition of molecular species of triacylglycerols in bovine milk fat. J. Dairy Sci. 76, 1850–1869.

Gunstone, F.D., 2004. The Chemistry of Oils and Fats. Blackwell Publishing Ltd., Coventry, UK.

Haighton, A.J., 1959. The measurement of the hardness of margarine and fats with cone penetrometers. J. Am. Oil. Chem. Soc. 36, 345–348.

Heertje, I., 1993. Microstructure studies in fat research. Food Microstruct. 12, 77–94.

Jensen, R.G., 1995. Handbook of Milk Composition. Academic Press, New York.

Marangoni, A.G., Acevedo, N., Maleky, F., Co, E., Peyronel, F., Mazzanti, G., Quinn, B., Pink, D., 2012. Structure and functionality of edible fats. Soft Matter 8, 1275–1300.

Peyronel, F., Ilavsky, J., Mazzanti, G., Marangoni, A.G., Pink, D.A., 2013. Edible oil structures at low and intermediate concentrations: II. Ultra-small angle X-ray scattering of in situ tristearin solids in triolein. J. Appl. Phys. 114, 234902.

Pink, D.A., Quinn, B., Peyronel, F., Marangoni, A.G., 2013. Edible oil structures at low and intermediate concentrations: I. Modelling, computer simulation and predictions for X-ray scattering. J. Appl. Phys. 114, 234901.

Rossel, J.B., 1967. Phase diagrams of triglyceride systems. Adv. Lipid Res. 5, 353–408.

Ramel, P., Co, E.D., Acevedo, N.A., Marangoni, A.G., 2016. Nanoscale structure and functionality of fats. Prog. Lipid Res. 64, 231–242.

Small, D.M., 1986. The Physical Chemistry of Lipids. Plenum Press, New York.

Timms, R.E., 1984. Phase behavior of fats and their mixtures. Prog. Lipid Res. 23, 1–38.

van den Tempel, M., 1961. Mechanical properties of plastic disperse systems at very small deformations. J. Colloid Sci. 16, 284–296.

CHAPTER 1

Characterization of the Nanostructure of Triacylglycerol Crystal Networks

Nuria C. Acevedo
University of Guelph, Guelph, ON, Canada

INTRODUCTION

The characteristics of a fat crystal network and the magnitude of the intercrystalline interactions created by its constituent crystallites contribute to many of the physical properties of food products. This structural network is the product of an aggregation process of triacylglyceride molecules into particles and particles into larger agglomerates, which continues until a space-filling three-dimensional assemblage is formed.

Throughout the years, much effort has been directed toward the development of specific characteristics in manufactured fat-rich materials for which it is crucial to understand their underlying structural organization and associated macroscopic properties.

Much work has been done on the analysis of modifications of the crystalline network induced by changes in processing conditions and the impact on the properties of fats (de Man and Beers, 1987; Heertje et al., 1987; Heertje et al., 1988; Shukla and Rizvi, 1996; Marangoni and Rousseau, 1996; Herrera and Hartel, 2000; Martini et al., 2002; Acevedo et al., 2012; Ribeiro et al., 2015). Furthermore, many microscopic techniques have been used to study the fat microstructure or mesoscale (within a scale range between 1 and 100 µm). A comprehensive review of the microscopic methods used with this purpose in the latest years has been undertaken by Tang and Marangoni (2006).

In 2010, work from our laboratory provided new insights into the structural organization of fat crystal networks at a scale never characterized before: the nanostructural scale. We found that what was previously assumed a fat "primary crystal" is in effect a fractal agglomerate of well-defined triacylglycerol (TAG) nanoplatelets. These nanoplatelets constitute the principal crystal entity of the network (Acevedo and Marangoni, 2010a). Moreover, we reported that fat nanostructure can be modified by using external fields or chemical transformations. We considered for the first time the possibility of engineering the nanostructure of such materials (Acevedo and Marangoni, 2010a, b; Maleky et al., 2011).

Therefore, the door to a new area of research of fats have been opened, where the applications of these new findings could eventually lead to the possibility of engineering the nanoscale of fats to enhance processing and formulate novel materials with tailored functionalities.

Structure-Function Analysis of Edible Fats, Second Edition
ISBN 978-0-12-814041-3, https://doi.org/10.1016/B978-0-12-814041-3.00001-0

The objective of this chapter is to provide a comprehensive description of the analytical methods used in the study of the nanostructural level of fat crystal networks. With the aim to provide the reader with a complete vision of the experimental techniques used in our laboratory, all the procedures involved, from sample preparation to data analysis along with some illustrative examples, will be included in this section.

NANOSTRUCTURE CHARACTERIZATION BY X-RAY POWDER DIFFRACTION

When fats crystallize, TAG molecules adopt a specific configuration and packing arrangement to enhance intra- and intermolecular interactions (Larsson, 1994). Polymorphism relates to the capability of molecules to arrange themselves in a crystal lattice in different ways of packing. Fats can crystallize in different polymorphic forms: α, β, and/or β' (occasionally different β' and/or β phases exit) (Simpson and Hagemann, 1982; Arishima and Sato, 1989).

Among the most commonly used techniques for studying nanocrystalline structure of fats is powder X-ray diffraction (XRD) because it is a well-known phase-selective method.

When fat crystals are analyzed XRD, two types of spacings can be recognized: long spacings in the small angle region of the pattern (1—15 degrees) and short spacings in the wide angle region (16—25 degrees) (Fig. 1.1).

The long spacings of the "unit cell" correspond to the thickness of the lamellae layer and depend on the length of TAG molecules and the angle of tilt of the chains relative to the normal plane (Lawler and Dimick, 1998). The unit cell is the smallest spatial unit of a fat crystal network whose repetition along the three axial directions results in the formation of a crystal lattice. On the other hand, the short spacings of the "subcell" refer to the cross-sectional packing of the hydrocarbon chains and are independent of the chain length. The subcell is the smaller repeating unit within the unit cell and along the acyl chains. Because each of the chain-packing subcells is characterized by a unique set of XRD lines, it is possible to identify the different polymorphic forms of fats (Fig. 1.1).

Furthermore, an advantage of XRD analysis at low-angle reflections is that this method provides the possibility of determining the average crystallite size of a nanomaterial from the broadening of the XRD reflections, by means of the so-called the Scherrer formula (West, 1984):

$$D = \frac{K\lambda}{\text{FWHM}\cos(\theta)}, \tag{1.1}$$

where θ is the diffraction angle, FWHM is the width in radians of the diffraction maximum measured at half-way height between background and peak, typically known as full width at half maximum (usually the first small angle reflection

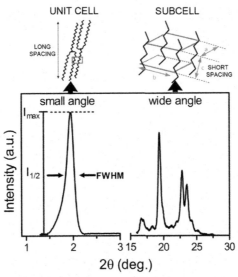

Figure 1.1 Example of a typical powder X-ray diffraction spectrum of a solid fat. In the small angle region, it is possible to observe the (001) reflection peak. The corresponding full width at half maximum (FWHM) is indicated with *arrows*. The peaks detected in the wide angle region determine the polymorphic form present in the sample.

corresponding to the (001) plane). λ is the X-ray wavelength; 1.54 Å for a copper anode. K (Scherrer constant) is a dimensionless number that provides information about the shape of the crystal and is typically in the order of the unit. Values for K can be between 0.62 and 2.08. For a spherical particle, the shape factor is 1, and in the absence of detailed shape information, $K = 0.9$ is a good approximation. For an excellent discussion of K, refer to the review by Langford and Wilson (1978).

It is worth noting that the Scherrer equation can only be used for mean sizes up to around 100–200 nm because peak broadening is inversely proportional to crystallite size. Thus, large sizes are subjected to large errors, as it is difficult to separate the peak broadening due to crystallite size from the broadening due to other factors, such as instrumental broadening.

Hence, for fat systems, D values obtained with this analysis represent only the thickness of the nanoplatelets, which are constituted by the stacking of several TAG lamellae. Nanoplatelets length and widths cannot be determined by the Scherrer analysis because they are usually much larger than the limit of 100–200 nm.

In fat systems, the Scherrer equation was used by Martini and Herrera (2002), although they ascribed the obtained results to a general crystal size without specifically differentiating between the three possible spatial dimensions of the fat nanocrystal. At the time this study was carried out, there was no information yet on the nanostructure of fat crystal networks.

In this section, we will focus on the use of the Scherrer analysis in fat matrices because XRD is a popular methodology and has been used for many years in the determination of short and long spacing of fat crystal networks.

Experimental Procedure

Sample Preparation

XRD patterns at the small angle region should be collected, scanning samples from approximately 0.9 to 8 degrees. To minimize background noise and obtain good-quality raw data, a low scanning rate is required, such as 0.02 degrees/min.

Data Analysis

Several software packages can be used for both processing and analysis of the raw XRD data. Jade (Materials Data Inc., Livermore, CA, USA) and PeakFit (Seasolve, Framingham, MA, USA) among others are the most commonly used software packages because of their versatility for spectroscopic techniques. In this section, the basic processing tools of PeakFit software used to fit XRD patterns and obtain FWHMs will be described. Furthermore, examples of the nanocrystal thickness calculation (D) through the Scherrer equation will be presented.

The first step is to launch PeakFit and load the XRD data in the program (Fig. 1.2A). Next, it is necessary to select the region of data to fit. The menu's *Prepare > section* option is used to activate (yellow points) or inactivate (gray points) specific data sections or points. Only active regions will be fitted because this tool affects the current state of the main data table. Therefore, it would be used to disable all elements of data that are not desired to be fitted under any circumstances. Fig. 1.2B shows the dual-coupled PeakFit graph that opens once this tool is selected. The upper graph is scaled for only active points (in this case, the peak corresponding to the (001) reflection); meanwhile the lower graph is scaled for all data points (in this case reflections from planes (002) and (003) as well). After sectioning, the proper data changes must be accepted. Then, only the chosen peak area/s will appear in the main window.

Peak-fitting procedures typically need to consider removal of background. Without correction for the background, much of the trace is fit with nonsensical peaks that do not provide a unique solution. When using the function *AutoFit and Substract Baseline* from the menu, a window like the one shown in Fig. 1.3A appears. This tool allows freely selecting specific points that the operator wants to include or exclude in a baseline fit. In our case, we chose to prefit the baseline using a *nonparametric* estimation procedure that can describe virtually any shape of baseline. Points used to determine the baseline will be active (yellow). All others will be grayed indicating that they are not being used in the baseline fit.

Once the baseline is fitted, PeakFit data table and graph will reflect the removal of the baseline (Fig. 1.3A, inset). For the sectioning and background tools, extreme care should

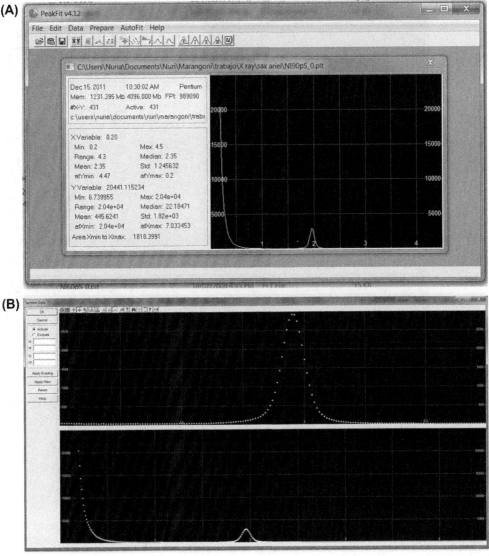

Figure 1.2 Dialog box used when opening X-ray diffraction data with PeakFit software (A). Dialog box used when selecting the specific data section to analyze (B).

be taken to avoid including or eliminating data points erroneously that could compromise the subsequent fitting. Finding and fitting the desired peak/s relies in the use of the *AutoFit Peaks* options, which are capable of finding hidden peaks if necessary. Fig. 1.3B shows the two PeakFit graphs displayed after selecting the function *AutoFit Peaks I Residuals*. This option should be used when each peak is identified by a local

(A)

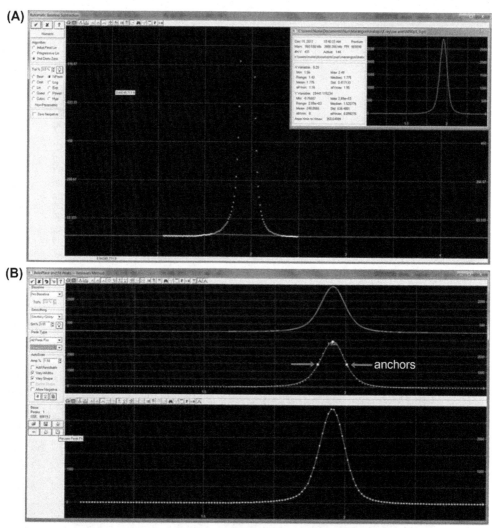

(B)

— anchors

Figure 1.3 Dialog box used to determine the baseline of the XRD pattern (A). Inset: resulting X-ray peak after baseline subtraction. PeakFit software window used to find and fit the desired peak/s (B).

maximum in the stream data; therefore there are no hidden peaks, such as in our case. Furthermore, the built-in function *Pearson IV* has been selected for peak fitting. It can be noted that the *Vary Widths* and *Vary Shape* options are enabled (Fig. 1.3B), and consequently, additional anchors appear on each peak. When the width or shape is allowed to vary, a peak refinement algorithm is used to characterize each peak as accurately as possible from the raw data. Once satisfied with the peak placement process, the automatic fitting can be performed through the *PeakFit with fast numerical update* and *PeakFit with full*

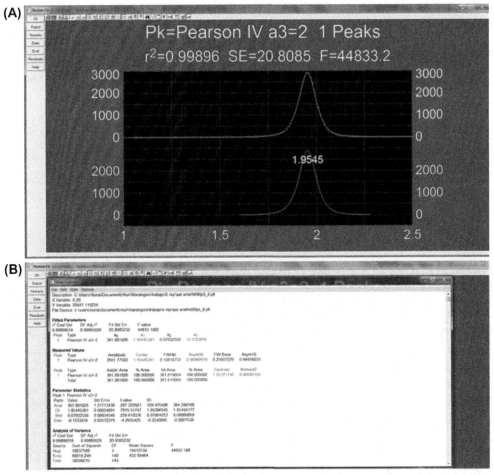

Figure 1.4 Screen capture of the PeakFit window used to review the state of the peak/s fitting (A). Numeric complete description of the fitting process used (B).

graphical update options. Finally, the *Review PeakFit* option will open a new window for evaluation of the fit (Fig. 1.4A) where the component peak functions, the data and the fitted curve are shown in the Y–axis plots. The obtained results, including a full numeric peak analysis can be obtained (Fig. 1.4B).

After the desired data from the fitting procedure is obtained, it is possible to proceed with the calculation of the nanoplatelet thickness by means of the Scherrer equation.

Table 1.1 shows the information extracted from the fitting process and values of nanoplatelets thickness (D) calculated using the Scherrer expression. Additional examples for other fat systems have been included as well.

Table 1.1 Examples of information obtained from the PeakFit analysis of the X-ray diffraction data (small angle region) and nanoplatelet thickness calculation using Scherrer equation

Sample	K	λ	FWHM (°)	FWHM (rad)	Peak center (2Φ)	Φ (°)	D (Å)
1. FHCO: HOSO[a] 30:70	0.9	1.5418	0.1202	2.0982×10^{-3}	1.9544	0.9772	**661.43**
2. FHSO:SO 40:60	0.9	1.5418	0.3670	6.4658×10^{-3}	2.2003	1.1001	**216.66**
3. Commercial shortening	0.9	1.5418	0.3721	6.4952×10^{-3}	2.27435	1.1372	**213.67**

FHCO, fully hydrogenated canola oil; *FHSO*, fully hydrogenated soybean oil; *FWHM*, full width at half maximum; *HOSO*, high oleic sunflower oil; *SO*, soybean oil.
[a]Example from Fig. 1.2−1.4.

NANOSTRUCTURE CHARACTERIZATION BY CRYO−TRANSMISSION ELECTRON MICROSCOPY

Transmission electron microscopy (TEM) is a tool that allows visualization of nanomaterial structures. This technique enables the resolution in the order of nanometers, and it is certainly a direct method, as it provides real images of the particles. For this reason, TEM has become a very attractive technique to study the nanostructure of fat crystal networks. However, several factors make it problematic, sometimes impossible, to apply TEM for the study of fat systems. First, one of the major challenges to overcome in the observation of fat matrices using this technique is that the presence of oil between fat crystals hinders the proper observation of nanoscale-sized structures. Additionally, TEM of a fat matrix must be carried out under cryogenic conditions (cryo−TEM) to avoid sample melting on exposure to the electron beam. What further complicates imaging such specimens is that under cryogenic temperatures the liquid oil trapped within the fat network freezes; consequently, the contrast between the vitrified oil and fat crystals is very poor, and it becomes very difficult to distinguish crystalline TAG particles with nanometer sizes.

In 2010, our group was able to successfully surpass these problems, leading to the visual characterization of fat nanoplatelets by cryo-TEM. We developed a sample treatment based on the removal of the entrapped oil phase using a cold solvent, disruption of the fat crystalline network, and extraction of crystal nanoplatelets. The main advantage of this new method is that it provides information that until the moment was impossible to obtain by other methods of investigation. A systematic understanding on the effects of composition and external fields on the morphology and structural characteristics of the resulting nanocrystals was gained by using this new technology (Acevedo and Marangoni, 2010b, 2014; Acevedo et al., 2012).

In this section, we will provide a comprehensive description of the experimental route to be followed in the systematic and quantitative characterization of the nanostructure of fats using cryo–TEM. We will focus in particular on sample preparation and data analysis.

Experimental Procedure

Sample Preparation

The structured fat material to be studied should be suspended in cold isobutanol in a ratio between 1:50 to 1:100 (w/w) according to the solid mass fraction present in the sample. Fat systems with higher solid mass fraction (and therefore less amount of liquid oil entrapped) may require a smaller amount of solvent to extract the oil fraction. Fig. 1.5 shows a schematic representation of the novel procedure developed and used in our laboratory. The new sample treatment is an adaptation of the work reported by Chawla and deMan (1990)

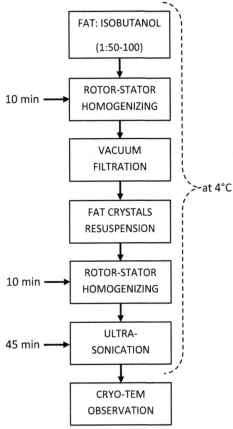

Figure 1.5 Experimental procedure for sample treatment used before the observation by cryo-TEM.

for the observation of fat mesoscale surfaces by scanning electron microscopy, and it is centered on the use of isobutanol for the separation of solid and liquid components of fats without a significant loss of solid phase.

To maximize results, fat sample treatment should be carried out at 4—10°C and as follows:

Once the fat—isobutanol mixtures are prepared, mix the suspension using a glass stirring rod to break up large fat aggregates and release them to the solvent. The well-mixed sample of fat plus isobutanol is mechanically homogenized using a rotor—stator at high rotor tip speed for 10 min. Assuming Couette-type flow at the rotor—stator tip (radius = 3 mm, gap size = 200 μm), we estimated a tip shear rate of $47,000 s^{-1}$, which has been proven to be sufficient to break low micrometer-scale objects and release fat nanocrystals from most of the samples.

Then, crystals and crystal aggregates are collected by vacuum filtration through a glass fiber filter of 1.0 μm pore size. After filtration, the recovered solid is placed in a clean beaker and resuspended with cold isobutanol. The resulting mixture is rehomogenized for 10 min using the rotor—stator until a suitable dispersion of crystals is obtained.

Immediately after the second homogenization process is complete, the suspension is transferred to a tube with sealing capacity to prevent solvent evaporation and the entry of dust or water. Finally, mixtures are sonicated for 30—45 min using an ultrasonic processor to complete the dispersion of the fat crystals before the observation by cryo-TEM.

It is important to mention that the sample treatment described above is critical when using cryo-TEM for characterization of the nanoscale because fat crystals tend to aggregate and form large agglomerates in relatively short periods of time. In fact, we found that even a few hours after finalizing sample treatment, fat elements formed larger structures, thus preventing the display of individual nanocrystals. Hence, it is recommended to carry out microscopy immediately after preparing the fat—solvent suspensions. Furthermore, the time intervals given here are merely a recommendation. When performing the sample treatment, the user might have to determine the appropriate length for each step of the experiment according to the effectiveness of the homogenization process for matrix disruption. Extended periods of rotor—stator or ultrasonication usage may be required for harder samples (avoiding the increase of temperature), as well as higher solvent proportion for oily samples.

Data Analysis

For the observation by cryo-TEM, the prepared suspension is placed on a perforated carbon-covered polymer membrane, which is, in turn, suspended on a copper microscopy grid. Therefore, because electrons must cross the grid with a thin film of the sample, this method is best suited for visualizing a size range from 4 to about 400 nm, which is roughly on the same order of the size of the fat nanocrystal being probed.

Here we will discuss some pertinent details for the understanding and interpretation of the images showed in this contribution. We will describe basic quantitative image analysis techniques performed with ImageJ (Wayne Rasband, NIH) to determine the size and number of features.

Fig. 1.6 shows some examples of the cryo-TEM micrographs resulting from the method previously described in Fig. 1.5. The extracted nanoplatelets are derived from fully hydrogenated canola oil (FHCO) (Fig. 1.6A), a 45:55 blend of fully hydrogenated soybean oil plus soybean oil (Fig. 1.6B) and two different roll in shortenings (Fig. 1.6C and D).

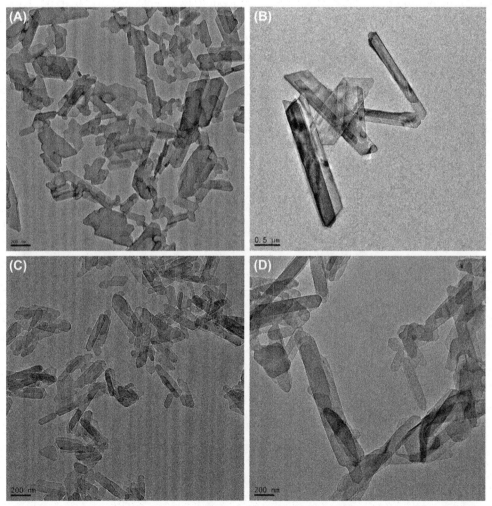

Figure 1.6 Examples of micrographs obtained by cryo-TEM using the new sample treatment developed by our group. These images correspond to fully hydrogenated canola oil (A), a 45:55 blend of fully hydrogenated soybean oil plus soybean oil (B) and two different roll in shortenings (C and D).

As can be observed in Fig. 1.6, cryo-TEM stands as a vital tool to get an impression of the homogeneity in nanoplatelet size and morphology of a given fat sample. Moreover, the acquired micrographs show that many of the nanoplatelets can be individually distinguished from the rest of particles and the background. Nevertheless, because of the lack of contrast or overlap of particles, image analysis could result challenging as nanocrystal borders might be difficult to differentiate. Thus, a semiautomatic procedure is necessary for image analysis, which relies on a manual particle selection by the observer. It is important to ensure that only crystals clearly discernible on the micrograph are selected. Fig. 1.7 depicts enlarged regions of cryo-TEM images where platelets, which longitudes and widths can be accurately quantified, are marked with arrows.

On the other hand, cryo-TEM images provide strong evidence that fat nanostructure does not consist of platelets of any unique size (see Fig. 1.6). Instead, it is composed of particles over a relatively wide size range. It is essential to describe this size range, thus size distribution curves must be constructed for data evaluation. A limitation or disadvantage is that the elaboration of these frequency size distribution curves is restricted to the sizing of typically many hundreds of nanostructural elements, which makes the procedure time-consuming.

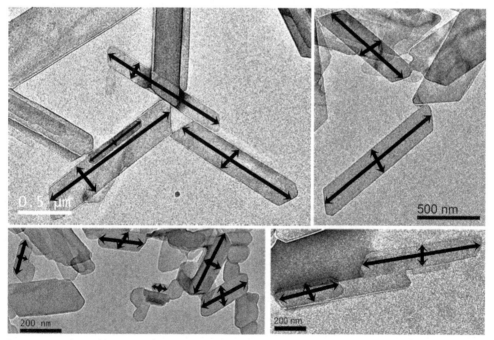

Figure 1.7 Enlarged sections of cryo-TEM images showing selected nanocrystals in which the particle boundaries can be clearly distinguished for further quantification by image analysis.

It should always be kept in mind that the images obtained using the cryo-TEM technique are two-dimensional projections of three-dimensional objects. In our case, a platelet gives rise to a dark rectangle with a black border, which derives from the scattering of electrons in the part of the beam that comes across the thickness of the platelet wall. If the platelets are small in longitude and width compared with their thickness, they could also be present in other orientations with respect to the direction of the electron beam and give rise to side-view projections. However, because platelet lengths and widths seem to be predominantly larger than the thickness, side views can only be obtained when nanoplatelets are piled with each other. Examples of side-view projections are seen in Fig. 1.8. When a cryo-TEM image shows platelet side views, the internal structure can be observed and even quantified. It consists of planar layers assembled into stacks that correspond to the piling of several TAG lamellae. As expected, the layers are oriented in parallel to the largest surface of the nanoplatelets.

ImageJ (Wayne Rasband, NIH) is a public domain Java image processing and analysis program that contains several useful tools among which the most important for our case are those used to measure distances and angles. Unlike most image processing programs, ImageJ does not have a main work area. Instead, ImageJ's main window is only a menu bar (shown at the top of Fig. 1.10) containing all the menu commands, a toolbar, a status bar, and a progress bar. Images, histograms, profiles, widgets, etc. are displayed in additional windows, which can be simultaneously shown on the screen.

When initiating the software, the desired image can be selected and opened from the menu *File > Open*. For the platelet dimensions measurement, the first step is to define a spatial scale from the active image; therefore, size results can be presented in calibrated units, such as nm or μm. Initially, the *Straight* line selection tool should be used to

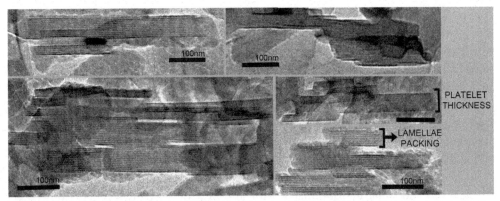

Figure 1.8 Side view of extracted nanoplatelets showing the layered internal structure given by the stacking of several triacylglycerol lamellae.

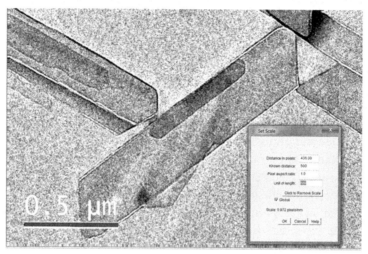

Figure 1.9 ImageJ software window displaying an enlarged region of the cryo-TEM micrograph used to set the scale before the nanocrystal size measurement process.

make a line selection that corresponds to a known distance, generally the scale bar in the image. Then, the *Analyze > Set scale* command will open a dialog box where the *Known distance* and *Unit of length* should be entered. In our example of Fig. 1.9, the scale corresponds to a length of 500 nm. As displayed in Fig. 1.9, ImageJ will fill automatically the distance in pixels field based on the length of the line selected and image resolution. In addition, a convenient function to use, especially when working with multiple images having equal magnification, is the option *Global*. When this option is checked, the scale defined in the set scale dialog is used for all opened images during the current session instead of just the active micrograph. Again, using the *Straight* line tool either the length or width of the nanoparticles to measure can be selected and then calculated from the menu *Analyze > Measure*. Measurement results will be displayed in the *Results table* as shown in Fig. 1.10 from which the data to be processed can be extracted. Additionally, when the *Angle tool* selection is chosen, it is possible to calculate the angles between platelets' corners, which makes possible comparisons, for instance, with structural crystallographic studies. However, it is noteworthy to mention that the angle column shown in Fig. 1.10 does not correspond to the angle formed by the platelet shape because the *Angle tool* has not been selected in the current example; instead, the registered value is the angle produced between the measuring drawn line and the normal plane.

In the same manner as described above, the user can measure the thickness of nanoplatelets and lamellae from images with side view of the particles (Fig. 1.8).

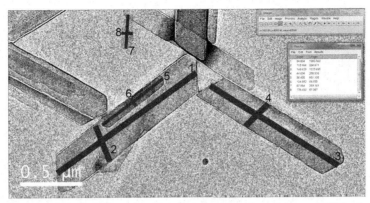

Figure 1.10 ImageJ software window showing an enlarged region of the cryo-TEM micrograph used during the nanoplatelet size measurement. The top right shows the ImageJ main menu bar along with the table of results with the obtained dimensions.

As recently reported by Acevedo and Marangoni (2010a), the measurement of the lamellae thicknesses yielded an average value of 4.23 ± 0.76 nm in samples of FHCO. This result has encountered good agreement with small angle powder X-ray diffraction data, which yielded a long spacing value (lamellae thickness) of 4.5 nm.

Regarding nanoplatelet thickness, it has also been demonstrated (Acevedo and Marangoni, 2010a,b) that values obtained with cryo-TEM and Scherrer's analyses are also in good agreement. For instance, for FHCO, the thickness values obtained are 31.32 ± 0.07 nm and 31.2 ± 2.3 nm using the Scherrer and cryo-TEM analysis.

Once image analysis is finalized, the next step is to transfer the data to a user's choice software suitable for data processing and statistics. Fig. 1.11 illustrates examples of particle size distribution for the platelet lengths and widths elucidated with the help of the GraphPad Prism 5 software (GraphPad Software, Inc., San Diego, CA, USA). The median values were calculated for about 1550 nanoplatelets in the case of the sample 1 (top of the figure) and around 724 particles in the case of the sample 2 (bottom of the Figure). It is important to emphasize that when analyzing fat nanostructure, all distributions can be approximated in this way where the median size corresponds to the maximum frequency and where no other maxima appear within the histogram (multimodal distributions).

From the size distributions, it is possible to obtain nanoplatelet lengths and widths and additional information on the fat nanocrystals, such as surface, area, and volume. However, occasionally, it is necessary to obtain information on particle size in terms of average diameter to describe structural elements using just one number. Thus, it is advisable to convert both nanoplatelets lengths and widths into equivalent diameter

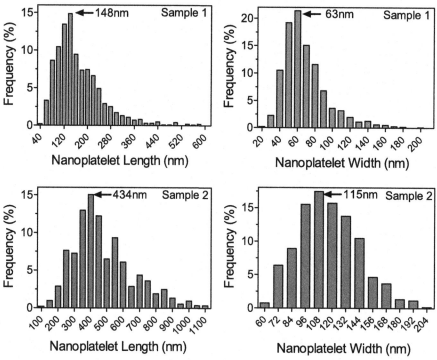

Figure 1.11 Size distributions of nanoplatelet longitudes and widths acquired after the analysis of the cryo-TEM images obtained for two different samples.

(ED). The expression of ED of a nonspherical particle is defined as the diameter of a sphere of equivalent cross-sectional area on the images and can be calculated from the projected lengths and widths on the cryo-TEM images as

$$ED = \sqrt{\frac{4 \times A}{\pi}}, \qquad (1.2)$$

where A is the platelet area (*length* × *width*) obtained from image analysis. For example, the experimental data showed in Fig. 1.11 yielded ED values of 109 and 252 nm for samples 1 and 2, respectively.

    To accurately interpreting the nanostructure of fat crystal networks, it is not only necessary to visualize and understand the arrangement of the primary crystals, but it is also important to understand the forces that keep these basic units together. It has been accepted that TAGs are flocculated solid particles stabilized together into the fat crystal network via Van der Waals forces (Van den Tempel, 1961; Nederveen, 1963). Van der Waals forces have been experimentally and theoretically studied for more than a century, in particular, for colloidal systems for interface energies and wetting (Parsegian, 2005; French et al., 2010). Marangoni et al. (2012) recently reviewed the approaches used to compute the energy and Hamaker coefficient describing the interaction between fat particles.

Based on the Lifshitz theory (Lifshitz, 1956), the Hamaker coefficient describes the interaction between two slabs of nanocrystalline fats (1 and 3) separated by an oily medium (2) and is defined as follows:

$$A = A_{\nu=0} + A_{\nu>0} = \frac{3}{4}kT\left(\frac{\xi_1 - \xi_3}{\xi_1 + \xi_3}\right)^2 + \frac{3h\nu}{16\sqrt{2}}\frac{\left(n_1^2 - n_3^2\right)^2}{\left(n_1^2 + n_3^2\right)^{3/2}}, \tag{1.3}$$

where ξ_1 and ξ_2 are the static dielectric constants for the slab and the medium, respectively; n_1, n_2 are the refractive index for the fat slab and the medium, respectively; k is Boltzman's constant ($1.38 \times 10{-23}$ J/K); T is the absolute temperature; h is Planck's constant ($6.62.10{-34}$ J s); and ν is the UV ionization frequency for the molecule (3.0 1015/s).

Several authors have calculated the Hamaker coefficient for TAGs using Eq. (1.3). For example, Johansson and Bergenståhl (1992) and Kloek (1998) calculated A for tristearin in a soybean oil medium obtaining values of $0.17\ 10^{-21}$ and $1.8\ 10^{-21}$ J., respectively. Furthermore, in a recent review, we calculated the Hamaker coefficient for FHCO and high oleic sunflower oil obtaining a value of $0.55\ 10^{-21}$ J (Acevedo et al. (2011).

To understand the degree to which nanocrystalline structure can influence London—van der Waals interaction and affect mechanical properties of fats, we can calculate, as an example, the elastic modulus G' for an FHCO sample using the expression (Narine and Marangoni, 1999; Marangoni, 2000):

$$G = \frac{A}{6\pi\gamma ad_0^2}\,\Phi^{\frac{1}{d-D}}. \tag{1.4}$$

In the case of FHCO, the nanoparticle diameter a is 109 nm (sample 1 in Fig. 1.11), $d_0 = 0.39$ nm (Nevderveen, 1963), the strain at the limit of linearity γ is $1.9\ 10^{-4}$ (Ahmadi et al., 2009), and the solid mass fraction Φ is ~ 1. Therefore, when using $A = 0.55 \times 10^{-21}$ J, a value of $G = 7.9 \times 10^6$ Pa is obtained, which agrees with the experimental value of 6.8×10^6 Pa reported by Ahmadi et al. (2009). Hence, mechanical properties can be calculated by a prior knowledge of the nanostructure of the fat crystal network.

This chapter offers a new approach in the characterization of nanostructure of fats. With the previously explained methods, the nanostructure of fat crystal networks can now be studied in a more systematic way, as it is possible to accurately quantify nanocrystalline elements. Reproducible results can be obtained by XRD and cryo-TEM with good agreement between techniques. Fig. 1.12 presents a schematic representation of the primary crystal of a TAG polycrystalline network, which larger dimensions can be characterized using the novel sample treatment developed by our

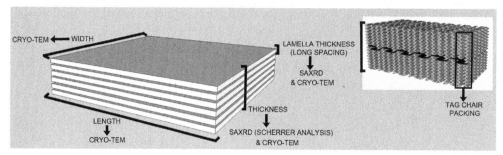

Figure 1.12 Schematic representation of a nanoplatelet and the organization of the triacylglyceride lattice within this nanoparticle and the preferential methods used to quantitatively determining their dimensions.

group followed by cryo-TEM. Cryo-TEM and the Scherrer analysis of XRD data give comparable results in the determination of nanoplatelet and lamella thicknesses. Therefore, both techniques can efficiently be applied in combination for thorough nanostructural studies.

REFERENCES

Arishima, T., Sato, K., 1989. Polymorphism of POP and SOS. III. Solvent crystallization of β2 and β1 polymorphs. J. Am. Oil Chem. Soc. 66, 1614—1617.

Acevedo, N.C., Block, J., Marangoni, A.G., 2012. Critical laminar shear-temperature effects on the nano- and mesoscale structure of a model fat and its relationship to oil binding and rheological properties. Faraday Discuss. 158, 171—194.

Acevedo, N.C., Marangoni, A.G., 2010a. Characterization of the nanoscale in triacylglycerol crystal networks. Cryst. Growth Des. 10, 3327—3333.

Acevedo, N.C., Marangoni, A.G., 2010b. Towards nanoscale engineering of triacylglycerol crystal networks. Cryst. Growth Des. 10, 3334—3339.

Acevedo, N.C., Marangoni, A.G., 2014. Functionalization of non-interesterified mixtures of fully hydroge- nated fats using shear processing. Food Bioprocess Technol. 7, 575—587.

Acevedo, N.C., Peyronel, M.F., Marangoni, A.G., 2011. Nanoscale structure intercrystalline interactions in fat crystal networks. Curr. Opin. Colloid Interface Sci. 16, 374—383.

Ahmadi, L., Wright, A.J., Marangoni, A.G., 2009. Structural and mechanical behavior of tristearin/triolein- rich mixtures and the modification achieved by interesterification. Food Biophy. 4, 64—76.

Chawla, P., deMan, J.M., 1990. Measurement of the size distribution of fat crystals using a laser particle counter. J. Am. Oil Chem. Soc. 67, 329—332.

deMan, J.M., Beers, A.M., 1987. Fat crystal networks: structure and rheological properties. J. Text. Stud. 18, 303—318.

French, R.H., Parsegian, V.A., Podgornik, R., Rajter, R.F., Jagota, A., Luo, J., Asthagiri, D., Chaudhury, M.K., Chiang, Y.-M., Granick, S., Kalinin, S., et al., 2010. Long range interactions in nanoscale science. Rev. Modern Phy. 82, 1887—1944.

Heertje, I., Leunis, M., van Zeyl, W.J.M., Berends, E., 1987. Product microscopy of fatty products. Food Microstruct. 6, 1—8.

Heertje, I., van Eendenburg, J., Cornelissen, J.M., Juriaanse, A.C., 1988. The effect of processing on some microstructural characteristics of fat spreads. Food Microstruct. 7, 189—193.

Herrera, M.L., Hartel, R.W., 2000. Effect of processing conditions on physical properties of a milk fat model system: microstructure. J. Am. Oil Chem. Soc. 77, 1197—1204.

Johansoon, D., Bergenståhl, B., 1992. The influence of food emulsifiers on fat and sugar dispersions in oils II. Rheology, colloidal forces. J. Am. Oil Chem. Soc. 69, 718–727.

Kloek, W., 1998. Mechanical Properties of Fats in Relation to Their Crystallization (Ph.D. dissertation). Wageningen Agric. Uni., Wageningen, The Netherlands.

Langford, J.I., Wilson, A.J.C., 1978. Scherrer after sixty years: a survey and some new results in the determination of crystallite size. J. Appl. Cryst. 11, 102–113.

Larsson, K., 1994. Lipids. Molecular Organization, Physical Functions and Technical Applications. The Oily Press LTD, Glasgow.

Lawler, P.J., Dimick, P.S., 1998. Crystallization and polymorphism of fats. In: Akoh, C.C., Min, D.B. (Eds.), Food Lipids: Chemistry, Nutrition and Biotechnology. Marcel Dekker, Inc, New York, pp. 229–250.

Lifshitz, E.M., 1956. The theory of molecular attractive forces between solids. Sov. Phys. JETP 2, 73–83.

Maleky, F., Smith, A., Marangoni, A.G., 2011. Laminar shear effects on crystalline alignments and nanostructure of a triacylglycerol crystal network. Cryst. Growth Des. 11, 2335–2345.

Marangoni, A.G., 2000. Elasticity of high volume-fraction fractal aggregates networks: a thermodynamic approach. Phys. Rev. B 62, 13951–13955.

Marangoni, A.G., Rousseau, D., 1996. Is plastic fat rheology governed by the fractal nature of the fat crystal network. J. Amer. Oil Chem. Soc. 73, 991–993.

Marangoni, A.G., Acevedo, N.C., Maleky, F., Co, E., Peyronel, F., Mazzanti, G., Quinn, B., Pink, D., 2012. Structure and functionality of edible fats. Soft Matter. https://doi.org/10.1039/C1SM06234D.

Martini, S., Herrera, M.L., Hartel, R.W., 2002. Effect of cooling rate on crystallization behavior of milk fat fraction/sunflower oil blends. J. Am. Oil Chem. Soc. 79, 1055–1062.

Martini, S., Herrera, M.L., 2002. X-ray Diffraction and crystal size. J. Am. Oil Chem. Soc. 79, 315–316.

Narine, S.S., Marangoni, A.G., 1999. Fractal nature of fat crystal networks. Physic Rev. 59, 1908–1920.

Nederveen, C.J., 1963. Dynamic mechanical behavior of suspensions of fat particles in oil. J. Colloid Sci. 18, 276–291.

Parsegian, V.A., 2005. Van Der Waals Forces. A Handbook for Biologists, Chemists, Engineers, and Physicists. Cambridge University Press, New York.

Ribeiro, A.P.B., Masuchi, M.H., Miyasaki, E.K., Domingues, M.A.F., Stroppa, V.L.Z., de Oliveira, G.M., Kieckbusch, T.G., 2015. Crystallization modifiers in lipid systems. J. Food Sci. and Tech. 52 (7), 3925–3946.

Simpson, T.D., Hagemann, J.W., 1982. Evidence of two β' phases in tristearin. J. Am. Oil Chem. Soc. 59, 169–171.

Shukla, A., Rizvi, S.S.H., 1996. Relationship among chemical composition, microstructure and rheological properties of butter. Milchwissenschaft 51 (3), 144–148.

Tang, D., Marangoni, A.G., 2006. Microstructure and fractal analysis of fat crystal networks. J. Am. Oil Chem. Soc. 83, 377–388.

Van den Tempel, M., 1961. Mechanical properties of plastic disperse systems at very small deformations. J. Colloid Sci. 16, 284–296.

West, A.R., 1984. Solid State Chemistry and its Applications. John Wiley & Sons, Chichester, West Sussex.

FURTHER READING

Marangoni, A.G., Rogers, M., 2003. Structural basis for the yield stress in plastic disperse systems. Applied Phys. Lett. 82, 3239–3241.

CHAPTER 2

Nucleation and Crystallization Kinetics of Fats

Keshra Sangwal[1], Kiyotaka Sato[2]

[1]Lublin University of Technology, Lublin, Poland; [2]Hiroshima University, Higashi Hiroshima, Japan

Fats and oils are complex systems composed primarily of mixtures of distinct triglycerides, which are esters of 1 mol of glycerol and 3 mol of fatty acids. Different fats and oils contain different proportions of chemically distinct triglycerides varying in carbon chain length and melting point. Therefore, instead of melting at a single temperature, a fat melts over a wide range of temperature, and high melting point glycerides are soluble in lower melting point glycerides. Consequently, the crystallization behavior of molten fats and oils is relatively complex in comparison with that of simple one-component systems. An additional difficulty associated with the crystallization of mixtures of triglycerides is the existence of different polymorphs of triglycerides.

In this chapter, the nucleation and crystallization behavior of fats and oils is presented and discussed. The discussion is confined to single-component systems, although fats and oils are composed of a major component and several other components, which may be considered as additives. The basic concepts of three-dimensional (3D) nucleation of crystals and transformation of polymorphic forms are presented first in Driving Force for Phase Transition, Three-dimensional Nucleation of Crystals, Nucleation and Transformation of Metastable Phases sections. Isothermal and nonisothermal crystallizations of fats are then described in Overall Crystallization and Nonisothermal Crystallization sections, respectively. Finally, induction period for nucleation is described in Induction Period for Nucleation section. Some of the topics described here have been reviewed previously (Marangoni, 2013; Rousset, 2002; Sato et al., 2013). Description of some of the topics in this chapter is based on the previous work of one of the authors (Sangwal, 2007). However, the theories of crystal growth are not described because they are not required specifically to understand the topics discussed here. For details on the kinetics and mechanism of crystal growth, the reader is referred to Chernov (1984), Mullin (2001) and Sangwal (2007).

DRIVING FORCE FOR PHASE TRANSITION

At a given temperature, all physical (e.g., condensation, crystallization, phase transitions in the solid phase, etc.) and chemical processes (e.g., heterogeneous reactions) in the

Structure-Function Analysis of Edible Fats, Second Edition
ISBN 978-0-12-814041-3, https://doi.org/10.1016/B978-0-12-814041-3.00002-2

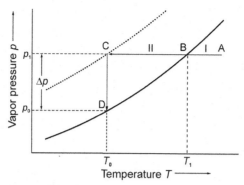

Figure 2.1 Schematic illustration of the dependence of vapor pressure p of a system on temperature. Solid curve shows the state when vapor phase I is in equilibrium with solid phase II, whereas dotted curve shows the upper limit of the metastable zone when precipitation of phase II occurs in the system.

vapor, melt, or solution phases of every system take place through the formation of 3D nuclei of a new phase and occur only when the medium is supersaturated. In the case of crystallization from vapor, liquid, or solid phase, nuclei, which can attain a size greater than that of a critical nucleus, develop into crystals of visible size by the attachment of growth species (i.e., molecules, atoms, or ions). To understand the conditions when nuclei of phase II are formed from a homogeneous phase I, such as vapor, melt, or solution, we refer to Fig. 2.1. The solid curve in the figure shows the situation when the two phases are in equilibrium conditions. A vapor—crystal phase change is characterized by a $p-T$ diagram, whereas a solution—crystal phase change by a $c-T$ diagram. Here p is the vapor pressure, c is the solute concentration, and p_0 and c_0 denote the saturation vapor pressure and the equilibrium solute concentration at temperature T_0, respectively.

We consider the change from vapor to crystalline phase during the cooling of the vapor, as shown by the line ABCD in the $p-T$ diagram of Fig. 2.1. Points B and D represent the equilibrium state in the two phases. At a constant vapor pressure p_1, a decrease in the temperature along the curve BC means its passing through the equilibrium curve, with phase I becoming metastable (i.e., supersaturated). The supersaturation of the vapor phase is represented by the difference $\Delta p = p_1 - p_0$. At a constant temperature T_0, the change may be considered to take place along the curve CD as a result of decreasing supersaturation to the point D, where Δp becomes zero. Thus, the supersaturated state of the vapor phase may be represented either with reference to a constant temperature T_0 by the difference Δp or with reference to a constant vapor pressure p_1 by the difference $\Delta T = T_1 - T_0$.

In the case of a $c-T$ diagram, at a constant temperature T, the supersaturation is represented by $\Delta c = c - c_0$, where c and c_0 denote the actual and the equilibrium concentrations of solute in the solution, respectively. The supersaturation σ is defined by the quotients $\Delta p/p_0$ and $\Delta c/c_0$, whereas the supersaturation ratio $S = p/p_0$ or c/c_0.

When $\Delta p = p - p_0$ or $\Delta c = c - c_0 > 0$, a system is said to become metastable. Under these conditions only nucleation, precipitation, deposition, and growth of new phase II are possible. Spontaneous nucleation (i.e., precipitation) of phase II occurs when Δp or Δc exceeds certain values Δp^* or Δc^*. These values of Δp^* or Δc^* characterize the width of the metastable zone width and are shown by the dotted curve in Fig. 2.1. When Δp or $\Delta c < 0$, the system is always stable. In this case, if phase II is present in the system, it undergoes evaporation or dissolution.

The driving force for the processes of crystallization (i.e., nucleation, precipitation, growth, deposition, etc.) is the difference in the Gibbs free energy G_I of supersaturated or supercooled mother phase I and the free energy G_{II} of the newly forming phase II and may be expressed by

$$\Delta G = G_{II} - G_I = \Delta H - T\Delta S, \tag{2.1}$$

where H and S denote the enthalpy and entropy of a phase, respectively, and $\Delta H = H_{II} - H_I$ and $\Delta S = S_{II} - S_I$. The dependence of the free energy G_I of vapor, melt, or solution phase I and G_{II} of solid phase II on temperature T and the free energy change ΔG with respect to temperature are schematically shown in Fig. 2.2. Obviously, the formation of phase II is possible when $\Delta G > 0$ and occurs for $T < T_0$.

In the state of thermodynamic equilibrium when $\Delta G = 0$ (and $T = T_0$), from Eq. (2.1) one has

$$\Delta S = \Delta H/T_0, \tag{2.2}$$

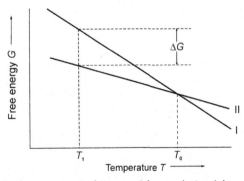

Figure 2.2 Change in Gibbs free energy G of vapor, melt, or solution (phase I) and solid (phase II) as a function of temperature of a system.

which, for nonequilibrium conditions, gives

$$\Delta G = \Delta H - T\Delta H/T_0 = \Delta H\Delta T/T_0, \tag{2.3}$$

where $\Delta T = T_0 - T$. Thus, the free energy change ΔG, given by Eq. (2.1), may be expressed in terms of temperature difference ΔT. In the case of a melt, ΔT is referred to as its supercooling and is denoted as $\Delta T = T_m - T$, where T is the temperature of the supercooled melt and T_m is the melting point of the solute (with $T < T_m$).

Using the relationship between the Gibbs free energy change ΔG and the chemical potential difference $\Delta \mu$, i.e.,

$$\Delta G = \sum_i n_i \, \Delta \mu_i, \tag{2.4}$$

where $\Delta \mu_i$ is the change in the chemical potential of n_i moles of component i, Eq. (2.3) may be given by

$$\Delta \mu = \Delta H \Delta T/T_0 \nu, \tag{2.5}$$

where ν is the number of particles produced by the dissociation of molecules in the liquid phase.

In the case of crystallization of molecular compounds, for which $\nu = 1$, with well-defined melting points T_m from the melt, Eq. (2.5) takes the form

$$\Delta \mu = \left(\frac{\Delta H_m}{T_m}\right)\Delta T, \tag{2.6}$$

with $\Delta T = (T_m - T)$ and ΔH_m equals to the enthalpy of melting. In the case of crystallization of molecular compounds at temperature T_c from solutions saturated at temperature T_0, the chemical potential difference for crystallization

$$\Delta \mu = \left(\frac{\Delta H_s}{T_0}\right)\Delta T, \tag{2.7}$$

where $\Delta T = (T_0 - T_c)$ and ΔH_s is the enthalpy of dissolution.

If c and c_0 are the concentrations of the solute in supersaturated and saturated solutions, respectively, the chemical potential difference may be given by

$$\Delta \mu = k_B T \ln(c/c_0), \tag{2.8}$$

where k_B is the Boltzmann constant. Denoting the supersaturation ratio c/c_0 by S and the supersaturation by $\sigma = (S - 1)$, Eq. (2.8) may be written in the form

$$\Delta \mu = k_B T \ln(1 + \sigma) \approx k_B T \sigma, \tag{2.9}$$

when $\ln(1 + \sigma) \approx \sigma$ for small values of σ.

The free energy change ΔG and the chemical potential difference $\Delta\mu$ of the above equations are the necessary driving forces for crystallization and the temperature difference ΔT and the supersaturation ratio S are measures of this driving force.

3D NUCLEATION OF CRYSTALS

The formation of 3D nuclei from atomic or molecular entities existing in the volume of a growth medium involves their aggregation in an ordered phase. The process is usually envisaged to occur as a result of collision of individual atoms or molecules C_1 in such a way that the collisions yield a sequence of aggregates of increasing size such as dimers C_2, trimers C_3, … i-mers C_i, according to the general reversible reaction:

$$i\, C_1 \Leftrightarrow C_i, \tag{2.10}$$

where C_i is the resulting aggregate formed by the addition of i monomers C_1. Aggregates such as C_2, C_3, …, C_i are usually called embryos, subnuclei, or clusters. The process of formation of subnuclei is in dynamic equilibrium, some of them grow larger, whereas others simply disintegrate with time. However, the statistical addition of individual atoms/molecules to some of the subnuclei leads to their development into a size when they no longer disintegrate. Such nuclei are said to attain a critical size and are called 3D stable nuclei.

Nucleation Barrier

The surface tension of a sphere is the lowest. Therefore, it is usually assumed that the nuclei prefer to attain a rounded shape. The reduction in the Gibbs free energy of the system due to the formation of a spherical nucleus of radius r is equal to the sum of the surface excess free energy ΔG_S and the volume excess free energy ΔG_V. The surface free energy ΔG_S is associated with the creation of the surface of the solid phase in the growth medium and is equal to the product of the surface area of the developing nucleus and the interfacial tension γ between the surface of the nucleus and the supersaturated medium surrounding it. Similarly, the volume excess free energy ΔG_V is a result of creation of the volume of the nucleus in the medium and is equal to the product of the volume of nucleus and the chemical potential difference $\Delta\mu$ per unit volume. Then one may write the free energy change as

$$\Delta G = \Delta G_S + \Delta G_V = 4\pi\, r^2\gamma - \frac{4}{3}\pi\, r^3\Delta\mu. \tag{2.11}$$

Because the creation of the new phase II from supersaturated medium is associated with a decrease in the chemical potential difference $\Delta\mu$, the ΔG_V term is a negative quantity. The two terms in the right-hand side of Eq. (2.11) depend differently on r. This behavior of ΔG associated with the formation of the nucleus is shown in Fig. 2.3 as a

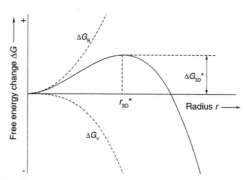

Figure 2.3 Change in Gibbs free energy ΔG as a function of radius r of nucleus formed in a supersaturated medium. Dashed curves show energy contributions ΔG_S and ΔG_V.

function of nucleus radius r. As may be seen from the figure, ΔG passes through a maximum and the maximum value ΔG^*_{3D} corresponds to the critical radius r^*_{3D}. The value of r^*_{3D} may be obtained by maximizing Eq. (2.11), taking $d\Delta G/dr = 0$, i.e.,

$$\frac{d\Delta G}{dr} = 8\pi\, r\gamma - 4\pi r^2\frac{\Delta\mu}{\Omega} = 0, \tag{2.12}$$

or

$$r^*_{3D} = \frac{2\gamma\Omega}{\Delta\mu}. \tag{2.13}$$

On substituting the value of r^*_{3D} from Eq. (2.13) into Eq. (2.11), one obtains

$$\Delta G^*_{3D} = \frac{16\pi}{3}\frac{\gamma^3\Omega^2}{(\Delta\mu)^2}$$

$$= \frac{16\pi}{3}\frac{\gamma^3\Omega^2}{(k_B T \ln S)^2} = \frac{16\pi}{3}\frac{\gamma^3\Omega^2}{(\Delta T)^2}\left(\frac{T_m}{\Delta H_m}\right)^2 = \frac{4\pi}{3}\gamma\, r^{*\,2}_{3D}. \tag{2.14}$$

Note that Eqs. (2.13) and (2.14) contain $\Delta\mu$, which is given by different relations defined in Driving Force for Phase Transition section, but Eq. (2.14) describes the dependence of ΔG^*_{3D} on supersaturation $\ln S$ and temperature difference ΔT as applied during crystallization from supersaturated solutions and supercooled melts (cf. Eqs. 2.9 and 2.6). The value of ΔG^*_{3D} is always a positive quantity because both γ and r^*_{3D} are positive quantities. Obviously, increasing supersaturation $\ln S$ and supercooling ΔT and decreasing interfacial energy γ facilitate the formation of 3D nuclei.

In Eq. (2.14), the factor $16\pi/3$ is a result of the geometry of a spherical nucleus. For nuclei of other geometries, this geometrical factor has other appropriate values. For a cube, for example, this factor has a value of 32.

The occurrence of a crystalline phase in a supersaturated medium depends on the radius r of the nuclei forming in it. When the radius r of the nuclei is smaller than r_{3D}^*, the nuclei dissolve. However, when $r > r_{3D}^*$, the nuclei are stable and grow. The critical radius r_{3D}^* is the minimum size of a stable nucleus. As seen from Fig. 2.3, only when the nucleus radius $r > r_{3D}^*$, the free energy ΔG for the formation of a nucleus decreases with an increase in r.

It should be noted that every phase transition is associated with two energy changes: the activation barrier ΔG_{3D}^*, given by Eq. (2.14), and the overall change in the free energy ΔG, given by Eq. (2.1). The value of the activation barrier ΔG_{3D}^* is associated with the process of formation of 3D nuclei, whereas the overall change in the free energy ΔG determines whether the 3D nuclei formed in the growth system remain stable after their formation. The higher the activation barrier ΔG_{3D}^*, the more difficult it is to attain stable nuclei. Similarly, the greater the change in ΔG, the higher the stability of the phase formed (nuclei). As seen from Eq. (2.1), the stability of the crystallized phase in the growth system at a given temperature T is related to changes in its enthalpy and entropy (i.e., ΔH and ΔS).

Nucleation Rate

Formation of subnuclei of large size is a statistical process involving a particular free energy change $\Delta G(i)$ of cluster formation and the formation of such subnuclei by the collision of a large number of atoms/molecules simultaneously is practically improbable. If C_0 is the concentration of possible sites for the formation of clusters of the new phase in the system, the equilibrium size distribution of clusters of size i may be described by (Kashchiev, 2000)

$$C(i) = C_0 \exp[- \Delta G(i)/k_B T], \qquad (2.15)$$

where $C(i)$ denotes the concentration of C_i clusters containing i atoms/molecules C_1. The equilibrium size distribution between the C_i and C_1 clusters may also be given by

$$C(i) = C_1 \exp\{ - [\Delta G(i) - \Delta G(1)]/k_B T\}, \qquad (2.16)$$

where $C_1 = C(1)$. The monomer concentration C_1 is related to C_0 by

$$C_1 = C_0 \exp[- \Delta G(1)/k_B T]. \qquad (2.17)$$

Because the equilibrium between C_i and C_1 clusters satisfies the law of mass action, using Eqs. (2.15) and (2.17), Eq. (2.16) may be written in the form

$$C(i) = C_0(C_1/C_0)^i \exp\{ - [\Delta G(i) - i\Delta G(1)]/k_B T\}. \qquad (2.18)$$

In this equation, the preexponential term essentially denotes the equilibrium constant of the reaction of Eq. (2.10), whereas $[\Delta G(i) - i\Delta G(1)]$ is the net free energy change involved in the formation of i-sized cluster from individual atoms/molecules.

At constant temperature and supersaturation conditions, the occurrence of 3D nucleation is described by the so-called stationary nucleation rate J_s, given by (Kashchiev, 2000)

$$J = zf^*C^*,$$
(2.19)

where z is the Zeldovich factor lying between 0.01 and 1, f^* is a kinetic factor associated with the frequency of attachment of monomers to the stable nucleus, and C^* is given by Eqs. (2.15)–(2.17) when the equilibrium concentration $C(i)$, the cluster formation free energy change $\Delta G(i)$, and the number i of atoms/molecules in the spherical critically sized nucleus of radius r^* are denoted by C^*, ΔG_{3D}^*, and i^*, respectively. Corresponding to the equilibrium concentration C^*, given by Eq. (2.15), one obtains

$$J = J_0 \exp\left(-\frac{\Delta G_{3D}^*}{k_B T}\right)$$

$$= J_0 \exp\left(-\kappa^*\frac{\gamma^3\Omega^2}{k_B T \,\Delta\mu^2}\right) = J_0 \exp\left(-\frac{B}{\ln^2 S}\right) = J_0 \exp\left(-\frac{B'}{T(\Delta T)^2}\right),$$
(2.20)

where the kinetic factor

$$J_0 = zf_e^* C_0 S,$$
(2.21)

and the constants (cf. Eq. 2.14)

$$B = \kappa^*\left(\frac{\gamma\Omega^{2/3}}{k_B T}\right)^3,$$
(2.22)

$$B' = \kappa^*\left(\frac{\gamma\Omega^{2/3}}{k_B}\right)^3\left[\frac{R_G T_m}{\Delta H_m}\right]^2.$$
(2.23)

In Eq. (2.23), the term in the square brackets is written in the form of a dimensionless quantity. Typically, $\Delta H_m/R_G T_m$ is about 2 for melts, 3 for anhydrous inorganic salts and 6 for organic compounds (Sangwal, 1989). Eq. (2.20) describes the nucleation rate J as a function of driving force for crystallization from supersaturated solutions and supercooled melts.

In the above expressions, f_e^* is the frequency of attachment of monomers to the nucleus at equilibrium, S is the supersaturation ratio (see Eqs. 2.8 and 2.9), ΔG_{3D}^* is given by Eq. (2.14), and κ^* is the shape factor for the nuclei. Typically, for melts and solutions, $0.01 < z < 1$, $C_0 \approx 10^{20}–10^{29}$ m^{-3}, $f_e^* \approx 1 – 10^{12}$ s^{-1} and $1 < S < 10$, $J_0 = 10^{15}–10^{40}$ m^{-1} s^{-1} (Kashchiev, 2000).

Eq. (2.20) represents the temperature dependence of nucleation rate J and is usually referred to as the classical theory of 3D nucleation. Obviously, it is an Arrhenius-type relation where the activation barrier ΔG_{3D}^* is essentially a measure of the "difficulty" for embryos to attain the radius r_{3D}^* of the stable nuclei in a growth medium.

The behavior of nucleation rate J on supersaturation ratio S (and on the related chemical potential difference $\Delta\mu$ defined in Driving Force for Phase Transition section) of systems is well described by Eq. (2.20). However, in the case of solutions at high concentrations and melts, the viscosity of the systems affects the nucleation rate J. The effect of viscosity η on J is due to its influence on the frequency f_e^* of attachment of monomers to the nucleus (see Eq. 2.10) associated with the diffusion of monomers to the nucleus and their subsequent integration. In this case, the nucleation rate is given by

$$J = J_0^* \exp\left(-\frac{B}{\ln^2 S} + \frac{E_\eta}{k_B T}\right),\tag{2.24}$$

where J_0^* is a new kinetic factor such that $J_0^* < J_0$ and E_η is the activation energy for viscous flow.

In the above type of systems, the nucleation rate J initially increases with increasing supersaturation $\ln S$, but later it begins to decrease such that the nucleation rate J passes through a maximum at a particular value of $\ln S$. This is due to the fact that, at a given temperature, at low values of $\ln S$ when the B term dominates, the rate increases with an increase in $\ln S$. However, at very high values of $\ln S$ when the B term becomes negligibly small in comparison with the E_η term, the latter term begins to dominate. Consequently, at very high values of $\ln S$, the nucleation rate J is decreased. Absence of nucleation means that the system is in the glassy state.

The derivation of Eq. (2.20) using Eq. (2.15) assumes that the interfacial tension γ of a critically sized nucleus is the same as that of a large bulk crystal. This assumption may hold in the case of large nuclei forming at low supersaturations but is certainly violated at high supersaturations when the cluster size approaches atomic/molecular dimensions (cf. Eq. 2.13). This may be understood better when the free energy change $\Delta G(i)$ for the formation of i-sized cluster is given by

$$\Delta G(i) = -i\Delta\mu + \kappa\gamma i^{2/3},\tag{2.25}$$

where κ is the shape factor for the cluster. For $i = 1$, Eq. (2.25) gives the free energy change

$$\Delta G(1) = -\Delta\mu + \kappa\gamma.\tag{2.26}$$

Using Eqs. (2.25) and (2.26), one obtains the net free energy change for the formation of the i-sized nucleus in the form

$$\Delta G_{3D} - i\Delta G(1) = \kappa\gamma\left(i^{2/3} - i\right). \tag{2.27}$$

This equation describes the net free energy change $[\Delta G_{3D} - i\Delta G(1)]$ involved in the formation of i-sized cluster according to Eq. (2.10) of the clustering reaction, and Eq. (2.18) of the equilibrium between C_i and C_1 clusters is said to be self-consistent because it reduces to Eq. (2.15) when $i = 1$.

When $i = i^*$, Eq. (2.27) represents the new activation barrier $\left[\Delta G_{3D}^* - i\Delta G(1)\right]$. The nucleation rate J in this case is also described by Eq. (2.20) with the kinetic factor J_0 given by Eq. (2.21). From Eq. (2.13), it may be noted that the interfacial energy of the stable nucleus is now size-dependent and decreases with an increase in $\Delta\mu$, and the nucleation rate J strongly increases with $\Delta\mu$.

With the new activation barrier defined by Eq. (2.27), Eq. (2.20) holds when the nuclei are of large size and the size decreases smoothly with increasing $\Delta\mu$. However, at very high $\Delta\mu$ when the nuclei contain only very few atoms/molecules, change in the number of atoms/molecules by a new atom/molecule is possible in succession of $\Delta\mu$ intervals. This feature of nucleus formation is not accounted for by the classical nucleation theory. In the limit $i^* \to 1$, the atomistic theory of nucleation based on self-consistency in $C(i)$, as required by the law of mass action (see Eq. 2.18), provides a more adequate description of the nucleation rate (Kashchiev, 2000)

$$J = J_0 \exp\left(\Phi^*/k_B T\right)\exp\left(i^*\Delta\mu/k_B T\right), \tag{2.28}$$

where Φ^* is the effective excess energy of the nucleus and is $\Delta\mu$-independent.

3D Heterogeneous Nucleation

Homogeneous 3D nucleation is possible when there is no external source that alters either the value of the kinetic parameter J_0 or the interfacial energy γ. It is well known that foreign particles present in bulk media as well as cracks and scratches on the walls of crystallizers frequently catalyze 3D nucleation. This type of nucleation is referred to as heterogeneous 3D nucleation.

Depending on the shape of the embryos forming on the foreign substrate, the embryos develop in two or three dimensions. For example, a cap-shaped embryo changes its size in 3D, whereas a disk-shaped embryo grows laterally when its height remains unchanged. In the former case, the nucleation is 3D, whereas in the latter case it is two-dimensional (2D). Therefore, in principle, nucleation on a foreign substrate can be both 2D and 3D.

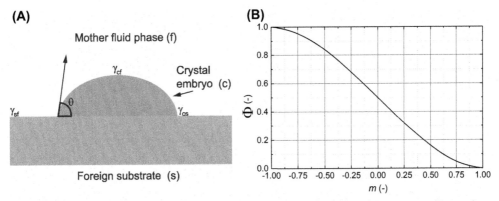

Figure 2.4 (A) Cross-section of a cap-shaped nucleus on the surface of a substrate; crystalline embryo (c), foreign substrate particle (s), and mother phase (f). (B) Dependence of the factor ϕ on $m = \cos\theta$ according to Eq. (2.30).

We consider here the case of 3D nucleation where the embryo forms on a flat substrate. In this case, the shape of the crystal embryo is essentially a spherical cap (see Fig. 2.4A), with the equilibrium wetting (contact) angle θ defined from 0 and π by

$$\cos\theta = (\gamma_{sf} - \gamma_{cs})/\gamma_{cf} = m, \tag{2.29}$$

where γ_{cs}, γ_{sf}, and γ_{cf} denote the interfacial energy between crystalline solid phase (c) and foreign particle surface (s), foreign particle surface (s) and mother fluid phase (f), and crystalline solid phase (c) and fluid phase (f), respectively. For a sufficiently large nucleus, $\gamma_{cf} = \gamma$. The contact angle θ defines a numerical factor ϕ, given by

$$\phi = (2 + \cos\theta)(1 - \cos\theta)^2/4. \tag{2.30}$$

The value of ϕ lies between 0 and 1, depending on the value of $m = \cos\theta$, as shown in Fig. 2.4B. Obviously, heterogeneous nucleation occurs when $-1 < m < 1$, and the situation $m = -1$ corresponds to homogeneous nucleation. When $\theta = 0$, $\phi = 0$; when $\theta = \pi/2$, $\phi = 1/2$; and when $\theta = \pi$, $\phi = 1$. The cases of $\theta = 0$ and $\theta = \pi$ are usually referred to as complete wetting and complete nonwetting, respectively.

Irrespective of whether nucleation is homogeneous or heterogeneous, the above treatment of 3D nucleation can be represented by unified formulas if the interfacial energy γ of nucleus formed by homogeneous nucleation is replaced by an effective interfacial energy γ_{eff}, given by

$$\gamma_{eff} = \Phi\gamma, \tag{2.31}$$

where the activity factor Φ is a number between 0 and 1 and is related to the numerical factor ϕ (see Eq. 2.30). For cap-shaped nuclei, $\Phi = \phi^{1/3}$.

The above treatment of heterogeneous 3D nucleation considers only changes in the interfacial energy of the nucleus and does not take into account the process of integration of growth species into the embryos during nucleation on foreign particles. For example, it is easy to realize that nucleation on a foreign substrate body (s) of radius r_s reduces the effective surface of collision of embryos, where growth species are incorporated into embryos (cf. Fig. 2.4A). This leads to a decrease in the value of the kinetic parameter J_0, thereby decreasing the nucleation rate J (see Eq. 2.20). This effect is opposite to that of the predicted decrease in the nucleation barrier ΔG^*_{3D} due to a decrease in the interfacial energy γ (see Eqs. 2.20 and 2.31). The decrease in the effective surface area for collision of embryos due to nucleation on foreign bodies is determined by the parameter (cf. Liu, 1999; Liu et al., 2000)

$$r' = r_s/r^*_{3D}, \tag{2.32}$$

where the radius of critical nucleus r^*_{3D} is given by Eq. (2.13). Because the nucleus radius r^*_{3D} decreases with an increase in supersaturation $\ln S$, for the foreign particle of a particular r_s, the ratio r' of radii increases with increasing $\ln S$. Therefore, the numerical factor ϕ now is a function of m and r'.

When the effective surface area for the collision of embryos is considered, the 3D nucleation is heterogeneous. Then the nucleation rate J_{het} may be given by an expression similar to Eq. (2.20) (cf. Liu, 1999; Liu et al., 2000)

$$J_{het} = J_{0(het)} \exp\left[-B_{het}/\ln^2 S\right] = J_{0(het)} \exp\left[-B'_{het}/T(\Delta T)^2\right], \tag{2.33}$$

where

$$J_{0(het)} = 4\pi a r_s^2 N_0 \phi'(m, r')[\phi(m, r')]^{1/2} J_0, \tag{2.34}$$

$$B_{het} = \phi(m, r')B, \quad B'_{het} = \phi(m, r')B'. \tag{2.35}$$

In Eqs. (2.34) and (2.35), N_0 is the number of nuclei per unit area on the surface, a is the dimension of growth species, $\phi'(m,r')$ is a parameter that depends on r' similar to $\phi(m,r')$ and lies between 0 and 1, whereas B and B' are given by Eqs. (2.22) and (2.23), respectively. We note that, for heterogeneous nucleation, $0 < \phi(m,r') < 1$, $0 < \phi'(m,r') < 1$, and $4\pi a r_s^2 N_0 < 1$. Consequently, $J_{0(het)} < J_0$, $B_{het} = \phi B < B$, and $B'_{het} = \phi B' < B'$. For homogeneous nucleation when $\phi(m,r') = \phi'(m,r') = 1$ and $4\pi a r_s^2 N_0 = 1$, Eq. (2.33) reduces to Eq. (2.20). In fact, when $r' \gg 1$, the nuclei may be considered to form on a planar substrate and Eq. (2.33) describes the nucleation rate J according to homogeneous 3D nucleation.

According to Eq. (2.33), the factors $\phi(m,r')$ and $\phi'(m,r')$ play different roles in different regimes of driving force. At low supersaturations $\ln S$, for example, when the nucleation barrier ΔG^*_{3D} is very high (see Eq. 2.20), the exponential term associated with the nucleation barrier is dominant over contribution from collisions of embryos due to the available effective surface area on the substrate, represented by the factors $\phi(m,r') = \phi'(m,r')$. Thus, nucleation rate is controlled by heterogeneous nucleation. However, at higher supersaturations, the preexponential term involving factors associated with effective collisions becomes dominant over the exponential term. Consequently, for any two sets of m and $\phi(m,r')$, there is always a critical value of supersaturation when heterogeneous nucleation occurring at low supersaturation undergoes transition to homogeneous nucleation. Obviously, for a given system, different foreign particles having distinct surface properties and/or different sizes control nucleation in different supersaturation regimes. Homogeneous nucleation occurs only when $\phi(m,r') = \phi'(m,r') = 1$, and this occurs always at very high supersaturations.

Impurities present in the medium can also lead to a decrease in the value of the kinetic parameter J_0 by physically blocking the existing active sites for the attachment of growth units. If θ_i is the fraction of blocked active sites of a nucleus, then the kinetic factor becomes $J_0(1-\theta_i)$. Assuming that the impurity adsorption on the active sites obeys the Langmuir adsorption isotherm, the relationship between heterogeneous nucleation rate J_{het} on impurity concentration c_i may be given by Eq. (2.33) with the preexponential factor $J_{0(het)} = J_0/(1 + K_L c_i)$, where K_L is the Langmuir constant.

NUCLEATION AND TRANSFORMATION OF METASTABLE PHASES

Practically, every solid is known to exist in different forms: amorphous phases, solvates, and polymorphs. For example, tripalmitoylglycerol (PPP), composed of three identical palmitic acid groups, shows three polymorphic forms: α, β', and β (Sato, 2001). They have different molecular structures characterized by the packing of their chains: disordered aliphatic chain conformation in the α polymorph, intermediate packing in the β' polymorph, and most dense packing in the β form. Therefore, the Gibbs free energy is the highest in the α form, intermediate in the β' form, and the lowest in the β polymorph. Other complex fats containing different fatty acid groups also exhibit a similar trend in their polymorphic properties, but in some cases, the β' form is more stable than the β form or two β forms are present (Arishima et al., 1991; Minato et al., 1996; Rousset et al., 1998; Rousset and Rappaz, 1996; Sato, 2001; Sato et al., 1989).

Apart from the difference in the molecular structures, polymorphic forms of a substance are also characterized by differences in their enthalpies of melting ΔH_m and melting points T_m (Arishima et al., 1991; Rousset and Rappaz, 1996; Sato et al., 1989). This behavior is illustrated in Fig. 2.5, which shows the dependence of ΔH_m on T_m for polymorphs of different fatty acid groups. There are two features of these polymorphs: (1) their melting points T_m and melting enthalpies ΔH_m increase with increasing stability of a polymorph and (2) the ratio $\Delta H_m/T_m$ increases with increasing T_m

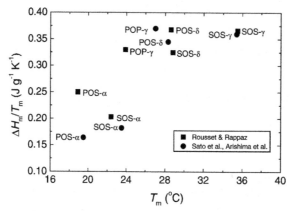

Figure 2.5 Relationship between $\Delta H_m/T_m$ and melting points T_m for the polymorphs of various fatty acids. *(Data from Sato et al. (1989); Arishima, T., N. Sagi, H. Mori, K. Sato., 1991. Polymorphism of POS: I. Occurrence of polymorphic transformation. J. Am. Oil Chem. Soc. 68, 710–715; Rousset, P., M. Rappaz., 1996. Crystallization kinetics of the pure triacylglycerols glycerol-1,3-dipalmitate-2-oleate, glycerol-1-palmitate-2-oleate-3-stearate, and glycerol-3,3-distearate-2-oleate. J. Am. Oil Chem. Soc. 73, 1051–1057.)*

(i.e., stability) of a polymorph. The increase in the ratio $\Delta H_m/T_m$ slowly decreases with increasing T_m of the polymorphs.

Under the same conditions, several phases can coexist, but, at constant volume and temperature conditions, the phase characterized by the lowest free energy $G = H - TS$ at a given temperature T is the stable one. The ability of formation of two forms, say α and β, from the melt may be explained in terms of the relative changes in their corresponding Gibbs free energy ΔG_α and ΔG_β. Fig. 2.6 shows the temperature dependence of the Gibbs free energy G for two phases α and β, crystallizing from, for example, the same melt or supersaturated solution. Fig. 2.6 is an extension of Fig. 2.2. The difference

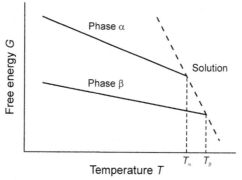

Figure 2.6 Dependence of Gibbs free energy G for two phases α and β, crystallizing from solution (or melt) on temperature T. *Dashed line* shows $G(T)$ dependence for the equilibrium concentration.

in the Gibbs free energy change for the formation of two polymorphs at a given temperature may be given by (cf. Eq. 2.3)

$$\Delta G_\beta - \Delta G_\alpha = \left(\frac{\Delta H_m}{T_m}\right)_\beta (T_{m\beta} - T) - \left(\frac{\Delta H_m}{T_m}\right)_\alpha (T_{m\alpha} - T)$$

$$= \left(\frac{\Delta H_m}{T_m}\right)_\beta \left\{ T_{m\beta}\left[1 - \delta\left(\frac{T_{m\alpha}}{T_{m\beta}}\right)\right] - T(1 - \delta)\right\}, \tag{2.36}$$

where the enthalpy of melting ΔH_m and the melting temperature T_m refer to a given polymorphic phase, the subscripts α and β refer to the values of $(\Delta H_m/T_m)$ and T_m of α and β forms, $(T_{m\alpha}/T_{m\beta}) < 1$, and $\delta = (\Delta H_m/T_m)_\alpha/(\Delta H_m/T_m)_\beta < 1$. Obviously, because the term in the curly brackets of Eq. (2.36) is always positive, $\Delta G_\beta - \Delta G_\alpha > 0$, implying that the formation of the stable β form is always favorable at $T < T_{m\beta}$.

At temperatures below T_β the higher stability of β phase does not mean that this should crystallize first. This may be explained by the dependence of nucleation rate J on the nucleation barrier ΔG_{3D}^* of each phase and the preexponential factor J_0 associated the kinetics of attachment of atoms/molecules to the developing embryos (see Eqs. 2.20 and 2.33). The higher stability of β phase than that of α phase implies that, at a particular temperature T, the interfacial tension γ of β phase is higher than that of α phase, and the concentration of atoms/molecules available for their attachment to developing nuclei is lower for β phase than that for α phase. The latter factor is easy to visualize in the case of supersaturated solutions. Both of these factors lead to an increase in the nucleation rate of α phase. Because the interfacial energy γ appears in the exponential terms of Eqs. (2.20) and (2.33), its effect is much pronounced at low supersaturations $\ln S$ when the exponential term dominates over the preexponential term. This explains why during the crystallization of various substances one frequently observes that the newly formed phase in the supersaturated mother phase is a metastable one rather than the thermodynamically most stable one.

After its formation, the metastable α phase undergoes transformation into the stable β phase because the system tends to attain the minimum free energy state by transformation of the metastable phase into the thermodynamically stable phase (Overall Crystallization section). The transformation occurs both in the solid state and in the solution. The former occurs when the two phases nucleate in the solid state, but in the solution–mediated transformation, the kinetics depend on the dissolution rates of the metastable phase and on the nucleation and growth rates of the stable phase. The process of transformation occurs by nucleation and growth of the thermodynamically stable crystallites in the metastable phase rather than in the old mother phase. This means that when direct formation of the stable phase is not possible, the overall crystallization becomes a two-stage process in which the formation of the first stage precedes the appearance of second stage. This process is described by the Ostwald step rule, which is also called the Ostwald rule of stages.

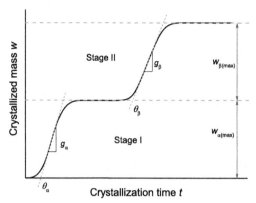

Figure 2.7 Schematic illustration of two-step model of crystallization in the bulk. Meanings of different symbols are defined. See text for details.

Fig. 2.7 illustrates schematically the two-stage process of crystallization of a stable form occurring in the bulk melt or supersaturated phase. The first stage is associated with the crystallization of the metastable α polymorph, whereas the second stage involves the transformation of the α polymorph into the stable β polymorph. Then the total crystallized mass

$$W_{\text{total}} = W_\alpha + W_\beta = w_{\alpha(\max)}\gamma_\alpha + w_{\beta(\max)}\gamma_\beta, \tag{2.37}$$

where W_α and W_β are the amounts (% solid or released heat as J/g) of metastable α and stable β forms, respectively, $\gamma_{\alpha,\beta}$ is the fraction the crystallized mass of the crystallized α or β form, and $w_{\alpha,\beta(\max)}$ is a parameter related to their maximum masses $w_0(\alpha,\beta)$ of the two forms (cf. Eq. 2.39). When the crystallized fraction γ_α of the metastable form is negligible, one observes only the second stage in the crystallization experiment. Physically, this situation is possible when the crystallizing unstable α phase transforms simultaneously into a more stable β phase. The crystallization behavior in this one-step mechanism is given by

$$W_{\text{total}} = w_{\beta(\max)}\gamma_\beta. \tag{2.38}$$

The quantities γ and w_0 of a crystallizing phase are related by Eq. (2.39). The quantity γ in Eqs. (2.37) and (2.38) is the fraction of crystalline phase α or β in a closed system, whereas w_0 is the maximum amount of the crystallized mass of a phase.

It should be mentioned that crystallization of fats is a complex phenomenon. It depends on crystallization temperature T_c, crystallization kinetics of different polymorphs, presence of impurities in the melt or supersaturated solution, and the concentration of impurity. An increase in crystallization temperature suppresses the crystallization of metastable phases.

OVERALL CRYSTALLIZATION

Continued nucleation and growth of crystallites of a material from its supersaturated solution or supercooled melt in a closed system leads to the crystallization of the entire mass of the material until the system reaches the equilibrium state. This type of crystallization is known as overall crystallization of a phase in a closed system. In terms of mass $w(t)$ crystallized in volume V at time t, the fraction y of crystalline phase in a closed system may be defined as

$$y = w(t)/w_0, \tag{2.39}$$

where w_0 is the maximum mass of the crystallizing phase.

Theoretical Interpretation of Isothermal Crystallization Kinetics

In the literature on the isothermal crystallization of fats, several models have been used. In this section, the basic equations of these approaches are presented, and their merits and demerits are pointed out.

Kolmogorov–Johnson–Mehl–Avrami Theory

Overall crystallization of a phase may occur by mononuclear (MN) or polynuclear (PN) mechanism involving the formation of one nucleus or many nuclei in the volume, respectively, and their subsequent growth. There are two ways of the formation of many nuclei in the volume. In the first case, the nuclei may be formed in the system at the initial moment $t = 0$ and thereafter they grow irreversibly until the completion of crystallization in the entire volume. In the second case, the nuclei form continuously during the crystallization process. These two processes are known as instantaneous and progressive nucleation mechanisms, respectively. However, during progressive nucleation the nucleation rate can be time-independent (stationary nucleation) as well as time-dependent (nonstationary nucleation).

Irrespective of whether nucleation is homogeneous or heterogeneous, nucleation always occurs on active centers. The total number N_0 of nuclei in the system cannot exceed the total number N_a of active centers present in it. The active centers thus determine both the maximum number of nuclei formed in the system and the duration of nucleation. During the nucleation process, the number of active centers is exhausted progressively during crystallization. Therefore, the time dependence of the kinetics of nucleation on active centers determines the overall crystallization kinetics. The theoretical expressions for MN or PN mechanism and for instantaneous or progressive nucleation are different from each other.

When nucleation occurs by progressive polynuclear mechanism, the crystallized mass during overall crystallization may be given by (Kashchiev, 2000)

$$w(t) = w_0\{1 - \exp[-(t/\Theta)^q]\},\tag{2.40}$$

where w_0 is the total mass for crystallization, the time constant

$$\Theta = \left(q\big/\kappa\, g_k^{q-1} J\right)^{1/q},\tag{2.41}$$

and the exponent

$$q = 1 + \nu d.\tag{2.42}$$

In the above equations, J is the rate of stationary nucleation, κ is the shape factor for nuclei (e.g., $4\pi/3$ for spherical nuclei); the growth constant g is defined as

$$r(t) = (gt)^\nu,\tag{2.43}$$

where ν is a number equal to $1/2$ and 1 for growth controlled by volume diffusion and mass transfer, respectively, and d is the dimensionality of growth. For the growth of needles, plates, and polyhedra, d is 1, 2, and 3, respectively. The above treatment is usually referred to as the Kolmogorov—Johnson—Mehl—Avrami (KJMA) theory. However, in the literature on kinetics of crystallization of fats and lipids, it is customarily called Johnson—Mehl—Avrami—Kolmogorov model.

In the case of instantaneous nucleation, the overall crystallization may also be given by Eq. (2.40) with the constant $q = \nu d$ (Kashchiev, 2000). Then the time constant Θ has a value different from that for progressive nucleation mechanism (see Eq. 2.41).

The nucleation and growth of the stable (β) phase within the metastable (α) phase makes the kinetics of formation of the former depend on those of the latter and may be described by (Kashchiev, 2000)

$$\gamma_\beta(t) = \gamma_\alpha(t)\left\{1 - \exp\left[-\left(\Theta_\alpha/\Theta_\beta\right)^\beta\right]\left[(t/\Theta_\alpha) - b'\right]^\beta\right\},\tag{2.44}$$

or, alternatively, in the form

$$\gamma_\beta(t) = \gamma_\alpha(t)\left(1 - \exp\left\{-\left[(t - \Theta_\alpha b')/\Theta_\beta\right]^\beta\right\}\right),\tag{2.45}$$

where $b' = (\alpha - 1/\alpha)^{1/\alpha}$, $\gamma_\alpha(t)$ is given by Eq. (2.40), α and β denote the value of q for the crystallization of metastable and stable form, respectively, whereas Θ_α and Θ_β are the corresponding time constants. We recall that $q = \nu d$ and $1 + \nu d$ for instantaneous and progressive nucleation mechanisms, respectively. Obviously, Eq. (2.45) is of the same form as Eq. (2.40), but crystallization of the stable β form occurs after a time period $\Theta_\alpha b'$ and a time constant different from Θ_α. In fact, when the quantity $\Theta_\alpha b' = 0$, Eq. (2.45) reduces to Eq. (2.40). However, as the ratio $\Theta_\alpha/\Theta_\beta > 0$ and the constant $b' > 0$, the dependence predicted by Eq. (2.44) or (2.45) differs significantly from that of Eq. (2.40).

It should be mentioned (Sangwal, 2007) that the data for the overall crystallization kinetics of involving transformation of a metastable α into a stable β form can be described both by Eqs. (2.40) and (2.45) with the exponents $q = \alpha = \beta = 4$. However, Eq. (2.40) describes the experimental data much better with $q > 4$.

The reproducibility of the experimental data by the best fit according to Eqs. (2.40) and (2.45) implies that

$$\left[(\Theta_\alpha/\Theta_\beta)^\beta\right]\left[(t/\Theta_\alpha) - b'\right]^\beta = \left[(t - \Theta_\alpha b')/\Theta_\beta\right]^\beta = (t/\Theta_\beta)^q, \tag{2.46}$$

where $\beta = 4$ and $q > 4$. From this equality, one finds

$$b' = (t/\Theta_\alpha)\left[1 - (t/\Theta_\beta)\right]^{(q-\beta)/\beta}, \tag{2.47}$$

where $q > 4$ and $\beta = 4$.

Eq. (2.45) describes the direct crystallization of a stable phase from melt or supersaturated solution as a two-stage process. It is based on the concept that the entire metastable form is simultaneously transformed into the stable form such that its presence is negligible in the crystallized mass. However, the KJMA theory can equally be extended to account for the simultaneous presence of both unstable and stable forms during isothermal crystallization. Then the total crystallized mass may be presented in two reaction stages in the form (cf. Eq. 2.37)

$$M_{\text{total}} = w_{\alpha(\max)}(1 - \exp\{-[(t/\Theta_\alpha)^\alpha]\}) + w_{\beta(\max)}$$
$$\left(1 - \exp\left\{-[(t - \Theta_\alpha b')/\Theta_\beta]^\beta\right\}\right), \tag{2.48}$$

where $w_{\alpha,\beta(\max)}$ is the maximum mass of the crystallized α or β form, observed as % solid or released heat as J/g.

In the above extended KJMA equation, the second time lag $\Theta_\alpha b'$ term is related to the time constant Θ_α and the exponent α of the crystallization of the metastable form α. Therefore, it is not an independent parameter, and Eq. (2.48), in fact, contains six parameters, which determine the shape of the two-step crystallization curve. The parameters $w_{\alpha(\max)}$ and $w_{\beta(\max)}$ determine the heights of the plateau of the first and the second steps, respectively. The higher the values of $w_{\alpha(\max)}$ and $w_{\beta(\max)}$, the more the heights of the plateau of the first and the second steps. The time constant $\Theta_{\alpha,\beta}$ determines the moment when the curve abruptly departs from the time axis. The higher the value of $\Theta_{\alpha,\beta}$, the more the delay when the curve leaves the time axis. The exponents α and β determine the slope of the straight-line part of the curve and its curvatures. The higher the values of these exponents, the higher the steepness of the crystallization curves.

Some Comments on Different Equations of the Kolmogorov–Johnson–Mehl–Avrami Theory

Eq. (2.40) of the KJMA theory contains two parameters, the time constant Θ and the exponent q, which describe isothermal crystallization in terms of nucleation and growth processes. However, in the literature on crystallization of fats, Eq. (2.40) is usually presented in the following forms:

$$w(t) = w_0[1 - \exp(- kt^q)], \tag{2.49}$$

$$w(t) = w_0\{1 - \exp[- (k't)^q]\}, \tag{2.50}$$

where the constants $k = \Theta^{-1/q}$ and $k' = \Theta^{-1}$. Eqs. (2.49) and (2.50) are generally known as the Avrami (or approximate Avrami) and the modified Avrami equations, respectively (Foubert et al., 2003; Padar et al., 2009). The difference between the two equations lies in the values of k and k'. In the former case, the constant k is a function of q; but in the latter case, k' is apparently independent of q. In fact, the modified Avrami equation is a reparameterized form of the original Avrami equation such that $k' = k^{1/q}$.

The derivation of Eq. (2.40) of the KJMA theory described above involves two basic parameters of crystallization (i.e., stationary nucleation rate J and growth rate g) and volume V_c of the crystallized mass per unit volume of the total mass V (Kashchiev, 2000), according to the relation

$$y = V_c(t)/V = 1 - \exp[- V_{ex}(t)/V], \tag{2.51}$$

where $V_{ex}(t)$ is the so-called extended volume, which is the total crystalline volume that would have been formed in the melt or supersaturated solution till time t, provided that the initial melt or saturated solution volume V is not exhausted by the growing crystallites and the crystallites do not contact with each other. Therefore, the problem of determining the crystallization kinetics is essentially to find V_{ex} for the particular case of a crystal shape. Initially, $V_c \approx V_{ex}$, but with increasing crystallization time $V_c < V_{ex}$ (Kashchiev, 2000). Consequently, V_{ex} is a function of time, but its value depends on the form of this dependence on time. This issue has been discussed by various authors (for example, see Foubert et al., 2003; Kashchiev, 2000; Padar et al., 2009), who derive the "precise" Avrami equation with the dependence of extended volume V_{ex} on the exact form of a dimensionless time. Finally, to define the solid fraction y, one replaces the volume V_c of the crystallized solid and V of the melt or supersaturated solution by masses $w(t)$ and w_0, respectively.

The above treatment predicts that the exponent q in Eq. (2.40) is an integer, but analysis of isothermal crystallization data often reveals that q is a noninteger. This noninteger value of q can result when (1) the dimensionality d of crystallite does not remain constant during crystallization and (2) growth of crystallites is controlled by diffusion of atoms/molecules in the volume.

The merits and demerits of the modified Avrami equation (2.50) have been discussed by various authors (Foubert et al., 2006; Khanna and Taylor, 1988; Marangoni, 1998; Padar et al., 2009). As pointed out by Marangoni (1998), the above reparameterization of the original Avrami equation is arbitrary and has no theoretical justification. This is due to the fact that the time constant Θ, which is inverse of the new parameter k', is a complicated function of nucleation rate J, growth constant g, and exponent q (see Eq. 2.42). Physically, the independence of k' on q is possible only in a narrow range of crystallization temperature T_c.

Finally, it should be noted that formally the traditional Avrami equation (2.49) can also be extended to describe two- or multistage processes. In fact, Narine et al. (2006) extended Avrami equation (2.49) to model the crystallization kinetics of lipids. In their extended Avrami equation, they used Avrami constants k_1, k_2, etc., and delay times τ_1, τ_2, etc. The physical meanings of constants k's and delay times τ's follow from Eq. (2.48), where, for example, for crystallization of the β phase in the second stage $k_2 = \Theta_\beta^{-\beta}$ and $\tau_2 = \Theta_\alpha b'$.

Model of Mazzanti, Marangoni, and Idziak

This model was advanced by Mazzanti et al. (2005) and later modified by them (2008). They called the model as the ordinary diffusion equation model. It deals with the modeling of individual contributions of unstable and stable forms of a crystallizing material. The model is based on the following postulates: (1) there are regions of liquid fractions A^* and B^* from which unstable α and stable β forms can be crystallized directly, but crystallization of the stable β phase can also occur from the region of fraction A^* by the transformation of the unstable α phase, (2) the transformation of the metastable α form into the stable β form is controlled by bulk diffusion, (3) the crystallization kinetics of both metastable and stable forms can be described by Eq. (2.42) of the KJMA theory, and (4) supersaturation σ available for crystallization is defined as the ratio of the uncrystallized mass in the liquid phase (melt or supersaturated solution) to the mass of the material that could crystallize. If w is the mass of a crystallized solid and w_0 is the value of w at equilibrium, the supersaturation

$$\sigma \approx (w_0 - w)/w_0 = 1 - w/w_0. \tag{2.52}$$

Denoting the untransformed material in the liquid phase by $L = w_0 - w$ and assuming that KJMA theory applies, the untransformed mass can be described by the differential form of the Avrami equation (2.49) in the form

$$\frac{\partial L}{\partial t} = -qkL\left[-\ln\left(\frac{L}{m_0}\right)\right]^{(q-1)/q}, \tag{2.53}$$

where k is the Avrami constant related to the time constant Θ ($k = \Theta^{-1/q}$; see Eqs. 2.49 and 2.50) and q is the growth mode exponent (see Eq. 2.42). Thus, the rate of phase transition is a function of the mass of the material left in the phase undergoing transformation. This phase can be the melt of the crystallizing material itself or its crystallized unstable α form.

During crystallization at a given temperature, if the material fraction w_0 that can crystallize into unstable α form is A^* and the additional liquid fraction B^* that can crystallize into stable β form, the total amount of the liquid fraction that can crystallize into the stable β form is $A^* + B^*$. As the unstable α phase grows, the material available from the region A^* becomes depleted, leaving behind an uncrystallized fraction A. This means that the amount of crystallized α form is $w_\alpha = A^* - A$. In the case of growth of the stable β phase for which the crystallized material is from both of the regions A^* and B^*, the uncrystallized portions of both A and B are reduced. The fractions of the stable β phase from A and B regions are $A/(A + B)$ and $B/(A + B)$, respectively. Following Mazzanti et al. (2008), the above concepts are used to obtain the equations given below.

The consumption of the liquid fraction A, as it is converted into unstable α phase and then stable β phase, after the occurrence of its formation at time $t_{\alpha\beta}$, is given by

$$\frac{\partial A}{\partial t} = -q_\alpha k_\alpha A \left[-\ln\left(\frac{A}{A^*}\right) \right]^{(q_\alpha - 1)/q_\alpha}$$
$$- \left\{ q_\beta k_\beta (A + B) \left[-\ln\left(\frac{A + B}{A^* + B^*}\right) \right]^{(q_\beta - 1)/q_\beta} \frac{A}{A + B} \right\}_{t > t_{\alpha\beta}}, \qquad (2.54)$$

where q_α and k_α are the Avrami constants describing the transformation of A to α form, whereas q_β and k_β refer to the constants describing that the β form is formed from the liquid $(A + B)$. The formation of unstable α phase from the liquid fraction A and its consumption due to transformation into β form is given by

$$\frac{\partial w_\alpha}{\partial t} = -q_\alpha k_\alpha A \left[-\ln\left(\frac{A}{A^*}\right) \right]^{(q_\alpha - 1)/q_\alpha}$$
$$- \left\{ q_c k_c w_\alpha \left[-\ln\left(\frac{w_\alpha}{w_\alpha(\max) - w_\alpha(\min)}\right) \right]^{(q_c - 1)/q_c} \right\}_{t > t_{\alpha\beta}}, \qquad (2.55)$$

where the constants q_c and k_c refer to the solid–solid transformation of α form into β form. The crystallization of the liquid fraction B into β phase is given by

$$\frac{\partial B}{\partial t} = -\left\{ q_\beta k_\beta (A + B) \left[-\ln\left(\frac{A + B}{A^* + B^*}\right) \right]^{(q_\beta - 1)/q_\beta} \frac{B}{A + B} \right\}_{t > t_{\alpha\beta}}. \qquad (2.56)$$

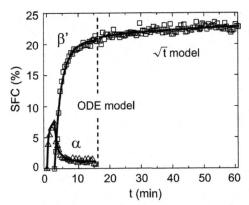

Figure 2.8 Plots of solid fat content (SFC) of α and β' polymorphs of anhydrous milk fat crystallized at 17°C and 90 s^{-1} shear as a function of time. Solid lines represent the fit obtained by using the model. Dashed line denotes the time $t_{\alpha\beta}$ when the growth of the β phase becomes diffusion controlled. *(Reproduced from Mazzanti, G., Marangoni, A.G., Idziak, S.H.J. Eur. Phys. J. E 27, 135–141. Copyright (2008), with permission from Springer.)*

The amount of β phase at time t can be computed from the mass balance equation

$$w_\beta = (A^* + B^*) - (A + B) - w_\alpha, \tag{2.57}$$

with the initial conditions: $A(t = 0) = A^*$, $B(t = 0) = B^*$, and $w_\alpha(t = 0) = 0$. The value of $w_\alpha(\text{min})$ is obtained from the procedure, such as integrated X-ray intensity profile from small-angle 2D X-ray diffraction pattern of the material, differential scanning calorimetry, or pulse nuclear magnetic resonance, followed to register the crystallized material content.

Using the above system of equations, their parameters are obtained by fitting the experimental time dependence of crystallization of unstable and stable phases. Fig. 2.8 shows plots of solid fat content of anhydrous milk fat crystallized at 17°C and 90 s^{-1}, shear as a function of time t. The lines represent the fit obtained by using the model. The dashed line denotes the time $t_{\alpha\beta}$ when the growth of the β phase becomes diffusion controlled, described by the relation

$$\frac{\partial w_\beta}{\partial t} = \frac{k_\beta}{2(t - t_{\alpha\beta})^{1/2}}, \tag{2.58}$$

where k_β is a kinetic constant (in min$^{-0.5}$), which is a function of temperature.

Gompertz's Model

Kloek et al. (2000) were the first to use Gompertz's empirical equation to describe the experimentally observed curves of isothermal fat crystallization. They were subsequently followed by Vanhoute et al. (2002) and Foubert et al. (2003). Following Foubert et al. (2003), we describe below the essential features of Gompertz's model.

In its original form Gompertz's equation is

$$Y = Y_0 \exp[-\exp(Y_1 - Y_2 t)], \tag{2.59}$$

where Y is the logarithm of the relative population size, and Y_0, Y_1, and Y_2 are empirical constants. To give specific meaning to the empirical parameters, they should be defined in terms of parameter w_{max} (the maximum value of Y), the growth rate g (tangent at the inflection point), and time lag θ (time axis intercept of the tangent), as illustrated schematically in Fig. 2.7. To find the inflection point at $t = t_{infl}$, one sets the second derivative of Eq. (2.59) to zero, which gives

$$t_{infl} = Y_1/Y_2. \tag{2.60}$$

The first derivative at t_{infl} gives the growth rate

$$G = (Y_0 - Y_2)/e, \tag{2.61}$$

where $e = 2.718$. The time lag θ may be calculated from the tangent line through the inflection point, i.e.,

$$\theta = (Y_1 - 1)/Y_2. \tag{2.62}$$

From Eqs. (2.61) and (2.62), one finds

$$Y_1 = \frac{ge}{Y_0}\theta + 1. \tag{2.63}$$

When $t \to \infty$, $Y \to Y_0$, and $Y_0 \to w_{max}$, then for the crystallization of stable β form from melt or supersaturated solution, Gompertz's empirical equation (2.59) may be written as

$$W_{total} = w_{\beta(max)} \exp\left\{ -\exp\left[\frac{2.718 g_\beta}{w_{\beta(max)}}(\theta_\beta - t) + 1\right]\right\}, \tag{2.64}$$

where $W_{total} (= Y)$ is the amount of the crystallized stable β form at time t, $w_{\beta(max)}$ is the maximum mass of the β form, and θ_β and g_β are, respectively, the time lag and growth rate of the stable β form. When crystallization of the stable β form is accompanied by simultaneous crystallization of metastable α form, the crystallization kinetics may be given by

$$W_{total} = w_{\alpha(max)} \exp\left\{ -\exp\left(\frac{2.718 g_\alpha}{w_{\alpha(max)}}(\theta_\alpha - t) + 1\right)\right\}$$
$$+ w_{\beta(max)} \exp\left\{ -\exp\left[\frac{2.718 g_\beta}{w_{\beta(max)}}(\theta_\beta - t) + 1\right]\right\}, \tag{2.65}$$

where θ_α and g_α are, respectively, the time lag and growth rate of the metastable α form, and $w_{\alpha(\mathrm{max})}$ is the maximum mass of the α form.

As in the case of KJMA theory, Eq. (2.65) contains six parameters. The parameters $w_{\alpha(\mathrm{max})}$ and $w_{\beta(\mathrm{max})}$ determine the heights of the plateau of the first and the second steps, respectively, whereas the time constants θ_α and θ_β determine the moment when the curve abruptly departs from the time axis. The growth rates g_α and g_β determine the slopes of the straight-line parts of the curve.

Model of Foubert, Dewettinck, Jansen, and Vanrolleghem

This approach was proposed by Foubert et al. (2006) and is an extension of their previous model (Foubert et al., 2003). It is based on consideration of (1) uncrystallized phase left in the crystallizing medium and (2) transformation of crystallized metastable phase α into stable phase β. The relative crystallizable phase $h(t)$ remaining in the medium is given by

$$h(t) = [w_{\mathrm{max}} - w(t)]/w_{\mathrm{max}} = 1 - \gamma(t), \qquad (2.66)$$

where w_{max} (% solid phase) is the maximum amount of crystallizing phase (i.e., solid phase) from that of the initial amount contained in the melt or supersaturated solution (see Fig. 2.7). Assuming that crystallization of solid phase is a first-order forward reaction and a reverse reaction of order n, the rate of relative crystallizable phase may be given by

$$\frac{dh(t)}{dt} = K(h^n - h), \qquad (2.67)$$

with

$$h(0) = [w_{\mathrm{max}} - w(0)]/w_{\mathrm{max}}, \qquad (2.68)$$

where K is the rate constant (defined as inverse of time t) and $w(0)$ is the amount of crystalline mass initially present in the melt or supersaturated solution. In view of difficulty in measuring $h(0)$ experimentally from the plots of crystallized mass x as a function of time t, Eq. (2.67) yields the relation

$$h(t) = \left\{1 + \left[(1-x)^{1-n} - 1\right]\exp\left[-(1-n)K(t - t_{\mathrm{in}}(x))\right]\right\}^{1/1-n}, \qquad (2.69)$$

where $t_{\mathrm{in}}(x)$ is the induction period required for the crystallization of x. According to Eq. (2.69), the induction period $t_{\mathrm{in}}(x)$ is related to the reference crystallized mass x by

$$t_{\mathrm{in}}(x) = \frac{1}{(1-n)K}\left(-\ln\left\{\frac{(1-x)^{1-n} - 1}{[1 - w(0)/w_{\mathrm{max}}]^{1-n} - 1}\right\}\right). \qquad (2.70)$$

Using Eqs. (2.66) and (2.67), the absolute amount $w(t)$ of crystallization at time t may be given by

$$\frac{dw(t)}{dt} = K(w_{max} - w) - Kw_{max}\left(\frac{w_{max} - w}{w_{max}}\right)^n. \tag{2.71}$$

Eq. (2.71) is the starting equation of the two-step model involving the transformation of metastable α phase into the stable β phase such that one α crystal transforms into one β crystal.

If g_α is the rate of formation of metastable α phase from the supersaturated solution or supercooled melt and g_β is the rate of transformation of the α phase into β phase, one may write

$$\frac{dy_\alpha(t)}{dt} = g_\alpha - g_\beta, \tag{2.72}$$

$$\frac{dy_\beta(t)}{dt} = g_\beta. \tag{2.73}$$

The rate g_β of formation of stable β phase may be given by

$$g_\beta = K_\beta(1 - y_\beta) - 1 \times K_\beta\left(\frac{1 - y_\beta}{1}\right)^{n_\beta}, \tag{2.74}$$

where K_β is the rate constant of β phase and n_β is the order of the reverse reaction. Similarly, the rate g_α of formation of α phase may be given by

$$g_\alpha = K_\alpha[1 - (y_\alpha + y_\beta)] - 1 \times K_\alpha\left[\frac{1 - (y_\alpha + y_\beta)}{1}\right]^{n_\alpha}, \tag{2.75}$$

where K_α is the rate constant of α phase and n_α is the order of the reverse reaction. The initial values of $y_\alpha(0)$ and $y_\beta(0)$ at $t = 0$ may be calculated from Eq. (2.75), i.e.,

$$y_\alpha(0) = 1 - 1 \times 1 + \left[\frac{(1 - x)^{1 - n_\alpha} - 1}{\exp(n_\alpha - 1)K_\alpha t_{in(\alpha)}}\right]^{1/(1 - n_\alpha)}, \tag{2.76}$$

$$y_\beta(0) = 1 - 1 \times 1 + \left[\frac{(1 - x)^{1 - n_\beta} - 1}{\exp(n_\beta - 1)K_\beta t_{in(\beta)}}\right]^{1/(1 - n_\beta)}, \tag{2.77}$$

where $t_{in(\alpha)}$ and $t_{in(\beta)}$ are the periods needed to reach x (taken arbitrarily as 1%) of crystallization of α and β phases, respectively.

To account for the fact that crystallization fractions y_α and y_β lead to different maximum values, the expressions for y_α and y_β are multiplied by the corresponding maximum values of $w_{\alpha(\max)}$ and $w_{\beta\,(\max)}$, observed as % solid or released heat as J/g. Then

$$W_\alpha = w_{\alpha(\max)} y_\alpha, \tag{2.78}$$

$$W_\beta = w_{\beta(\max)} y_\beta, \tag{2.79}$$

where W_α and W_β are the amounts of crystallized α and β phases, respectively. The total amount W_{total} of crystallization is then given as the sum of W_α and W_β, i.e., $W_{\text{total}} = W_\alpha + W_\beta$; see Eq. (2.38). The experimental data on W_{total} and W_β are fitted simultaneously.

The fitting procedure involves eight parameters. Foubert et al. (2006) examined the influence of these parameters on the overall crystallization curves. It was shown that the parameters $w_{\alpha(\max)}$ and $w_{\beta(\max)}$ determine the heights of the plateau of the first and the second steps, respectively. The time constants $t_{\text{in}(\alpha)}$ and $t_{\text{in}(\beta)}$ determine the moment when the curve abruptly departs from the time axis. The higher the value of $t_{\text{in}(\alpha)}$ and $t_{\text{in}(\beta)}$, the more the delay when the curve leaves the time axis. However, the rate constants K_α and K_β as well as n_α and n_β determine the slopes of the straight-line parts of the curve and their curvatures. The higher the values of these parameters, the higher the steepness of the crystallization curves.

Comparison of Different Models of Isothermal Crystallization Kinetics

As mentioned in Kolmogorov–Johnson–Mehl–Avrami Theory section, the KJMA theory provides information on the dimensionality d of growing crystallites as well as on the nucleation and growth processes. Therefore, it is not surprising that this theory, as presented by Eq. (2.49), has been most widely used in the isothermal crystallization kinetics of fats. The main advantage of this theory is that all of its parameters have well-defined physical significance. Gompertz's empirical equation, on the other hand, contains empirically fitting parameters, which have no physical significance. In the Foubert model, although the physical meaning can be assigned to practically all of the parameters, yet the weakest point is the arbitrary choice of the induction period $t_{\text{in}}(x)$ required for the crystallization of $x = 1\%$.

Two typical examples of the influence of temperature and time on the overall crystallization of different materials from melts are illustrated in Figs. 2.9 and 2.10. Figs. 2.9 and 2.10 show the experimental data obtained at different temperatures T_c on isothermal crystallization of polypropylene and cocoa butter (CB), respectively. The data on polypropylene were reported by Lopez-Manchado et al. (2000) and later analyzed by Padar et al. (2009) using the "precise" Avrami equation, with dimensionality $d = 3$, based on Eq. (2.51) and approximate Avrami equation (2.49). The crystallization kinetics of polypropylene was investigated by differential scanning calorimetry. The data on CB were

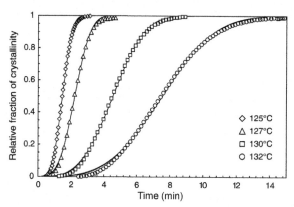

Figure 2.9 Dependence of overall crystallization fraction y of polypropylene on time t at different temperature T_c. The data, shown by symbols, are from Lopez-Manchado et al. (2000). Solid curves are drawn by using Avrami's "precise" equation, whereas dotted curves are drawn by using Eq. (2.49). *(Adapted from Padar, S., Jeelani, S.A.K., Windhab, E.J., 2008. J. Am. Oil Chem. Soc. 85, 1115–1126. Copyright (2008), with permission from Springer.)*

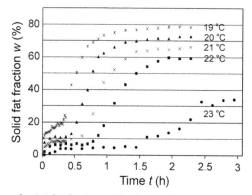

Figure 2.10 Dependence of solid fat fraction (%) of cocoa butter on time t at different crystallizing temperature T_c. *(Adapted from Foubert, I., Vanrolleghem, P.A., Dewettinck, K. Eur. J. Lipid Sci. Technol. 107, 660–672. Copyright (2005), with permission from Wiley.)*

reported by Foubert et al. (2005) and were obtained by pulsed nuclear magnetic resonance method.

The following general features of the overall isothermal crystallization may be summarized from Fig. 2.9:

1. Crystallization of polypropylene is a single-stage process.
2. During the single-stage crystallization of polypropylene, there is a small time lag, or induction period $t_{in(\beta)}$, when detectable fraction of the crystallizing phase is formed. This time lag increases with an increase in crystallizing temperature T_c.
3. The slope g_β of the linear parts of the crystallization curves of polypropylene involving single stage decreases with an increase in crystallization temperature T_c.

4. Both the "precise" and the "approximate" Avrami equations describe the experimental data on polypropylene reasonably well, and there is practically insignificant difference between the best fit of the experimental data by the two equations. The best-fit curves by the two equations are indistinguishable in the crystallized fraction between about 0.2 and 0.8, and the deviation between the experimental data and the best-fit curves increases with increasing T_c.

In contrast to the above case, from Fig. 2.10 the following features may be noted:

1. Crystallization of CB occurs in two stages. During this two-stage crystallization, there is practically no induction period $t_{in(\alpha)}$ for the onset of crystallization of the metastable α phase, but the induction period $t_{in(\beta)}$ for the crystallization of the stable β phase, as defined by extrapolating the linear parts of the curves to the time axis, increases with increasing T_c.

2. The slopes g_α and g_β of the linear parts of the crystallization curves in the region of the crystallization of the metastable and stable forms decrease with an increase in T_c.

3. Between the two linear parts defining g_α and g_β, there is a plateau region along the time axis. The width of this plateau region increases with an increase in T_c.

Most of the above features may be explained by the models described above, but features (2) and (4) concerning crystallization of polypropylene and feature (3) concerning crystallization of CB deserve particular attention. To interpret the dependence of the width of the plateau region on crystallization temperature T_c, we define the plateau region as $\Delta t_{in} = [t_{in(\beta)} - t_{f(\alpha)}]$, where $t_{f(\alpha)}$ is the value of the finish time up to $w_{\alpha(max)}$ of α form and $t_{in(\beta)}$ is the time when β form appears in the overall crystallization curve (see Fig. 2.7). Then from Fig. 2.10, we consider the trajectories of $t_{f(\alpha)}$ and $t_{in(\beta)}$ obtained by joining their values corresponding to various T_c. If the solid fraction scale is represented by crystallization temperature T_c, it is easy to recognize that this procedure yields two linear plots of temperature T_c against time t such that there are two extreme temperatures T_0 and T_{max} when the plateau width Δt_{in} is zero and maximum, respectively, as shown in Fig. 2.11. Here T_0 is the crystallization temperature when plateau is not observed. Denoting the time corresponding to T_0 by t_0, from Fig. 2.11, one obtains

$$\frac{T_c - T_0}{t_{in(\beta)} - t_0} = -b_\beta, \qquad \frac{T_c - T_0}{t_{f(\alpha)} - t_0} = -b_\alpha, \qquad (2.80)$$

where b_α and b_β are the slopes of $T(t)$ plots corresponding to the end and beginning of crystallization of unstable α and stable β forms, respectively, $T_c < T_0$ and $b_\beta < b_\alpha$. From these relations, one finds the relationship between plateau width Δt_{in} and crystallization temperature T_c:

$$\Delta t_{in} = t_{in(\beta)} - t_{f(\alpha)} = (T_c - T_0)\left(\frac{1}{b_\beta} - \frac{1}{b_\alpha}\right). \qquad (2.81)$$

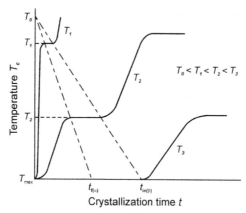

Figure 2.11 Schematic illustration of dependence of crystallization temperature T_c on $t_{f(\alpha)}$ and $t_{in(\beta)}$. It is assumed that T_c decreases linearly with increasing both $t_{f(\alpha)}$ and $t_{in(\beta)}$ with slopes b_α and b_β, respectively. See text for details.

According to Eq. (2.81), Δt_{in} increases with an increase in T_c, as observed in the case of CB (Fig. 2.10). When $b_\alpha \rightarrow \infty$, crystallization of β phase occurs in single stage. Crystallization of polypropylene is an example of this special case (Fig. 2.9). However, instead of intermediate plateau regions in the time dependence of crystallization curves, accelerated transformation regions have also been observed (see Padar et al., 2009). The above explanation can equally be applied to such cases.

Now we consider feature (4) of crystallization of polypropylene. For this purpose, we reexamine, as an example, the experimental data on the crystallization kinetics of CB (Fig. 2.12) followed by differential scanning calorimetry (Foubert et al., 2003).

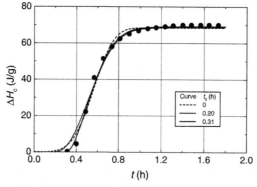

Figure 2.12 Example of fit of Avrami Eq. (2.40) to the crystallization kinetics of cocoa butter with corrections t_0 in time. Dashed curve is for $t = 0$, but dotted and continuous curves are drawn with $t_0 = 0.20$ and 0.31 h. *(Original data from Foubert, I., Dewittinck, K., Vanrolleghem, P.A., 2003. Modelling of the crystallization kinetics of fats, Trends Food Sci. Technol. 14, 79–92.)*

Foubert et al. (2003) compared the quality of fit of Avrami equation (2.49), Gompertz equation (2.65), and Foubert equations (2.76) and (2.77) for the above data. These authors concluded that the Gompertz equation provides a better fit than the Avrami equation, but Foubert equation gives the best fit. However, it may be noted that when one makes an allowance for the time t_0 needed by a system to attain the crystallization temperature T_c, the fit of Avrami equation for the data of crystallization kinetics improves remarkably. Fig. 2.12 illustrates an example of this type of fit with corrected time for the data on CB analyzed by Foubert et al. (2003). The dashed curve represents the crystallized mass, measured as crystallization heat ΔH_c, as a function of recorded time t. This curve shows the original fit reported by Foubert et al. (2003), with $t_0 = 0$. The dotted and continuous curves are drawn for the fit with the rescaled time $t^* = t - t_0$, where $t_0 = 0.20$ and 0.31 h, respectively, are the corrections used. Obviously, the fit improves on the introduction of the correction time t_0 and the best fit in the entire time interval is obtained when $t_0 = 0.31$ h in the case of the analyzed data.

Measurement of the time period t_{in} required for the detection of crystallization by different technique is based on the cooling of a saturated medium to a predefined crystallization temperature T_c as rapidly as possible, followed by continuous monitoring of the crystallization process. However, it is well known (Wright et al., 2000) that the detection of the initial stages of crystallization depends on the sensitivity of experimental techniques and the rapidness with which the preset isothermal crystallization temperature T_c of the sample in the experimental setup is reached. The latter factor mainly depends on the heat transfer between the sample and its surroundings used for cooling, whereas the former is related to the time t_g required for the growth of stable critically sized 3D nuclei to achieve sizes for the detection of the crystallites by the measurement technique. Consequently, the time period recorded by different techniques for the occurrence of crystallization is always more than that required for the induction period t_{in}, which is the period required for the formation of stable nuclei in a system supersaturated at apparent T_c. In fact, these are main errors that lead to erroneous conclusion about the validity of crystallization models.

Finally, it should be mentioned that another period not accounted for in the measured induction period t_{in} is the time t_{reorg} required for the reorganization and orientation of growth species before the formation of stable 3D nuclei, but this period is also included in the definition of the induction period t_{in}, given by

$$t_{in} = t_{reorg} + t_N + t_g, \tag{2.82}$$

where t_N is the actual time required for the formation of critically sized nucleus at a given value of the driving force as determined by supersaturation $\ln S$ or supercooling ΔT. The classical 3D nucleation theory accounts for this t_N alone. This means that, strictly spoken, the relationship between induction period t_{in} and driving force for 3D nucleation predicted by the classical nucleation theory hold when t_{reorg} and t_g are negligibly small in comparison with t_N.

NONISOTHERMAL CRYSTALLIZATION

Solubility and Supersolubility of Solutions

The temperature dependence of solubility c_0 of a solute in a solvent forming an ideal solution at temperature T_0 is usually given by the relation

$$\ln c_0 = \frac{\Delta H_m}{R_G T_m}\left(1 - \frac{T_m}{T_0}\right),\qquad(2.83)$$

where ΔH_m is the enthalpy of melting of the solute, T_m is its melting point, and R_G is the gas constant. A similar relation holds in the case of real systems, where one obtains the enthalpy of dissolution ΔH_s instead of ΔH_m. The value of ΔH_s depends on the solvent used. This relation is usually referred to as the solubility for regular or real systems. Deviations from the ideal solubility behavior are due to the nature of interactions between solute ions and solvent molecules, and the solutions are then referred to as regular solutions.

Using Eq. (2.83), one may define supersaturation $\ln S$ from the relationship between the ratio of solution concentrations c_1 and c_2 corresponding to temperatures T_1 and T_2, respectively, and supercooling ΔT in the form (see Fig. 2.13)

$$\ln S = \ln\left(\frac{c_2}{c_1}\right) = \frac{\Delta H_s}{R_G}\frac{\Delta T}{T_1 T_2} = \frac{\lambda u}{1-u},\qquad(2.84)$$

where the dimensionless enthalpy of dissolution

$$\lambda = \Delta H_s / R_G T_2,\qquad(2.85)$$

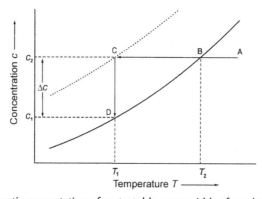

Figure 2.13 Schematic presentation of metastable zone width of a solute–solvent system.

$u = \Delta T/T_2$ with $\Delta T = (T_2 - T_1)$ such that $T_2 > T_1$ and $c_2 > c_1$. As mentioned in Driving Force for Phase Transition section, there is a level of excess solute Δc defined with respect to the temperature T_1, which is achieved by cooling the system from temperature T_2 to T_1 when the new phase nucleates in the system. Therefore, the supersaturation $\ln S$, given by Eq. (2.84), can be defined in terms of either the excess solute concentration $\Delta c = (c_2 - c_1)$ with respect to solubility c_1 at T_1 or the supercooling $\Delta T = (T_2 - T_1)$ defined with respect to T_2 corresponding to the solubility c_2. However, there is always a critical level of the excess solute concentration in the supersaturated solution when nucleation of the new phase occurs instantaneously, i.e., the new phase precipitates. This excess solute level corresponds to the upper limit of the state of metastable equilibrium and defines the width of the metastable zone width in terms of supersaturation $\ln S$ or supercooling ΔT.

Metastable zone width of a solute—solvent system can be determined experimentally either by polythermal (nonisothermal) or isothermal method. The nonisothermal method of metastable zone width is based on the determination of the maximum supercooling ΔT_{max}. A solution of known saturation temperature T_0 is cooled from a temperature higher than T_0 at a constant cooling rate R to a temperature T_{lim} at which first crystals are detected in the solution. The maximum supercooling ΔT_{max} is defined as the difference between the saturation temperature T_0 and the temperature T_{lim}, i.e., $\Delta T_{max} = T_0 - T_{lim}$. The isothermal method, on the other hand, consists of attaining solution supersaturation as fast as possible to attain a predefined temperature for nucleation, followed by the measurement of time period for the appearance of first detectable nuclei in the solution. When the appearance of detectable nuclei occurs immediately after attaining the predefined supersaturation corresponding to the limiting temperature T_{lim}, one considers this excess solute concentration Δc_{max} as the width of the metastable zone width with reference to solute concentration c_{lim} corresponding T_{lim}.

During the last four decades, voluminous literature has emerged on the metastable zone width, as determined by the maximum supercooling ΔT_{max} using the conventional polythermal method, of various compounds in different solvents (Nývlt et al., 1985; Sangwal, 2011b). It is found (Sangwal, 2010, 2011a) that with an increase in T_0, the value of ΔT_{max} decreases for some solute—solute systems; it increases for some other solute—solvent systems, whereas it remains practically constant for the remaining systems. The experimental data of maximum supercooling ΔT_{max} for solute—solvent systems have traditionally been analyzed until now as a function of cooling rate R using Nývlt's equation (2.90), proposed in 1968, which contains two empirical parameters (i.e., nucleation order m and nucleation constant k_m). However, the physical significance of these parameters in Nývlt's equation has remained obscure until now. Since 2008, several papers have been devoted to the understanding of the effect of various experimental factors on the metastable zone width.

Theoretical Interpretations of Metastable Zone Width

There are two types of approaches to explain the dependence of metastable zone width on various factors. The first type of approaches assumes the formation of critically sized 3D nuclei during the cooling of a solution below saturation temperature T_0 (Kubota, 2008; Nyvlt, 1968; Sangwal, 2009a,b), but they differ in the way the nucleation rate J depends on the developed solution supersaturation $\ln S$. Most of these approaches (Kubota, 2008; Nyvlt, 1968; Sangwal, 2009b) assume that the nucleation rate J is related to the maximum supersaturation $\ln S_{max}$ by simple power law i.e., $J \propto (\ln S_{max})^m$, whereas another approach (Sangwal, 2009a,c, 2010) assumes that the nucleation rate J is described by the classical 3D nucleation theory. The second type of approaches is based on the concept of overall crystallization involving progressive and instantaneous nucleation mechanisms (Kashchiev et al., 2010a,b).

Nývlt's Approach

This approach assumes (Nývlt, 1968; Nývlt et al., 1985) that the excess solute concentration Δc, which is a measure of supersaturation $\ln S$, is related to supercooling ΔT by

$$\Delta c = \left(\frac{dc_0}{dT}\right)_T \Delta T,$$ (2.86)

and that, in the vicinity of metastability, the nucleation rate J is related to excess solute concentration Δc by the power law relation

$$J = k_m (\Delta c)^m,$$ (2.87)

and with the cooling rate $R = \Delta T / \Delta t$ by

$$J = \left(\frac{dc_0}{dT}\right)_T R,$$ (2.88)

where m is the apparent nucleation order, $(dc_0/dT)_T$ is the temperature coefficient of solubility at temperature T and k_m is the nucleation constant. The value of k_m depends on the processes of formation and growth of stable nuclei into visible entities and the experimental method used for the measurement of metastable zone width.

On substituting the value of $\Delta c = \Delta c_{max}$, corresponding to the maximum supercooling ΔT_{max}, from Eq. (2.86) in Eq. (2.87) and equating the nucleation rate J given by Eqs. (2.87) and (2.88), one obtains

$$\Delta T_{max} = \left(\frac{dc_0}{dT}\right)_T^{(1-m)/m} \left(\frac{R}{k_m}\right)^{1/m}.$$ (2.89)

On taking logarithms on both sides of Eq. (2.89) and rearrangement, one obtains

$$\ln \Delta T_{max} = \frac{1-m}{m}\ln\left(\frac{dc_0}{dT}\right)_T - \frac{1}{m}\ln k_m + \frac{1}{m}\ln R. \qquad (2.90)$$

Eq. (2.90) predicts a linear dependence of $\ln \Delta T_{max}$ on $\ln R$. This linear dependence enables to calculate the values of m and k_m because $(dc_0/dT)_T$ can be determined from solubility data.

It is usually found that the quantity $(dc_0/dT)_T$ strongly depends on temperature T, but the quantity $(dc_0/dT)/T$ is practically temperature-independent. Therefore, it is expected that, in contrast to the original Nývlt's Eq. (2.90), an expression relating dimensionless maximum supercooling $u_{max} = \Delta T_{max}/T_0$ with the cooling rate R describes the experimental $\Delta T_{max}(R)$ data better. Then from Eq. (2.89), one gets the relation (Sangwal, 2009b)

$$\ln u_{max} = \left[\frac{1-m}{m}\ln\left(\frac{dc_0}{T_0\,dT}\right) - \frac{1}{m}\{\ln k_m + (2m-1)\ln T_0\}\right] + \frac{1}{m}\ln R. \qquad (2.91)$$

This expression is similar to the original Nývlt's Eq. (2.90). According to this equation, plots of $\ln u_{max}$ against $\ln R$ give slope $1/m$ and intercept equal to the term contained in the square brackets.

Self-Consistent Nývlt-Like Equation of Metastable Zone Width (Sangwal, 2009b)

This approach involves a redefinition of nucleation rate J as a function of dimensionless supersaturation ratio S by Eq. (2.84) and, as in Nývlt's approach, assumes that the nucleation rate J may be given by a power law relation of the form

$$J = K_m(\ln S)^m, \qquad (2.92)$$

where m is the so-called apparent nucleation order and K_m is a new nucleation constant related to the number of stable nuclei forming per unit volume per unit time. On substituting the value of $\ln S$ from Eq. (2.84) in (2.92), one obtains

$$J = K_m[\lambda u/(1-u)]^m. \qquad (2.93)$$

Because the nucleation rate J is proportional to the rate of change of solution supersaturation $(\Delta c/c_1)/\Delta t$, one may write the relationship between the nucleation rate J and cooling rate $R = \Delta T/\Delta t$ in the form

$$J = f\frac{\Delta c}{c_1\Delta t} = f\frac{\Delta c}{c_1\Delta T}\frac{\Delta T}{\Delta t} = f\left(\frac{\lambda}{1-u}\right)\left(\frac{R}{T_2}\right), \qquad (2.94)$$

where the proportionality constant f has units as number of entities (i.e., particles) per unit volume.

For $u = u_{max}$ and $T_2 = T_0$, on equating Eqs. (2.93) and (2.94) to eliminate J, one gets

$$u_{max} = \left(\frac{f}{K_m T_0}\right)^{1/m} \left(\frac{\lambda}{1 - u_{max}}\right)^{(1-m)/m} R^{1/m}.$$ (2.95)

On taking logarithm on both sides of Eq. (2.95), one obtains

$$\ln u_{max} = \Phi + \beta \ln R,$$ (2.96)

where Φ is the value of $\ln u_{max}$ when $\ln R = 0$, $\beta = 1/m$, and Φ is given by

$$\Phi = \frac{1-m}{m} \ln\left(\frac{\lambda}{1 - u_{max}}\right) + \frac{1}{m} \ln\left(\frac{f}{K_m T_0}\right).$$ (2.97)

Eq. (2.96) predicts a linear dependence of $\ln(\Delta T_{max}/T_0)$ on $\ln R$, with slope β and intercept Φ. The linear dependence enables to calculate the values of nucleation order m from slope β and the term $\ln(f/K_m T_0)$ from the intercept because, for $u_{max} \leq 0.1$, $\ln[\lambda/(1 - u_{max})] \approx \ln \lambda$ and can be calculated from solubility data of the investigated compound. The factor f may be calculated independently from the equilibrium solute concentration at temperature T_0.

Eq. (2.96) is similar in form to the traditional Nývlt's equation (90). However, it has three advantages over Nývlt's equation: (1) there is a consistency in the units of the left-hand and right-hand sides, (2) the new nucleation constant K_m has the same units as the nucleation rate J (i.e., nuclei per unit volume per unit time), and (3) nucleation constant K_m and nucleation order m are related to constant B of the classical 3D nucleation theory by the empirical relations (Sangwal, 2009a,b)

$$m = 2B/(\ln S_{eff})^2, \quad K_m = B/(\ln S_{eff})^{0.565},$$ (2.98)

where $\ln S_{eff}$ is the effective supersaturation for nucleation, and B is given by Eq. (2.22). The relationship between K_m and B holds for $B \leq 4$. The effective supersaturation $\ln S_{eff} < \ln S_{max}$.

Approach Based on Classical Theory of 3D Nucleation (Sangwal, 2009a)

This approach is based on the following two concepts: (1) the dependence of 3D nucleation rate J on supersaturation $\ln S$ is described by the classical 3D nucleation theory (Eq. 2.20) and (2) the nucleation rate J is proportional to the rate of change of solution supersaturation $(\Delta c/c_1)/\Delta t$ and is given by Eq. (2.94). When $u = u_{max}$, $T_2 = T_0$, and $T_1 = T_{lim}$, from Eqs. (2.94) and (2.20), one obtains

$$\exp\left[-\kappa \frac{\omega^3}{\lambda^2(1 - u_{max})u_{max}^2}\right] = f\left(\frac{\lambda}{1 - u_{max}}\right)\frac{R}{J_0 T_0},$$ (2.99)

where

$$\omega = \frac{\gamma \Omega^{2/3}}{k_B T_0},$$ (2.100)

J_0 is the preexponential factor and all other parameters have the same meaning as above. On taking logarithm on both sides of Eq. (2.99) and rearrangement, one obtains

$$\frac{1}{(1 - u_{max})u_{max}^2} \approx u_{max}^{-2} = \frac{X}{b} - \frac{1}{b}\ln R = F(1 - Z \ln R),$$ (2.101)

where $F = X/b = 1/Zb$, $Z = 1/X$,

$$b = \kappa \omega^3 / \lambda^2,$$ (2.102)

$$Z = \frac{1}{X} = \ln\left\{\frac{f}{AT_0}\frac{\lambda}{(1 - u_{max})}\right\}.$$ (2.103)

Eq. (2.101) predicts that, at a given saturation temperature T_0, the quantity u_{max}^{-2} decreases linearly with an increase in $\ln R$, with slope $1/b$ and intercept F.

The main advantage of this approach is that the effects of different experimental parameters can be explained satisfactorily in terms of two parameters of the classical nucleation theory (see Eq. 2.20): (1) parameter J_0 associated with the kinetics of formation of nuclei in growth medium and (2) changes in the solid—liquid interfacial energy γ. However, it should be noted that the constant Z is not a dimensionless quantity. Instead, its value depends on the units of cooling rate R, but the term $Z\ln R$ is a constant quantity for a crystallizing system.

Approach Based on Progressive 3D Nucleation

Kashchiev et al. (2010a) considered the early stages of crystallization occurring by the formation of stable 3D nuclei progressively and their subsequent growth during cooling of solution at a constant rate and derived the relation between u_{max} and R of the form

$$\ln u_{max} = \left[\frac{a_2}{a_1(1 - u_{max})u_{max}^2} - \frac{1}{a_1}\ln R_0\right] + \frac{1}{a_1}\ln R,$$ (2.104)

where the constants $a_1 > 0$, $a_2 > 0$, and $R_0 > 0$. For u_{max} determined by fraction y of crystallized mass, a_1 and a_2 are given by

$$a_1 = 3 + \frac{3nqd}{qd + 1}, \quad a_2 = \frac{b}{qd + 1},$$ (2.105)

and R_0 is a "threshold value" of cooling rate R and is determined, among others, by the detection limit for the mass fraction y of crystallites by a particular experimental technique. In Eq. (2.105), n is the growth exponent lying between 1 and 2 when crystallites grow by normal growth or spiral growth mechanism, and d is the growth dimensionality.

It may be noted from relations (2.105) that the values of a_1 and a_2 depend on the values of n, q, d, and b. From Eq. (2.105), it also follows that, irrespective of the mechanism by which crystallites attain visible dimensions, a_1 increases at the expense of a decrease in a_2. However, it may be seen that this approach predicts $3 \leq a_1 \leq 7.5$ when crystallites grow by the normal growth and spiral growth mechanisms. The lowest value equal to 3 corresponds to the situation when the crystallites or aggregates of nuclei do not grow, whereas the highest value corresponds to the situation when $n = 2$, $d = 3$, and $q = 1$. When crystallites grow by 2D nucleation mechanism, the growth exponent n can attain much higher values. Then a_1 can have values exceeding 7.5.

Depending on whether the term $\ln u_{max}$ or the term in the square brackets on the left-hand side of Eq. (2.104) is constant in a particular range of cooling rate R, two cases arise. When the second term in the square brackets is constant, Eq. (2.104) takes the form of the self-consistent Nývlt-like Eq. (2.96) with $a_1 = m$. However, when the term $\ln u_{max}$ is constant, one has

$$\frac{1}{(1 - u_{max})u_{max}^2} \approx \frac{1}{u_{max}^2} = \frac{1}{a_2}[(a_1 \ln u_{max} + \ln R_0)] - \frac{1}{a_2}\ln R. \qquad (2.106)$$

This equation is similar to Eq. (2.101), with $a_2 = b$.

Approach Based on Instantaneous 3D Nucleation

Kashchiev et al. (2010b) also considered the case of instantaneous nucleation in which all crystallites appear at a particular time and subsequently grow in the absence of newly developing nuclei. These authors derived the relation

$$\ln u_{max} = \frac{1}{n+1}\ln R - \frac{1}{n+1}\ln R_0^*, \qquad (2.107)$$

where the new "threshold" cooling rate R_0^* also depends, among others, on the detection limit for mass crystallization by a measurement technique.

Because $1 < n < 2$ when crystallites grow by normal growth and spiral growth mechanisms, one obtains the traditional $m = n + 1$ in Nývlt's equation to lie between 2 and 3.

Experimental Results on Metastable Zone Width of Solute–Solvent Systems

In the studies of the metastable zone width as determined by the nonisothermal methods, there are three issues of concern: (1) the dependence of dimensionless cooling

$u_{max} = \Delta T_{max}/T_0$ on cooling rate R, (2) the nature of the nucleation order m reported for different systems, and (3) role of saturation temperature T_0. These issues are described below.

Dependence of Dimensionless Supercooling u_{max} on Cooling Rate R

It is observed (Sangwal, 2009a,b; 2011a,b) that u_{max} increases practically linearly with increasing R in some solute—solvent systems, but u_{max} increases steeply first, followed by a progressively slow increase with an increase in R in other systems. Typical examples of the former and latter systems are potassium tetraborate tetrahydrate (KTB)—water (Sahin et al., 2007) and 3-nitro-1,2,4-triazol-5-one (NTO)—water systems (Kim and Kim, 2001), which are characterized by high and low values of $\beta = 1/m$ and both Eqs. (2.96) and (2.101) applied when $u_{max} < 0.1$ (Sangwal, 2011b).

Fig. 2.14A and B presents the plots of $\ln(\Delta T_{max}/T_0)$ against $\ln R$ according to Eq. (2.96) for POP and PPP, respectively, whereas Fig. 2.15A and B shows the above data for the two compounds as plots of $(T_0/\Delta T_{max})^2$ against $\ln R$ according to Eq. (2.101). The experimental data were obtained using turbidimetry and growth exotherms. These techniques are denoted in the figure by TBT and CRY, respectively.

It was observed (Sangwal, 2009b, 2011a; b) that in several cases the plots of $(T_0/\Delta T_{max})^2$ against $\ln R$ according to Eq. (2.101) deviate from the linear dependence. However, when the applied cooling rate R for a system is corrected such that the corrected cooling rate $R^* = (R - R_c)$, where R_c is the threshold value of cooling rate R, the data exhibit a better fit with a higher regression coefficient RC. This threshold cooling rate R_c corresponds to the lower limit of the cooling rate R when relation (2.101) applies. Physically, R_c corresponds to the situation when a system begins to respond to the cooling procedure and is associated with the setting up of a thermal equilibrium between the solution and the environment.

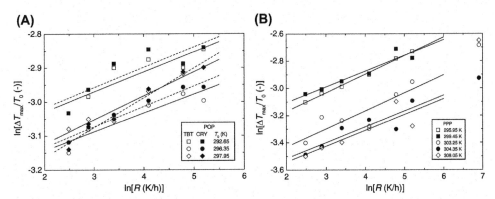

Figure 2.14 Plots of $\ln(\Delta T_{max}/T_0)$ against $\ln R$ for (A) POP and (B) PPP solutions at various saturation temperature T_0 according to Eq. (2.96). Linear plots represent the best fit of data. *(Reproduced from Sangwal, K., Smith, K.W. Cryst. Growth Des. 10, 640–647. Copyright (2010), with permission from American Chemical Society.)*

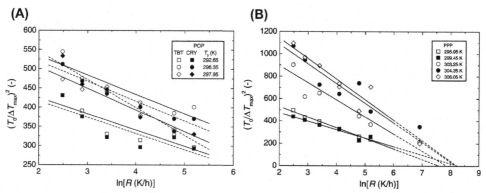

Figure 2.15 Plots of $(T_0/\Delta T_{max})^2$ against $\ln R$ for (A) POP and (B) PPP solutions against $\ln R$ at various saturation temperature T_0 according to Eq. (2.101). Linear plots represent the best fit of data. *(Reproduced from Sangwal, K., Smith, K.W. Cryst. Growth Des. 10, 640–647. Copyright (2010), with permission from American Chemical Society.)*

It is found that the value of $m = 1/\beta$ is relatively insensitive to saturation temperature T_0 (Sangwal, 2009a, 2011a; Sangwal and Smith, 2010) but depends on the investigated compound and the technique used for the measurement of maximum supercooling (Sangwal and Smith, 2010). In fact, the value of $m = 1/\beta$ highly depends on the investigated compound. For example, as reported by Sangwal and Smith (2010), the values of $m = 1/\beta$ calculated from data obtained using turbidimetry are 16.2 and 8.1 for POP and PPP, respectively, but the value of β calculated from data obtained by growth exotherms gives m as 14.5 for POP. Obviously, the different values of m for POP are associated with the sensitivity of the two techniques.

The parameter Z shows trends similar to those of β and $\beta \approx Z$ for different solute–solvent systems (Sangwal, 2009a,c; 2011a,b; Sangwal and Smith, 2010). This equality means that $\ln S_{eff} \approx 2(\Delta H_s/R_G T_{lim})^2$; cf. Eqs. (2.8) and (2.98). Because $\ln S_{eff} < \ln S_{max}$ (see Self-consistent Nývlt-like Equation of Metastable Zone Width section (Sangwal, 2009b)) and heat of dissolution ΔH_s is related to solubility c_0 of a solute in a solvent (cf. Eq. 2.83), the equality between β and Z implies that these parameters are connected with the solubility of a solute in the solvent. However, it may be noted that in the case of PPP and POP the higher values of β and Z for a compound obtained from data using growth exotherms than those obtained from data using turbidimetry are caused by the growth of stable nuclei to detectable dimensions.

It is well known (Sangwal, 1989) that the solubility of different compounds increases with a decrease in the solid–liquid interfacial energy γ. Therefore, a decrease in γ also means a decrease in the factor B of Eqs. (2.22) and (2.98) and an increase in the values of β and Z. The higher the solubility of different compounds in a given solvent, the lower the values of β and Z. Thus, the values of β and Z for a compound are intimately connected with its solubility in a solvent.

The Nucleation Order m

A survey of the data on m for aqueous solutions of various solutes at different temperatures for unseeded and seeded systems without and with intentionally added foreign substances (i.e., additives) reveals the following features (Sangwal, 2011b):

1. In general, $1.5 < m < 10.5$ but the frequency of occurrence is 51% for $2.5 < m < 3.5$. Values of m as low as 1.41 for $(NH_4)_2HPO_4$ and as high as 20.05 for unfiltered KH_2PO_4 solutions have also been found.

2. No specific trends are found for the values of m for a solute at a given T_0 for unseeded and seeded systems nor at different T_0 for seeded or unseeded systems.

3. Enormously different values of m even for the same solute have been observed. For example, for unseeded $(NH_4)_2SO_4$ system, the value of m equals to 8.26 and 2.45 at 40 and 60°C, respectively, and for unseeded $Ca(NO_3)_2 \cdot 4H_2O$ system, the value of m equals to 2.29 and 13.64 at 19.5 and 38.7°C, respectively, have been reported (Nývlt et al., 1985).

The above features are consistent with the progressive nucleation mechanism. The values of m exceeding about 3 suggest that crystallization in aqueous solutions takes place by progressive nucleation during metastable zone width measurements, whereas the observation $m \geq 7.5$ for progressive nucleation is expected when crystallites grow by 2D nucleation mechanism. These enormously different values of m even for the same system by progressive nucleation are also possible when crystallites grow by different growth mechanisms.

It is observed (Sangwal, 2009a,b; 2011a,b; Sangwal and Smith, 2010) that the value of m for a solute–solvent system is relatively insensitive to saturation temperature T_0. This is indeed predicted by the progressive nucleation mechanism when the crystallites of a substance have the same dimensionality and grow by the same mechanism (i.e., n, q, and d), and the interfacial energy γ of the system remains unchanged; see Eq. (2.105). However, the value of m depends on the investigated compound and the technique used for the measurement of maximum supercooling. For example, the values of m obtained from experimental $u_{max}(R)$ data using turbidimetry are 16.2 and 8.1 for POP and PPP in acetone solutions, respectively, and the value of m from data obtained by growth exotherms is 14.5 for POP. The different values of m are due to different interfacial energies γ for PPP–acetone and POP–acetone solutions, whereas the different values of m for POP–acetone system are associated with the sensitivity of the two techniques. The fact that the value of m obtained from the data obtained from growth exotherms is higher than that from turbidimetry implies that detection techniques, which record delayed onset of crystallization nucleation event developing nuclei lead to lower nucleation order m.

Role of Saturation Temperature T_0

The values of $-\Phi$ obtained from the fit of the experimental $u_{max}(R)$ data by Eq. (2.96) is expected to be equal to the values of $\ln(F^{1/2})$ obtained by using Eq. (2.101). It is found

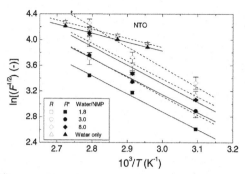

Figure 2.16 Plots of $\ln(F^{1/2})$ against $1/T_0$ for NTO in different solvents. Dashed and solid lines present data for uncorrected R and corrected R^*, respectively. *(Reproduced from Sangwal, K. Cryst. Eng. Comm. 13, 489–501. Copyright (2011), with permission from Royal Chemical Society.)*

(Sangwal, 2009a, 2010; 2011a; b) that $-\Phi$ and $\ln(F^{1/2})$ are usually different, although $-\Phi = \ln(F^{1/2})$ have also been observed. The former discrepancy occurs when $\Delta T_{\max}/T_0$ is relatively high. In the case of seeded phosphoric acid aqueous solutions, for example, where $\Delta T_{\max}/T_0 \approx 0.011$, the values of $-\Phi$ are in good agreement with those of $\ln(F^{1/2})$ at a particular T_0 (Sangwal, 2010).

The data of $\ln(F^{1/2})$ as a function of saturation temperature T_0 may be described by an Arrhenius-type relation (Sangwal, 2009a, 2010; 2011a),

$$\ln\left(F^{1/2}\right) = \left[\ln\left(F^{1/2}\right)\right]_0 - E_{sat}/R_G T_0, \qquad (2.108)$$

where $[\ln(F^{1/2})]_0$ denotes the extrapolated value of $\ln(F^{1/2})$ corresponding to $T_0 = \infty$ and E_{sat} is the activation energy associated with the process of formation of crystallites in the solution. The data of $-\Phi$ as a function of saturation temperature T_0 can also described by Eq. (2.108) (Sangwal, 2009b). Fig. 2.16 illustrates the plots of $\ln(F^{1/2})$ against $1/T_0$ for NTO in different solvents according to Eq. (2.108). Dashed and solid lines represent the best fit of the data with as-applied and corrected cooling rates indicated by R and R^*, respectively.

It should be mentioned that the values of the activation energy E_{sat} for different compounds can be positive (Sangwal, 2009a–c, 2011a) as well as negative (Sangwal, 2010). When E_{sat} is equal to the activation energy for diffusion E_D, $\Delta T_{\max}/T_0$ is independent of T_0. When $E_{sat} > E_D$, $\Delta T_{\max}/T_0$ decreases with an increase in T_0. However, when $E_{sat} < E_D$, $\Delta T_{\max}/T_0$ increases with an increase in T_0.

Nonisothermal Crystallization Kinetics

We recall here that nonisothermal crystallization is based on the cooling of a liquid at a particular cooling rate R from the equilibrium state at temperature T_0 to a lower temperature T_c where crystallization takes place and that this notion is used in the derivation of all of the theoretical relationships, described earlier, between normalized temperature difference $\Delta T/T_0$ and cooling rate R. Therefore, so far as the understanding of the metastable zone width has been concerned, all of the equations are equally applicable in the case of melts, with essentially the same physical quantities and parameters.

Evolution of overall nonisothermal crystallization kinetics of fatty acid melts at different cooling rates R is similar to isothermal crystallization kinetics. At a given cooling rate R, the time dependence of the overall crystallization curve shows an initial period t_{in} without observable crystallized mass and then there is a sudden increase in the crystallized mass, which subsequently approaches a constant value, as shown in Fig. 2.17. As in the case of isothermal crystallization, nonisothermal crystallization also comprises two processes: (1) an initial period, usually referred to as induction period t_{in}, which involves the nucleation of a new phase and (2) its subsequent overall crystallization γ during the cooling of supersaturated solution or melt at a constant cooling rate R. For a given system, both of these processes depend on the cooling rate R.

It is observed that the induction period t_{in} and the duration of the overall crystallization γ for a crystallizing phase decrease with increasing cooling rate R (see Fig. 2.18), whereas an increase in R induces the formation of more metastable phases such that the crystallization of the stable phase occurring at low R is completely replaced by the crystallization of the metastable phase with an increasing R (Fig. 2.17). The nature of the nonisothermal crystallization curves is similar to that of isothermal crystallization curves. Therefore, basically the mechanisms proposed by isothermal crystallization are also valid here, but the effect of cooling rate has to be included in the explanation of

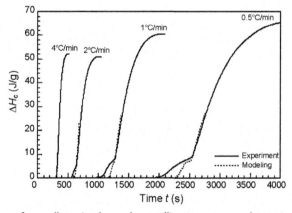

Figure 2.17 Evolution of overall nonisothermal crystallization mass as determined by latent heat ΔH released by cocoa butter samples for different cooling rate R. *(Reproduced from Rousset, P., 2002. Modeling crystallization kinetics of triacylglycerols. In: Marangoni, A.G., Narine, S.S. (Eds.), Physical Properties of Lipids, pp. 1—33. Copyright (2002) by Marcel Dekker, New York. With permission from Taylor and Francis.)*

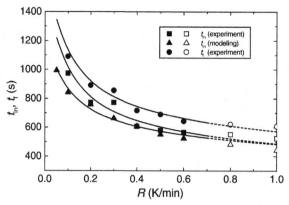

Figure 2.18 Dependence of the times t_{in} and t_f for the nonisothermal crystallization kinetics of SOS on cooling rate R. Plots are drawn according to Eq. (2.111) with constants given in Table 2.1. *(Original data from Rousset, P., Rappaz, M., 1996. See text for details.)*

nonisothermal crystallization (for example, see: Narine et al., 2006; Rousset and Rappaz, 1996; Rousset et al., 1998; Smith et al., 2005).

Rousset and Rappaz (1996) and Rousset et al. (1998) analyzed the time dependence of volume fraction of crystallized SOS and POS–SOS mixture by splitting the cooling curves into small time steps of constant temperature and fitting the data with Avrami equation (2.49) considering k and q as temperature-dependent parameters, time–temperature–transformation diagrams and fictitious times between two successive time steps. This procedure enabled to calculate time t_s (denoted here as t_{in}) of start and time t_f of finish of crystallization as a function of cooling rate R, as shown in Fig. 2.18 by squares and circles, respectively, for the crystallization of SOS by cooling the melt from 100°C. In the figure, triangles represent data obtained by modeling the process using additivity principle based on decomposition of the $t_{in}(R)$ curve into small steps. The figure also shows the data on t_{in} and t_f for SOS melts initially quenched between 50 and 30°C and then allowed to cool at cooling rates of 0.8 and 1 K/min. These data are presented as open squares, circles, and triangles.

It may be seen from Fig. 2.18 that simulation predicts the trend of the curves but gives somewhat lower values of t_{in} and t_f than the experimental data. In any case, these trends of the dependence of t_{in} and t_f on R for nonisothermal crystallization of SOS demonstrate that nonisothermal crystallization can also be described by Avrami theory. Smith et al. (2005) also found that the KJMA theory best describes the experimental data on the nonisothermal crystallization kinetics of POP, whereas the temperature dependence of the Avrami rate constant k_m is best explained by a Vogel–Fulcher relationship where melting point of the crystallizing species is taken as the reference temperature. Narine et al. (2006) extended the Avrami theory to describe the crystallization kinetics of high-volume fraction lipids and suggested that lipids crystallize and grow into networks via very specific growth modes.

Following Kim and Mersmann (2006), from the maximum supercooling ΔT_{max} attained in the system at a constant cooling rate R, we define the induction period t_{in} by the relation

$$t_{in} = \Delta T_{max}(R)/R, \qquad (2.109)$$

where the time when the first crystal is detected in the system is taken as the induction period t_{in}, and we have denoted ΔT_{max} as a function of R because, in a cooling crystallization experiment from a given equilibrium temperature T_0, the value of ΔT_{max} somewhat increases with increasing R. Assuming that $\Delta T_{max}(R)$ increases with R following the simple relation

$$\frac{\Delta T_{max}(R)}{\Delta T_{max}(R_{lim})} = \left(\frac{R}{R_{lim}}\right)^p, \qquad (2.110)$$

where $\Delta T_{max}(R_{lim})$ is the limiting value of ΔT_{max} when R approaches R_{lim}, and the exponent $P < 1$, Eq. (2.109) may be written as

$$t_{in} = \frac{\Delta T_{max}(R_{lim})}{R_{lim}^p R^{1-p}}. \qquad (2.111)$$

According to Eq. (2.111), t_{in} is strictly not inversely proportional to R because of the dependence of ΔT_{max} on R. The validity of Eq. (2.111) is demonstrated in Fig. 2.18, which shows the dependence of the times t_{in} and t_f for the nonisothermal crystallization kinetics of SOS on cooling rate R. The continuous and the dashed curves are drawn with the best-fit constants of the data up to 0.6 K/min (Table 2.1). The reported values of simulated t_{in} as a function of R, shown in Fig. 2.18, are somewhat lower than the experimental values, but they can also be fitted according to Eq. (2.111) with somewhat different values of the constants.

Table 20.1 Values of constants of Eq. (111)

Sample	Data	$\Delta T_{max}(R_{lim})$ (K)	$1/R_{lim}$ (min/K)	P
SOS	$t_{in}(R)^a$	2.397	$4.7 \cdot 10^{-4}$	.693
	$t_{in}(R)^c$	1.478	$4.7 \cdot 10^{-4}$	.755
	$t_f(R)^a$	2.40	$4.7 \cdot 10^{-4}$	.714
POS-SOS	$t_{in}(R)^b$	3.553	$8.09 \cdot 10^{-7}$	.393
	$t_f(R)^b$	3.050	$1.13 \cdot 10^{-6}$	.402

[a]Experimental data by Rousset and Rappaz (1996).
[b]Experimental data by Rousset et al. (1998).
[c]Modeling data by Rousset and Rappaz (1996).

A behavior similar to that of SOS was also observed in the case of nonisothermal crystallization kinetics of a 25:75 w/w POS−SOS mixture on cooling rate R (Rousset et al., 1998). The values of the constants of Eq. (2.111) for the two samples are listed in Table 2.1.

INDUCTION PERIOD FOR NUCLEATION

According to the 3D nucleation theories, the nucleation rate J depends on supersaturation $\ln S$, and there is a threshold value of supersaturation S_{max} when the metastable phase precipitates spontaneously into the new stable phase. This means that after the initial moment $t = 0$ of attaining supersaturation in the old phase, formation of an appreciable amount of the new phase requires a certain time t_{in}, called induction period or induction time. This time is an experimentally measurable quantity and is a measure of the ability of the system to remain in the metastable state; also see Section: Comparison of Different Models of Isothermal Crystallization Kinetics, and Eq. (2.82).

Theoretical Background

As described in Section: Theoretical Interpretation of Isothermal Crystallization Kinetics, crystallization of a solid phase in the mother phase is possible by MN and PN mechanisms. These two mechanisms describe two extreme cases of formation of detectable nuclei. The induction period t_{in} according to the two mechanisms may be represented by single expression (Kashchiev, 2000; Kashchiev et al., 1991):

$$t_{in} = t_{MN} + t_{PN} = \frac{1}{JV} + \left(\frac{q\gamma}{\kappa J g^{q-1}} \right)^{1/q}, \tag{2.112}$$

where γ is the smallest experimentally detectable fraction of overall crystallization (see Eq. 2.39), κ is the shape factor, V is the volume of the parent phase, the growth rate g is related to supersaturation $\ln S$ by: $g = A(\ln S)^n$ (with A as growth rate constant and $1 < n < 2$), and the exponent q is given by Eq. (2.42).

When

$$V < \left(\frac{\kappa A^{q-1}}{q\gamma J^{q-1}} \right)^{1/q}, \tag{2.113}$$

MN mechanism holds, which applies to supersaturated phases involving sufficiently small volumes. Then

$$t_{in} = t_{MN} = \frac{1}{J_0 V} \exp\left(\frac{B}{\ln^2 S} \right), \tag{2.114}$$

where J_0 is the kinetic factor and B is given by Eq. (2.22). In the other case involving large volume, PN mechanism is applicable. Then

$$t_{in} = t_{PN} = \left(\frac{q\gamma}{\kappa J_0 A^{q-1}}\right)^{1/q} \left[S(S-1)^{q-1}\right]^{-1/q} \exp\left(\frac{B}{q \ln^2 S}\right). \qquad (2.115)$$

In Eq. (2.115), the term $S(S-1)^{q-1} \approx S$. Therefore, the above equations may be expressed in the form

$$\ln t_{in} = -\ln(J_0 V) + \frac{B}{\ln^2 S}, \qquad (2.116)$$

$$\ln t_{in} = \frac{1}{q} \ln\left(\frac{q\gamma}{\kappa J_0 S A^{q-1}}\right) + \frac{B}{q \ln^2 S}. \qquad (2.117)$$

It may be noted that Eq. (2.117) reduces to Eq. (2.116) of MN mechanism when $q = 1$. Then $\kappa = V_0$, $\gamma = 1$, and $J_0 S$ now includes S, which was omitted in Eq. (2.21). According to these dependencies, the plots of $\ln t_{in}$ against $\ln^2 S$ are straight lines with slopes B and B/q, respectively.

When the atomistic model of nucleation applies, from Eq. (2.28), one obtains

$$\ln t_{in} = [-\ln J_0 + \Phi^*/k_B T] - i^* \ln S. \qquad (2.118)$$

A plot of $\ln t_{in}$ against $\ln S$ gives a slope equals to the number i of molecules in the critical nucleus.

The above equations describe the nucleation kinetics data for supersaturated solutions. However, using the relation between $\ln S$ and $\Delta T = T_m - T_c$, given by (cf. Eq. 2.84), the corresponding relations according to the PN mechanism and atomic models for nucleation from melts may be written in the form

$$\ln t_{in} = K_1 + \frac{B'}{q T_c (\Delta T)^2}, \qquad (2.119)$$

$$\ln t_{in} = K_2 - i^* \lambda (\Delta T/T_c), \qquad (2.120)$$

where K_1 and K_2 are constants denoting the first term in Eqs. (2.117) and (2.118), respectively, B' is given by Eq. (2.23), and the dimensionless melting enthalpy λ is given by Eq. (2.85), with $\Delta H_s = \Delta H_m$ and $T_2 = T_0$.

From the experimental data on the induction period t_{in} for nucleation as a function of relative supersaturation $\ln S$ or temperature difference ΔT for systems with or without an additive, one can verify the validity of the above models. These equations apply both to 3D homogeneous and heterogeneous nucleation. However, it should be noted that the value of the interfacial energy γ predicted by Eq. (2.119) of the PN model depends on the value of q and increases as $q^{1/3}$. If the dimensionality d of the nuclei is ignored (i.e., when $d = 0$ such that $q = 1$), Eq. (2.119) takes the same form of the dependence following from the inverse relationship between nucleation rate J and t_{in}. Eq. (2.119) then represents the traditional Fisher–Turnbull relationship (for example, see Marangoni 2005).

The nucleation may equally be homogeneous or heterogeneous, depending on the value of supersaturation and solvent composition. In heterogeneous nucleation, the induction period corresponding to a given value of supersaturation of a system is reduced due to a decrease in the interfacial energy and kinetics of attachment of growth entities (see Eq. 2.33).

Typical Examples of Experimental Data

In this section, some features of the dependence of induction period t_{in} on ΔT are described using examples of different polymorphs of PPP and CB. The data on the measurement of induction period t_{in} for different polymorphs of PPP were reported by Sato and Kuroda (1987) as the time when the occurrence of a crystal was detected optically after attaining a predefined temperature T_c of the melt during isothermal crystallization between 35 and 65°C. These authors reported that t_{in} is the shortest for the α form, intermediate for the β' form, and the longest for the β form. The data on t_{in} as a function of crystallization temperature T_c in the range between 19 and 23°C for the α and β' polymorphs of CB were reported by Foubert et al. (2005). These $t_{in}(T_c)$ data were obtained by analysis of the isothermal crystallization kinetics using the model of Foubert et al. (2003); see Eq. (2.69). The reported melting point T_m of CB is 36°C, whereas the starting CB material for the crystallization of the β' polymorphs were from two different sources and are indicated as CB-A and CB-B. In both cases, supercooling $\Delta T = T_m - T_c$ was calculated by taking into account their corresponding T_m.

Fig. 2.19A shows the experimental $t_{in}(T_c)$ data, reported by Sato and Kuroda (1987), in the form plots of $\ln t_{in}$ on $1/T_c(\Delta T)^2$ according to Eq. (2.120) based on the classical nucleation theory. While plotting the data, the melting points T_m of α, β', and β forms, as read off from the plots of original $t_{in}(T_c)$ data reported by Sato and Kuroda (1987), were taken as 45, 57, and 66°C, respectively. It may be seen that the plots for all of the three polymorphs can be represented by two straight lines with transition values of $1/T_c(\Delta T)^2$ corresponding to T_c equal to about 42, 53, and 55°C for the α, β', and β forms, respectively. Fig. 2.19 presents the above experimental $t_{in}(T)$ data on the crystallization of different polymorphs of PPP as plots of $\ln t_{in}$ against $\Delta T/T_c$, according to Eq. (2.120) based on the atomistic theory. Here, there are linear dependencies of the $\ln t_{in}(\Delta T)$ data in the entire range of $\Delta T/T_c$ for different polymorphs.

Fig. 2.20A and B shows another example of the plots of $\ln t_{in}$ against $1/T_c(\Delta T)^2$ for the $t_{in}(T_c)$ data for the α and β' polymorphs of CB according to Eq. (2.119) of the classical nucleation theory. In the calculations of supercooling $\Delta T = T - T_m$, $T_m = 36$°C, as reported by Foubert et al. (2005), was taken. It may be seen from the plots that the data in the entire investigated temperature interval for the two polymorphs follow

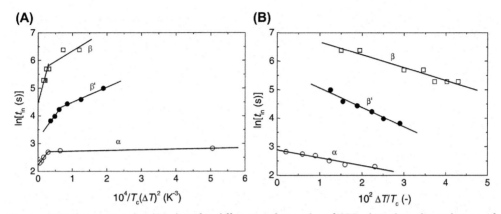

Figure 2.19 Experimental $t_{in}(T_c)$ data for different polymorphs of PPP plotted as dependences of (A) lnt_{in} on $1/T_c(\Delta T)^2$ and according to Eq. (2.119) of classical nucleation theory and (B) lnt_{in} on $\Delta T/T_c$ Eq. (2.120) of atomistic theory. *(Original data from Sato, K., Kuroda, T., 1987. Kinetics of melt crystallization and transformation of tripalmitin polymorphs. J. Am. Oil Chem. Soc. 64, 124–127.)*

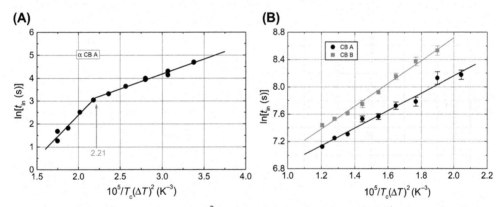

Figure 2.20 Plots of lnt_{in} against $1/T_c(\Delta T)^2$ for the $t_{in}(T_c)$ data for (A) α and (B) β' polymorphs of cocoa butter according to Eq. (2.120). Note that there are two different slopes for the data in (A). *(Original data from Foubert, I., Vanrolleghem, P. A, Dewittinck, K., 2005. Insight in model parameters by studying temperature influence on isothermal cocoa butter crystallization. Eur. J. Lipid Sci. Technol. 107, 660–672.)*

different trends. The plots of lnt_{in} against $1/T_c(\Delta T)^2$ for the β' form of two different samples CB-A and CB-B exhibit linear dependences in the entire T_c range. In contrast to the β' form, the slope of the plot of the data of lnt_{in} against $1/T_c(\Delta T)^2$ for the α form consists of at least two parts below and above $1/T_c(\Delta T)^2$ equal to $2.21 \cdot 10^{-5}$ K^{-3}. This transition temperature corresponds to $T_c = 293.5$ K.

It should be mentioned that two different linear regions of different slopes in the plots of lnt_{in} against $1/T_c(\Delta T)^2$ for the $t_{in}(T_c)$ data from melts and lnt_{in} against $1/\ln^2 S$

from solutions are frequently encountered in isothermal crystallization of a variety of compounds. This difference is associated with different types of nucleation occurring at low and high driving force as defined by ΔT or $\ln S$ and the transition occurring at particular values of the driving force. A lower slope of the plot of $\ln t_{in}$ against $1/T_c(\Delta T)^2$ or $1/\ln^2 S$ above the transition value of the driving force than a higher slope of the plot below this transition value of the driving force implies that nucleation is heterogeneous predominantly at low driving force ΔT in comparison with that at high ΔT when homogeneous nucleation occurs. The trends of the $\ln t_{in}[1/T_c(\Delta T)^2]$ plots of Figs. 2.19 and 2.20 suggest that heterogeneous and homogeneous nucleation mechanisms occur during the crystallization of the three polymorphs of PPP and the α form of CB in the range of small and high supercooling ΔT, respectively. The occurrence of heterogeneous nucleation at low values of the driving force is associated with a decrease in the interfacial energy γ of a solute–solvent system due to the presence of impurities inherently present in them.

It is of interest to note that the slopes of the plots of $\ln t_{in}$ against $1/T_c(\Delta T)^2$ for CB in the region of high $1/T_c(\Delta T)^2$ (i.e., at $T_c < 293.5°C$ corresponding to low driving force) for the α form and in the entire range of $1/T_c(\Delta T)^2$ for the β' form of different two samples are comparable. This means that the values of the interfacial tension γ in these cases are comparable, and the nucleation is mainly heterogeneous in the three cases.

The linear dependencies of the plots of $\ln t_{in}$ against $\Delta T/T_c$, according to Eq. (2.120) based on the atomistic theory in the entire range of $\Delta T/T_c$ for the different polymorphs of PPP in Fig. 2.19B, suggest that the atomistic theory of nucleation is applicable. The slopes of the plots are 27.5, 68.4, and 45.1 for α, β', and β forms, respectively. Because the slope is equal to $i^*\lambda$, the molecules in the critical nucleus i^* are about 4, 10, and 7 for α, β', and β forms, respectively, for a typical value of $\lambda = 6$. A small number of molecules in the critical nuclei are indeed expected according to the atomistic theory.

The data of Fig. 2.20A and B for CB were also examined using Eq. (2.120). It was found that the plots of $\ln t_{in}$ against $\Delta T/T_c$ were linear in the entire T_c range for the α form and for the β' form of the two samples and that the slope $i^*\lambda$ is 222 ± 9 for the α form and 75 ± 5 and 91 ± 6, respectively for the β' form of CB-A and CB-B samples. Assuming that $\lambda = 6$ for different polymorphs, one finds that the critical nuclei of the α form contain i^* equal to about 37 molecules, whereas those of the β' form of CB-A and CB-B samples about 12 and 15 molecules, respectively. These nuclei are relatively large, and the linear dependence between $\ln t_{in}$ against $\Delta T/Tc$ does not guarantee that the atomistic theory represented in the form of Eq. (2.120) is always valid because the effective excess energy Φ^* of the nucleus is also a function of the driving force (see Nucleation Rate section).

ACKNOWLEDGMENT

The authors express their gratitude to Dr K. Wójcik for his assistance with the preparation of the figures.

REFERENCES

Arishima, T., Sagi, N., Mori, H., Sato, K., 1991. Polymorphism of POS: I. Occurrence of polymorphic transformation. J. Am. Oil Chem. Soc. 68, 710–715.

Chernov, A.A., 1984. Modern Crystallography III: Crystal Growth. Springer, Berlin.

Foubert, I., A Vanrolleghem, P., Dewittinck, K., 2005. Insight in model parameters by studying temperature influence on isothermal cocoa butter crystallization. Eur. J. Lipid Sci. Technol. 107, 660–672.

Foubert, I., Dewittinck, K., Vanrolleghem, P.A., 2003. Modelling of the crystallization kinetics of fats. Trends Food Sci. Technol. 14, 79–92.

Foubert, I., Dewittinck, K., Janssen, G., Vanrolleghem, P.A., 2006. Modelling two-step isothermal fat crystallization. J. Food Eng. 75, 551–559.

Kashchiev, D., 2000. Nucleation: Basic Theory with Applications. Butterworth-Heinemann, Oxford.

Kashchiev, D., Borissova, A., Hammond, R.B., Roberts, K.J., 2010a. Effect of cooling rate on the critical undercooling for crystallization. J. Cryst. Growth 312, 698–705.

Kashchiev, D., Borissova, A., Hammond, R.B., Roberts, K.J., 2010b. Dependence of the critical undercooling for crystallization on the cooling rate. J. Phys. Chem. 114, 5441–5446.

Kashchiev, D., Verdoes, D., van Rosmalen, G.M., 1991. Induction time and metastability limit in new phase formation. J. Cryst. Growth 110, 373–380.

Khanna, Y.P., Taylor, T.J., 1988. Comments and recommendations on the use of the Avrami equation for physico-chemical kinetics. Polym. Eng. Sci. 28, 1042–1045.

Kim, K.-J., Kim, K.-M., 2001. Nucleation kinetics in spherulitic crystallization of explosive compound: 3-nitro-1 2,4-triazol-5-one. Powder Technol. 119, 109–116.

Kim, K.-J., Mersmann, A., 2006. Estimation of metastable zone width in different nucleation processes. Chem. Eng. Sci. 56, 2315–2324.

Kloek, W., Walstra, P., van Vliet, T., 2000. Crystallization kinetics of fully hydrogenated palm oil in sunflower oil mixtures. J. Am. Oil Chem. Soc. 77, 389–398.

Kubota, N., 2008. A new interpretation of metastable zone widths measured for unseeded solutions. J. Cryst. Growth 310, 629–634.

Liu, X.Y., 1999. A new kinetic model for three-dimensional heterogeneous nucleation. J. Chem. Phys. 111, 1628–1635.

Liu, X.Y., Tsukamoto, K., Sorai, M., 2000. New kinetics of $CaCO_3$ nucleation and microgravity effect. Langmuir 16, 5499–5502.

Lopez-Manchado, M.A., Blaglotti, J., Torre, L., Kenny, J.M., 2000. Effects of reinforcing fibers on the crystallzation of polypropylene. Polym. Eng. Sci. 40, 2194–2204.

Marangoni, A.G., 1998. On the use and misuse of the Avrami equation in the characterization of the kinetics of fat crystallization. J. Am. Oil Chem. Soc. 75, 1465–1467.

Marangoni, A.G., 2013. Nucleation and crystal growth kinetics, pp. 27–99. In: Structure and Properties of Fat Crystal Networks, second ed. CRC Press, Boca Raton, FLA.

Mazzanti, G., Marangoni, A.G., Idziak, S.H.J., 2005. Modelling phase transitions during the crystallization of a multicomponent fat under shear. Phys. Rev. E 74, 041607.

Mazzanti, G., Marangoni, A.G., Idziak, S.H.J., 2008. Modelling of two-regime crystallization in a multicomponent lipid system under shear flow. Eur. Phys. J. E 27, 135–141.

Minato, A., Ueno, S., Yano, J., Wang, Z.H., Seto, H., Amemiya, Y., Sato, K., 1996. Synchrotron radiation X-ray diffraction study on phase behavior of PPP-POP binary mixture. J. Am. Oil Chem. Soc. 73, 1567–1572.

Mullin, J.W., 2001. Crystallization, fourth ed. Butterworth-Heinemann, Oxford (Chap. 5).

Narine, S.S., Humphrey, K.L., Bouzidi, L., 2006. Modification of the Avrami model for application of the melt crystallization of lipids. J. Am. Oil Chem. Soc. 83, 913–921.

Nývlt, J., 1968. Kinetics of nucleation from solutions. J. Cryst. Growth 3/4, 377–383.

Nývlt, J., Söhnel, O., Matuchova, M., Broul, M., 1985. The Kinetics of Industrial Crystallization. Academia, Prague.

Padar, S., Jeelani, S.A.K., Windhab, E.J., 2009. Crystallization kinetics of cocoa fat systems: experiments and modeling. J. Am. Oil Chem. Soc. 85, 1115—1126.

Rousset, P., 2002. Modeling crystallization kinetics of triacylglycerols. In: Marangoni, A.G., Narine, S.S. (Eds.), Physical Properties of Lipids. Marcel Dekker, New York, pp. 1—36.

Rousset, P., Rappaz, M., 1996. Crystallization kinetics of the pure triacylglycerols glycerol-1,3-dipalmitate-2-oleate, glycerol-1-palmitate-2-oleate-3-stearate, and glycerol-3,3-distearate-2-oleate. J. Am. Oil Chem. Soc. 73, 1051—1057.

Rousset, P., Rappaz, M., Minner, E., 1998. Polymorphism and solidification kinetics of the binary system POS-SOS. J. Am. Oil Chem. Soc. 75, 857—864.

Sahin, O., Dolas, H., Demir, H., 2007. Determination of nucleation kinetics of potassium tetraborate tetrahydrate. Cryst. Res. Technol. 42, 766—772.

Sangwal, K., 1989. On the estimation of surface entropy factor, interfacial-tension, dissolution enthalpy and metastable zone-width for substances crystallizing from solution. J. Cryst. Growth 97, 393—405.

Sangwal, K., 2007. Additives and Crystallization Processes: From Fundamentals to Applications. Wiley, Chichester.

Sangwal, K., 2009a. Novel approach to analyze metastable zone width determined by the polythermal method: physical interpretation of various parameters. Cryst. Growth Des. 9, 942—950.

Sangwal, K., 2009b. A novel self-consistent Nývlt-like equation for metastable zone width determined by the polythermal method. Cryst. Res. Technol. 44, 231—247.

Sangwal, K., 2009c. Effect of impurities on the metastable zone width of solute-solvent systems. J. Cryst. Growth 311, 4050—4061.

Sangwal, K., 2010. On the effect of impurities on the metastable zone width of phosphoric acid. J. Cryst. Growth 312, 3316—3325.

Sangwal, K., 2011a. Some features of metastable zone width of various systems determined by polythermal method. Cryst. Eng. Comm 13, 489—501.

Sangwal, K., 2011b. Recent developments in understanding of the metastable zone width of different solute—solvent systems. J. Cryst. Growth 318, 103—109.

Sangwal, K., Smith, K.W., 2010. On the metastable zone width of 1,3-dipalmitoyl-2-oleoylglycerol, tripalmitoylglycerol and their mixtures in acetone solutions. Cryst. Growth Des. 10, 640—647.

Sato, K., 2001. Crystallization behaviour of fats and lipids: a review. Chem. Eng. Sci. 56, 2255—2265.

Sato, K., Arishima, T., Wang, Z.H., Ojima, K., Sagi, N., Mori, H., 1989. Polymorphism of POP and SOS: I. Occurrence and polymorphic transformation. J. Am. Oil Chem. Soc. 66, 664—672.

Sato, K., Bayes-Garcia, L., Calvet, T., Cuevas-Diarte, M.A., Ueno, S., 2013. External factors affecting polymorphic crystallization of lipids. Eur. J. Lipid Sci. Technol. 115, 1224—1238.

Sato, K., Kuroda, T., 1987. Kinetics of melt crystallization and transformation of tripalmitin polymorphs. J. Am. Oil Chem. Soc. 64, 124—127.

Smith, K.W., Cain, F.W., Talbot, G., 2005. Kinetic analysis of nonisothermal differential scanning calorimetry of 1,3-dipalmitoyl-2-oleoylglycerol. J. Agric. Food Chem. 53, 3031—3040.

Vanhoute, B., Dewettinck, K., Foubert, I., Vanlerberghe, B., Huyghebaert, A., 2002. The effect of phospholipids and water on the isothermal crystallization of milk fat. Eur. J. Lipid Sci. Technol. 104, 490—495.

Wright, A.J., Narine, S.S., Marangoni, A.G., 2000. Comparison of experimental techniques used in lipid crystallization studies. J. Am. Oil Chem. Soc. 77, 1239—1242.

CHAPTER 3

Powder X-ray Diffraction of Triglycerides in the Study of Polymorphism

Stefan H.J. Idziak
University of Waterloo, Waterloo, ON, Canada

INTRODUCTION

The use of X-ray diffraction in the study of polymorphism has been an essential tool for decades, with very early work using photographic film laying the foundation for our understanding of the polymorphic nature of triglycerides (TAGs). This chapter will provide a reasonably complete description of the theoretical basis for X-ray diffraction and will also describe how to simulate X-ray diffraction patterns of some TAGs. A short description of the different polymorphic forms seen in TAGs will be given followed by some comments on the usefulness of powder diffraction experiments for studying TAG polymorphism.

CRYSTAL STRUCTURE

Crystal Lattice

TAGs can form crystalline solids, which melt when exposed to higher temperatures. In the crystalline form, the molecules are arranged in a periodic fashion and in principle, if the location of a single molecule is known, then the location of all the other molecules in the crystal can be determined mathematically. This section will review crystalline structures and develop some of the methodologies needed to describe them.

The fundamental building block in a crystal is referred to as a unit cell, and the perfect crystal can then be described as an infinite repetition of the unit cells in space. In two dimensions, this is simple to visualize as seen in Fig. 3.1A, where a typical rectangular unit cell is shown.

The two-dimensional plane can be completely filled with these rectangles, with no empty spaces as shown in the figure. This requirement of completely filling space places a restriction on the number of different unit cells that are in possible in two dimensions at five. They are known as square, rectangular, centered rectangular, hexagonal, and oblique lattices, all of which correspond to different symmetries. These unit cells can

Structure-Function Analysis of Edible Fats, Second Edition
ISBN 978-0-12-814041-3, https://doi.org/10.1016/B978-0-12-814041-3.00003-4

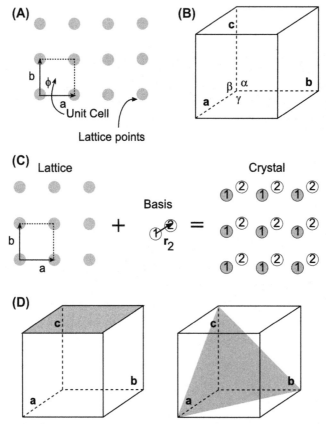

Figure 3.1 (A) Sketch of a lattice in two dimensions with unit cell vectors specified. (B) A three-dimensional unit cell. (C) Lattice and basis model showing how to construct a crystal in two dimensions. (D) Drawing of two different crystal planes specified by their Miller indices. (001) on the right and (111) on the left.

be characterized by two lattice vectors **a** and **b**, or by the lattice parameters, given by the lengths of the vectors, a and b, and the angle ϕ between them. Other rotational symmetries, such as fivefold, are forbidden (it is impossible to completely fill a plane with pentagons, for example).

The five two-dimensional unit cells can be characterized by different values of a, b, and φ. The square lattice, for example, requires a = b and $\varphi = 90$ degrees. The crystalline lattice can then be described by all the points that can be reached by adding different combinations of **a** and **b** vectors; in other words, the points of the lattice can be described by $u\mathbf{a} + v\mathbf{b}$, where u and v are integers.

In three dimensions (3D), which is how crystals are constructed, the unit cell is now defined by either three lattice vectors, **a**, **b**, and **c**, or by the six lattice parameters a, b, c,

Table 3.1 Bravais lattices in three dimensions

System	Lattice parameters	Types of Bravais lattice
Cubic	$a = b = c$ $\alpha = \beta = \gamma = 90$	Simple Body-centered Face-centered
Tetragonal	$a = b \neq c$ $\alpha = \beta = \gamma = 90$	Simple Body-centered
Orthorhombic	$a \neq b \neq c$ $\alpha = \beta = \gamma = 90$	Simple Body-centered Face-centered Base-centered
Trigonal	$a = b = c$ $\alpha = \beta = \gamma \neq 90$	Simple
Hexagonal	$a = b \neq c$ $\alpha = \beta = 90$ $\gamma = 120$	Simple
Monoclinic	$a \neq b \neq c$ $\alpha = \gamma = 90 \neq \beta$	Simple Base-centered
Triclinic	$a \neq b \neq c$ $\alpha \neq \beta \neq \gamma \neq 90$	Simple

α, β, γ as shown in Fig. 3.1B, 14 different lattices, known as Bravais lattices, are possible in 3D (Cullity and Stock, 2001; Hahn, 2006), all with different constraints on the lattice parameters. These constraints are described in Table 3.1. As in the 2D case, all lattice points can be reached by $\mathbf{T} = u\mathbf{a} + v\mathbf{b} + w\mathbf{c}$, where u, v, and w are all integers.

Several of these lattices have additional constraints, such as containing an additional lattice point in the center of the unit cell or on two or more faces. The interested reader should consult a textbook dealing with X-ray diffraction, such as the excellent one by Cullity (Cullity and Stock, 2001) for more details.

Of course, a real crystal has to be made of physical objects, such as atoms or molecules and not simply of lattice points. These different atoms form the basis of the crystal, with different atoms located in different positions within the unit cell. The crystal is then formed by placing the atoms of the basis in each unit cell with atom "n" located a distance $\mathbf{r}_n$ from the origin of each unit cell. This is illustrated in Fig. 3.1C where the two-dimensional unit cell is described by the lattice vectors $\mathbf{a}$ and $\mathbf{b}$ and the two atoms are located at positions $\mathbf{r}_1$ and $\mathbf{r}_2$, respectively (with $\mathbf{r}_1 = 0$ in this example). These two atoms are located in each unit cell and are always positioned $\mathbf{r}_1$ and $\mathbf{r}_2$ from the origin of the particular unit cell. In a perfect crystal then, knowing the location of the atoms within a single unit cell can be used to determine the location of all the atoms in the crystal. Therefore, knowledge of the crystal lattice vectors and the position of all the atoms in a single unit cell is all that is necessary to fully describe the crystalline structure of the material. If the lattice vectors are given by **a**, **b**, and **c** and the positions of all the atoms in a unit cell are given by $\mathbf{r}_i$, then the position of all the atoms in the crystal can be described by

$$u\mathbf{a} + v\mathbf{b} + w\mathbf{c} + \mathbf{r}_i,$$

where u, v, and w are integers.

Crystallographic Planes

As shall be shown later, X-rays diffract off of different planes in the crystallographic lattice, and the intensity of the diffracted beams depends on the location and type of atoms within the unit cell. These parallel planes intersect all the lattice points in the crystal with two sample planes shown in Fig. 3.1D. The (001) plane is shown on the left, and the (111) plane is shown on the right. These planes, identified using traditional Miller indices, are referred to by the notation (*hkl*) and are described as follows:

1. Determine where the plane intersect the a, b, and c axes of the unit cell. In the case of the (001) plane, the plane is parallel to the a axis; therefore, its intercept is at infinite (∞). Its intercept with the b axis is again at infinite, whereas its intercept with the c axis is at $1c$.

2. The plane is labeled by calculating the reciprocals of the intercepts. Therefore, again for the (001) plane, the fractional intercepts are at $\infty, \infty, 1$, and the reciprocals will be $\left(\frac{1}{\infty}, \frac{1}{\infty}, \frac{1}{1}\right)$ or (001). The commas separating the different coordinates are omitted by convention. If one of the intercepts was negative, this would be denoted by a bar on top of the reciprocal, such as the $(00\bar{1})$.

X-RAY DIFFRACTION

Geometrical Origin of Scattering

X-rays can be described both as electromagnetic waves, typically with wavelength around 1 Å (or 0.1 nm) and as particles called photons, which have a particular energy, usually around 10 keV. A typical in-house diffractometer can use a copper X-ray tube, which produces X-rays with a wavelength of 1.54 Å. This wavelength range is important, as it corresponds roughly to the size of the crystalline lattice, which makes interference effects important. X-rays are scattered by the electrons in the material (Baym, 1973) with the resulting scattered waves either destructively or constructively interfering. An X-ray diffraction peak can be detected when the waves constructively interfere; no peaks are seen when destructive interference occurs.

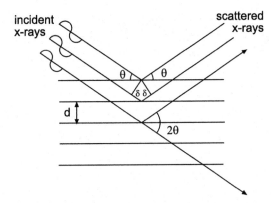

Figure 3.2 Diagram representing the geometrical approach to X-ray diffraction, with incident X-rays reflecting off of different crystal planes.

Fig. 3.2 shows a geometrical description of X-rays scattering from a set of planes in a crystalline material (these correspond to the (001) plane). Please note that a quantitative description based on scattering theory will be developed later.

Incident X-rays strike the crystal at an angle θ with respect to the crystalline planes. The X-rays are sufficiently energetic that they can penetrate the crystal to a reasonable depth and will interact with a large number of parallel planes (a typical X-ray beam produced using a copper source, producing 1.54 Å wavelength X-rays, will lose about half its intensity to absorption after passing through about 1 mm of a typical fat). Each plane will tend to reflect some small amount of the incident X-ray beam (as light is reflected by a mirror) with the reflected X-rays also leaving at an angle θ with respect to the planes as shown in the figure. Because the X-rays incident on the crystal are in phase, constructive interference in the reflected beam can only occur if the difference in path length that the X-rays travel is one full wavelength (or an integer number of wavelengths). As seen in the figure, X-rays reflecting off-the-top plane travel some distance from the source to the detector, whereas X-rays reflecting off the next plane travel the same distance, plus a small extra length, given by 2δ. Constructive interference then occurs when $n\lambda = 2\delta$, where n is an integer and λ is the wavelength of the incident X-rays.

If the distance between the adjacent planes is given by d (the d-spacing of the planes), then simple geometry can be used to show that $\delta = d \sin\theta$; therefore, the condition necessary for constructive interference is

$$n\lambda = 2d \sin \theta,$$

Which is known as Bragg's law where n is an integer (to reflect the possibility that the path length difference can be equal to one or more full wavelengths to get constructive interference).

In a typical X-ray diffraction experiment, the angle between the incident and reflected X-ray beams (2θ in the figure) is measured. This readily gives the d-spacing between the planes responsible for the particular reflection, with

$$d = \frac{\lambda}{2 \sin \theta}.$$

Given enough different measured diffraction angles, a set of d-spacings can be determined, which can be used to determine the crystallographic lattice of the material.

Single-Crystal X-ray Diffraction

Single-crystal X-ray diffraction is the gold standard for structure determination, with structures determined with great certainty. The primary technical difficulty with this method (and why it is not used for all structure determinations) is the difficulty associated with growing the needed single crystal in a size large enough for diffraction experiments. With access to modern synchrotron techniques (Dauter et al., 2010), this constraint has been reduced, but some materials still do not readily grow single crystals of reasonable size. These single crystals are required so that the entire piece of material inserted into the X-ray beam can be characterized by a single set of three crystal lattice vectors. This single set gives the location of all the lattice points within the entire crystalline material.

In these experiments, the angle between the crystal and the incident X-ray beam must be adjusted for each different set of diffracting lattice planes, as well as the angle between the detector and the transmitted beam. The relatively recent widespread use of 2D X-ray detectors has made this type of measurement relatively straightforward. A complete set of data includes the intensity of the diffracted X-ray beams as well as the angles describing the rotation of the crystal and the position of the diffracted beams. Given a reasonably large number of measured diffracted beams (also known as reflections), with the number required depending on the complexity of the crystalline structure, relatively straightforward analysis can be performed to yield the position of all the atoms in the unit cell, which fully describes the crystalline structure.

Powder X-ray Diffraction

In contrast, powder X-ray diffraction requires minimal sample preparation, but this is countered by the difficulty in obtaining structural information.

Most crystalline materials we experience in our day-to-day life, for example, the metal on a car or the crystallized fat in a chocolate bar, are not giant single crystals. Instead they are formed by an extremely large collection of very small crystals or crystallites that

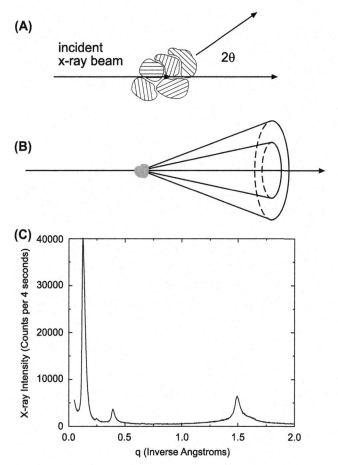

Figure 3.3 (A) X-rays scattering off a powder sample containing a very large number of randomly oriented crystallites. (B) Cones of scattering observed when scattering from a good powder sample. (C) X-ray powder diffraction pattern of cocoa butter in the α form.

are fused together with defects between the individual crystallites. This polycrystalline arrangement is actually responsible for some of the mechanical properties of these materials.

From an X-ray diffraction perspective, this collection of randomly oriented crystallites makes life very simple as illustrated in Fig. 3.3A. Assuming a very large number of randomly oriented crystallites, all with identical crystal lattices, an incident X-ray beam will sample all the crystalline orientations simultaneously.

When an X-ray beam is incident on the powder sample, X-rays are diffracted off in cones, with the opening angle of the cones corresponding to the various Bragg angles 2θ, as seen in Fig. 3.3B. Ideally, these cones of scattering are uniform, as there will always be a crystallite in an appropriate orientation for each Bragg angle and there is no need to rotate the sample. Although the cone of scattering appears to be continuous, it is important to realize that each part of the cone corresponds to scattering from different crystallites in the sample; the cone is actually made of numerous X-ray beams, all originating from diffraction by different crystallites. For this reason, it is essential to ensure that the sample has a large number of randomly oriented crystallites: too small a number and the cones will not be continuous and will look more like collections of individual diffracted beams. If the crystallite orientations are not random (say they have been oriented by a shear field (MacMillan et al., 2002; Mazzanti et al., 2003)), the cones will tend to have preferred orientations, which will also complicate further data analysis.

In the early days of structural characterization of fats, powder diffraction experiments were conducted by wrapping a piece of film around the studied material. The film would intersect with the cones of scattering and yield narrow bright circular segments (called lines) at positions corresponding to diffraction peaks at the various Bragg angles. Depending on the intensity of the lines, the diffraction peaks were characterized by intensities ranging from very strong (vs) to very weak (vw). A more modern powder diffractometer will consist of an X-ray detector, typically one that can count single photons, that revolves around the sample. The scattering intensity can then be measured quantitatively as a function of scattering angle, yielding a typical powder diffraction pattern for the α form of cocoa butter as seen in Fig. 3.3C. One- and two-dimensional X-ray detectors can also be used to perform powder diffraction measurements. These make data acquisition very fast with the detector intersecting many diffraction cones simultaneously, producing a diffraction pattern in one exposure. Image analysis can be done with the 2D images to extract intensity as a function of diffraction angle.

Unfortunately, although data collection is straightforward, a unique solution for the structure cannot be derived from a powder X-ray diffraction data set, as will be discussed later. Although the lattice can usually be readily found, the position of all the atoms in the basis is more difficult to elucidate, particularly for very large molecular systems, such as TAGs.

Calculation of D-Spacings and the Reciprocal Lattice

The positions of the diffraction peaks observed in the typical powder diffraction pattern shown in Fig. 3.3C correspond to the different d-spacings, or interplanar spacings, of the crystal lattice. Given a particular Bravais lattice, geometry can be used to calculate the spacing between all the different parallel planes (hkl), as seen in a variety of textbooks, such as that by Cullity (Cullity and Stock, 2001).

There is, however, an alternate way to calculate the d-spacings (and hence the Bragg angles) that will prove to be useful for calculations of the actual diffracted beam intensities. If the real-space unit cell (the one describing the real crystal lattice) is defined by the three lattice vectors, **a**, **b**, and **c**, we can calculate the three reciprocal lattice vectors[1]:

$$\mathbf{A} = 2\pi \frac{\mathbf{b} \times \mathbf{c}}{\mathbf{a} \cdot (\mathbf{b} \times \mathbf{c})}$$

$$\mathbf{B} = 2\pi \frac{\mathbf{c} \times \mathbf{a}}{\mathbf{a} \cdot (\mathbf{b} \times \mathbf{c})}$$

$$\mathbf{C} = 2\pi \frac{\mathbf{a} \times \mathbf{b}}{\mathbf{a} \cdot (\mathbf{b} \times \mathbf{c})},$$

where the $\times$ refers to the vector cross product and the $\cdot$ to the vector dot product. These three vectors can be used to generate a 3D lattice, known as the reciprocal lattice, with lattice points at location $\mathbf{G}_{hkl} = h\mathbf{A} + k\mathbf{B} + l\mathbf{C}$. The length of this vector $G_{hkl} = |\mathbf{G}_{hkl}|$[2] can be related to the spacing between real-space lattice planes, with spacing between planes (*hkl*) given by

$$d_{hkl} = \frac{2\pi}{G_{hkl}}$$

The vectors $\mathbf{G}_{hkl}$ also turn out to be perpendicular to the planes (*hkl*).

Quantitative Description of X-ray Diffraction

X-rays incident on a material interact predominantly with the electron charges in the material; this interaction is well described using quantum mechanics (Baym, 1973). After the analysis presented there, it can be shown (Als-Nielsen and McMorrow, 2001; Cullity and Stock, 2001) that the intensity of the X-ray beam scattered from a material can be written as

$$I(\mathbf{q}) = S(\mathbf{q})S^*(\mathbf{q}),$$

where $S(\mathbf{q})$ refers to the structure factor of the crystalline material, $S^*(\mathbf{q})$ refers to its complex conjugate, and **q** is the scattering vector, which corresponds to an arbitrary position in reciprocal space.

---

[1] Crystallographers tend to use an alternate definition of the reciprocal lattice vectors, which omits the 2π factor. This will affect how all the following equations look (factors of 2π can be missing).

[2] The length of a vector can be calculate by computing the square root of the dot product of the vector with itself, i.e., $G = \sqrt{\mathbf{G} \cdot \mathbf{G}}$.

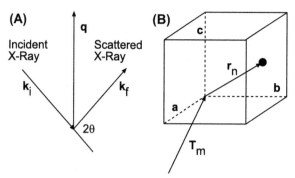

Figure 3.4 (A) Alternate view of X-ray diffraction, see text for details. (B) Sketch representing the location of an atom in a typical unit cell, with **T** representing the location of the unit cell origin and $\mathbf{r}_n$ the location of the atom in the unit cell.

The structure factor can then be written as the Fourier transform of the electron charge density in the material ($\rho(\mathbf{r})$)

$$S(\mathbf{q}) = \int \rho(\mathbf{r})\, e^{i\mathbf{q}\cdot\mathbf{r}} d^3 r$$

remembering that $e^{i\mathbf{q}\cdot\mathbf{r}}$ can also be written as $\cos(\mathbf{q}\cdot\mathbf{r}) + i\sin(\mathbf{q}\cdot\mathbf{r})$, where $i = \sqrt{-1}$, and the integral is carried out over the entire volume of the crystal. This is important as knowledge of the electron charge density implies knowledge of the crystal structure (if you know where all the electrons are, then you know where the atoms are located). This also demonstrates that the different X-ray intensity measured is dependent on the position of the individual atoms in the material.

This expression describes the structure factor (and hence the intensity) measured as a function of scattering vector $\mathbf{q}$ (also referred to as the momentum transfer), while most powder diffraction experiments give data as a function of scattering angle, or 2θ. An alternate view of X-ray scattering (Als-Nielsen and McMorrow, 2001) is shown in Fig. 3.4A, where the incident and scattered X-ray beams are shown. The lengths of these vectors, $\mathbf{k}_i$ and $\mathbf{k}_f$ correspond to $2\pi/\lambda$, where λ is the X-ray wavelength of the incident ($\mathbf{k}_i$) and reflected ($\mathbf{k}_f$) beams. Assuming elastic scattering, where no energy is lost, the incident and reflected wavelengths are the same; therefore, the magnitudes of $\mathbf{k}_i$ and $\mathbf{k}_f$ are the same. The scattering vector $\mathbf{q}$ can then be shown to be

$$\mathbf{q} = \mathbf{k}_f - \mathbf{k}_i,$$

in other words, the vector difference between the reflected and incident X-ray beam wave vectors. The reciprocal lattice vectors are perpendicular to the real-space lattice planes; therefore, it will turn out that $\mathbf{q}$ also describes the orientation of the crystal needed to get a particular lattice plane into the diffraction condition. Therefore, $\mathbf{q}$ can be related to the detector angle (2θ) and the angular orientation of the crystal. In a powder diffraction experiment, all of the crystalline orientations are averaged out; therefore, the reflected X-ray intensity is no longer measured as a function of $\mathbf{q}$ but by its magnitude q. It is fairly straightforward to show that

$$q = \frac{4\pi}{\lambda}\sin\theta = \frac{2\pi}{d},$$

where θ is half the Bragg scattering angle 2θ and d is the d-spacing of the lattice planes giving the reflection.

The periodicity of the crystal can be used to simplify the structure factor expression. Each lattice point can be labeled by a vector $\mathbf{T}_m$ as seen in Fig. 3.4B, where $\mathbf{T}_m = u\mathbf{a} + v\mathbf{b} + w\mathbf{c}$ and u, v, and w are integers; and we will assume that there are N atoms in each unit cell (in other words the basis consists of N atoms) with each atom located at a position $\mathbf{r}_n$ with respect to the origin of an individual unit cell. Then the structure factor can be rewritten as

$$S(\mathbf{q}) = \sum_{m}\sum_{n=1}^{N} \int \rho_n(\mathbf{r})\, e^{i\mathbf{q}\cdot(\mathbf{T}_m+\mathbf{r}_n+\mathbf{r})}\, d^3r,$$

where the first sum is over all the lattice points and the second sum is over all the atoms in a single unit cell. This can be rewritten as

$$S(\mathbf{q}) = \sum_{m} e^{i\mathbf{q}\cdot\mathbf{T}_m} \sum_{n=1}^{N} e^{i\mathbf{q}\cdot\mathbf{r}_n} \int \rho_n(\mathbf{r})\, e^{i\mathbf{q}\cdot\mathbf{r}}\, d^3r.$$

The integral (or Fourier transform of the charge distribution of an individual atom) is known as the atomic form factor (or atomic scattering factor) and describes how an individual atom interacts with an X-ray photon. These form factors have been calculated and can be found in various tables (Prince, 2006). The structure factor can then be written as

$$S(\mathbf{q}) = \sum_{m} e^{i\mathbf{q}\cdot\mathbf{T}_m} \sum_{n=1}^{N} e^{i\mathbf{q}\cdot\mathbf{r}_n} f_n(q),$$

where $f_n(q)$ is the atomic form factor of the nth atom in the unit cell.

These sums are independent of each other, and it turns out that the structure factor is only nonzero when $\sum_{m} e^{i\mathbf{q}\cdot\mathbf{T}_m} \neq 0$, which implies $\mathbf{q}\cdot\mathbf{T}_m = 2\pi j$, where j is an integer. This only occurs when the scattering vector $\mathbf{q}$ is exactly a reciprocal lattice vector $\mathbf{G}$. Thus scattering intensity only occurs (is measurable) at locations in reciprocal space corresponding to reciprocal lattice vectors, i.e., $\mathbf{q} = \mathbf{G}_{hkl} = h\mathbf{A} + k\mathbf{B} + l\mathbf{C}$ and is zero everywhere else. In other words, scattering only occurs at discrete locations in reciprocal space. The structure factor is therefore proportional to

$$S(hkl) \propto \sum_{n=1}^{N} f_n(q) e^{i(h\mathbf{A}+k\mathbf{B}+l\mathbf{C})\cdot\mathbf{r}_n}$$

where the form factor still depends on the magnitude of the scattering vector, which is defined by h, k, and l. In practice, form factors in tables are given as a function of $\frac{\sin\theta}{\lambda}$, which can easily be related to q by noting that $\frac{\sin\theta}{\lambda} = \frac{q}{4\pi}$. If $\mathbf{r}_n$ is expressed in fractional format

$$\mathbf{r}_n = u_n\mathbf{a} + v_n\mathbf{b} + w_n\mathbf{c}.$$

Then the structure factor can be rewritten as

$$S(hkl) \propto \sum_{n=1}^{N} f_n(q) e^{i2\pi(hu_n+kv_n+lw_n)}.$$

It is important to comment that the structure factor is not measured in a diffraction experiment but the square of the structure factor (the intensity). If the structure factor itself was measured, a single Fourier transform of the structure factor would yield the crystal structure $\rho(\mathbf{r})$ directly. Unfortunately, the intensity is measured and the structure factor is not equal to $\sqrt{I(\mathbf{q})}$ as the relevant phase information cannot be determined. Therefore, elucidation of the real-space crystal structure is not trivial, even with a complete 3D description of the scattering intensity coming from the crystalline material.

The Quick Summary

In short, what this means is that diffraction peaks corresponding to the (hkl) planes will be observed and will have intensity proportional to the square of the structure factor $\sum_{n=1}^{N} f_n e^{i2\pi(hu_n+kv_n+lw_n)}$. This sum is performed over all the atoms in a single unit cell, where (u_n,v_n,w_n) is the fractional position of the nth atom in the unit cell, i.e., the real position of the atom is at

$$u_n\mathbf{a} + v_n\mathbf{b} + w_n\mathbf{c}.$$

The measured X-ray intensity will be

$$I(\mathbf{q}) = S(\mathbf{q})S^*(\mathbf{q}) = |S(\mathbf{q})|^2.$$

The peaks will be observed at reciprocal lattice positions $\mathbf{G}_{hkl}$, which can be related to the d-spacing of the diffracting planes

$$d_{hkl} = \frac{2\pi}{|\mathbf{G}_{hkl}|} = \frac{2\pi}{|h\mathbf{A} + k\mathbf{B} + l\mathbf{C}|}.$$

In a powder sample, these peaks will be observed at angles 2θ that satisfy Bragg's law

$$\lambda = 2d_{hkl} \sin(\theta_{hkl})$$

or

$$2\theta_{hkl} = 2 \sin^{-1}\left(\frac{\lambda}{2d_{hkl}}\right).$$

This can also be written using the (simpler) notation of reciprocal space:

$$q = \frac{4\pi}{\lambda} \sin \theta \quad \text{or} \quad 2\theta = 2 \sin^{-1}\left(\frac{q\lambda}{4\pi}\right).$$

Sample Calculation

A simple calculation of the X-ray scattering intensity of common table salt, NaCl, will now be done to bring the concepts together. NaCl forms a face-centered cubic structure with one Na and one Cl in the basis. However, we will treat this as a simple cubic structure with eight atoms (four Na and four Cl) in the basis, as it makes a more interesting example. A cubic unit cell is characterized by a single length (the edge of the cube), which is 5.639 Å for NaCl (Cullity and Stock, 2001). The positions of each atom in the unit cell are shown in Table 3.2.

Therefore, the real-space lattice vectors describing the structure are

$$\mathbf{a} = 5.639\hat{x} \text{ Å}$$

$$\mathbf{b} = 5.639\hat{y} \text{ Å}$$

$$\mathbf{c} = 5.639\hat{z} \text{ Å},$$

Table 3.2 Positions of atoms in NaCl unit cell

Atom type	Fractional position		
	u_n	v_n	w_n
Na	0	0	0
Na	0.5	0.5	0
Na	0.5	0	0.5
Na	0	0.5	0.5
Cl	0.5	0.5	0.5
Cl	0	0	0.5
Cl	0	0.5	0
Cl	0.5	0	0

where $\widehat{x}$, $\widehat{y}$, and $\widehat{z}$ are the unit vectors in the x, y, and z Cartesian directions. The reciprocal lattice vectors can then be calculated as

$$\mathbf{A} = 2\pi \frac{\mathbf{b} \times \mathbf{c}}{\mathbf{a} \cdot (\mathbf{b} \times \mathbf{c})} = 2\pi \frac{5.639\widehat{y} \times 5.639\widehat{z}}{5.639\widehat{x} \cdot (5.639\widehat{y} \times 5.639\widehat{z})} = \frac{2\pi}{5.639}\widehat{x} = 1.114\widehat{x}\ \mathring{A}^{-1}$$

$$\mathbf{B} = 1.114\widehat{y}\ \mathring{A}^{-1}$$

$$\mathbf{C} = 1.114\widehat{z}\ \mathring{A}^{-1}$$

Diffraction peaks will then be observed at reciprocal lattice positions:

$$\mathbf{G}_{hkl} = h\mathbf{A} + k\mathbf{B} + l\mathbf{C}.$$

For powder diffraction, peaks will be seen at scattering vectors

$$q_{hkl} = |\mathbf{G}_{hkl}| = 1.114\sqrt{h^2 + k^2 + l^2}.$$

The structure factor for each peak can then be calculated (for example, for the (111) peak)

$$q_{111} = |\mathbf{G}_{111}| = 1.114\sqrt{1^2 + 1^2 + 1^2} = 1.930\ \mathring{A}^{-1}$$

$$S(hkl) \propto \sum_{n=1}^{N} f_n(q)e^{i2\pi(hu_n + kv_n + lw_n)}$$

$$S(111) \propto f_{Na}\left(e^{i2\pi(1\cdot0+1\cdot0+1\cdot0)} + e^{i2\pi\left(1\cdot\frac{1}{2}+1\cdot\frac{1}{2}+1\cdot0\right)} + e^{i2\pi\left(1\cdot\frac{1}{2}+1\cdot0+1\cdot\frac{1}{2}\right)} + e^{i2\pi\left(1\cdot0+1\cdot\frac{1}{2}+1\cdot\frac{1}{2}\right)}\right)$$

$$+ f_{Cl}\left(e^{i2\pi\left(1\cdot\frac{1}{2}+1\cdot\frac{1}{2}+1\cdot\frac{1}{2}\right)} + e^{i2\pi\left(1\cdot0+1\cdot0+1\cdot\frac{1}{2}\right)} + e^{i2\pi\left(1\cdot0+1\cdot\frac{1}{2}+1\cdot0\right)} + e^{i2\pi\left(1\cdot\frac{1}{2}+1\cdot0+1\cdot0\right)}\right),$$

where $f_{Na} = 8.984$ and $f_{Cl} = 13.608$ (Prince, 2006) and the structure factor is proportional to 15.04. The intensity (structure factor squared) is then proportional to 226.2. Similarly, the intensity of the (200) peak can be calculated as 7345.9. These can both be normalized to the intensity of the (200) peak so that the (200) peak has relative intensity 100, and the (111) has relative intensity 3.1. This is then repeated for each of the possible scattering vectors so that intensities for all planes (hkl) can be calculated. These are not absolute intensities but can be compared relative with each other. The intensities are typically normalized to a strong peak; therefore, they can be compared with real measurements. The calculated intensities for several sets of planes are shown in Table 3.3.

It turns out that diffraction peaks that satisfy hkl not being all even or all odd have intensity exactly equal to zero, which tells us that the structure is not simple cubic but

Table 3.3 Measured and calculated X-ray diffraction intensities for NaCl. See text for details

Plane	Bragg angle (2θ, degrees)	Measured peak position (Å^{-1})	Calculated peak position (Å^{-1})	Measured intensity (normalized to (200) peak)	Calculated intensity	Multiplicity factor	Lorentz polarization factor	Corrected calculated peak intensity
(111)	27.4	1.929	1.930	10.8	3.1	8	15.5	5.7
(200)	31.7	2.227	2.228	100	100	6	11.1	100
(220)	45.4	3.149	3.151	65.2	72.2	12	4.6	59.8
(311)	53.9	3.694	3.695	2.9	1.5	24	2.9	1.6
(222)	56.5	3.857	3.859	15.6	57.1	8	2.5	17.5

is in fact face-centered cubic. Similar measurements of missing diffraction peaks are used to differentiate between all the different space groups associated with the different Bravais lattices (see Hahn, 2006 for a more detailed explanation).

Corrections to Calculated X-ray Intensities

A typical powder diffraction pattern for NaCl is shown in Fig. 3.5, with intensities normalized to the (200) diffraction peak. The measured and calculated peak positions and intensities are presented in Table 3.3. The differences between the measured and calculated intensities are large for some of the peaks but can be explained by taking a look at the actual diffraction experiment. A detailed description of these corrections can be found in the several textbooks (Als-Nielsen and McMorrow, 2001; Cullity and Stock, 2001).

Two effects will tend to broaden the calculated diffraction peaks. The crystallites in the powder are not perfect, which will tend to broaden the calculated diffraction peaks and make them wider, not linelike. The calculated diffraction pattern also must be convolved with some kind of peak function, like a Gaussian or pseudo-Voigt, to accommodate for the intrinsic resolution of the X-ray diffractometer. Both these will change the shape of the peak but not the intensity.

The calculated intensities must be corrected for a multiplicity factor (for example, in a cube, diffraction from the (100), (010), (001), ($\bar{1}$00), (0$\bar{1}$0), and (00$\bar{1}$) planes) all occurs at the same magnitude of scattering vector q; therefore, the calculated intensity of the (100) peak must be multiplied by 6 to compare with the measured intensity (an alternative is to index this peak as all six of the listed peaks above and then add the calculated intensity from each of these reflections to get the total intensity). The measured scattering intensity is also affected by real experimental phenomena as well. X-rays with different polarizations scatter with different intensities from the atoms, giving rise to a polarization

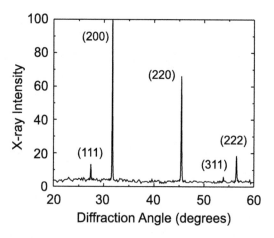

Figure 3.5 X-ray powder diffraction of table salt (NaCl) with peaks labeled by their crystal planes.

factor, which differs depending on the X-ray source and scattering geometry (for example, powder diffraction at the synchrotron will typically have a different polarization factor than an in-house diffractometer). The scattering intensity must also be corrected for geometrical effects such as that arising as a result of a real, finite-sized, detector passing through the conical diffraction rings coming from the sample. These last two corrections are often combined into a Lorentz polarization factor, which depends on the exact experimental details. A typical Lorentz polarization factor used for in-house diffractometers is

$$I \propto \frac{1 + \cos^2 2\theta}{\sin^2 \theta \cos \theta}.$$

The corrected calculated intensities are also shown in Table 3.3 and agree quite well with measured values.[3]

TYPICAL TRIGLYCERIDE STRUCTURES

TAG polymorphism has been studied for a long time, with early studies by Duffy (Duffy, 1853) clearly showing different forms with different melting temperatures. Historically, TAGs have been described as crystallizing into different crystalline forms (Chapman, 1962; Clarkson and Malkin, 1934; Ferguson and Lutton, 1941; Filer et al., 1946; Abrahamsson et al., 1978) (not phases as many of the structures are nonequilibrium) that can broadly be grouped by how the hydrocarbon chains are thought to be arranged. These forms can now loosely be described as γ, α, β′, and β, in order of increasing stability. Little is known about the γ form; therefore, we will not be discussing it. This nomenclature (α, β′, and β) is historical and is based on how long-chained molecules (not necessarily TAGs) typically crystallize into different polymorphic forms (Abrahamsson et al., 1978; Bunn, 1939).

Several authors have grouped the different forms by characteristic d-spacings seen in their powder X-ray diffraction patterns (Chapman, 1962; Clarkson and Malkin, 1934; Ferguson and Lutton, 1941; Lutton, 1950; D'souza, Deman, and Deman, 1990). Table 3.4 shows the different short d-spacings (larger q-values) associated with the different forms with values found in D'souza et al. (1990). In addition, all these forms also have diffraction peaks at smaller angles (or smaller q), which are referred to as long spacing diffraction peaks (observed at smaller angles as can be seen from Bragg's law). The long d-spacing is typically associated with the length of the molecule and had been generally thought to describe how the molecules stack on themselves in say

---

[3] We will not discuss corrections due to absorption or thermal motion of the atoms in the crystal. The resulting Debye—Waller factor is described in many X-ray diffraction texts (Cullity and Stock, 2001; Als-Nielsen and McMorrow, 2001).

Table 3.4 Characteristic d-spacings associated with different polymorphs derived from D'souza et al. (1990)

Polymorph	Characteristic short spacings (Å)	Chain-packing subcell
α	4.15	Hexagonal
β′	4.2 and 3.8 or 4.27, 3.97, and 3.71	Orthorhombic
β	4.6 (often strongest reflection)	Triclinic

the 2L or 3L conformations shown in Fig. 3.6A. Of course, molecular tilting with respect to the planes can complicate this description as shown in Fig. 3.6B, where a measured d-spacing would look more like that coming from a 2L structure. Harmonics of this peak can also be seen, for example, if a peak is seen at some scattering vector q, then peaks may also be seen at 2q, 3q, etc. In other words, if a particular *d*-spacing is observed, peaks can be seen for *d*-spacings corresponding to $^1/_2$, 1/3, $^1/_4$, etc. of the original d-spacing (these come from the different values of n in Bragg's law).

The characteristic short spacings for each form are said to be associated with the subcell of the structure arising from the packing of the hydrocarbon chains in the unit cell. Typical sketches of how the chains pack are shown in Fig. 3.6C for the α, β′, and β forms. Please note that there is strong single-crystal diffraction evidence supporting the description of the β form in TAGs. There is no definitive evidence supporting the hexagonal α form, and single-crystal studies of the β′ form have demonstrated that different subcells can exist in this form. Typical diffraction patterns of cocoa butter in these three forms can be seen in Fig. 3.7.

The Subcell

The subcell requires some consideration. TAG molecules (let us look at fully saturated molecules) have three alkyl chains, which presumably crystallize in the typical zigzag fashion of the pure alkane. Therefore, the overall crystal structure can be thought of as being comprised by the stacking of the molecules in the c direction of the unit cell and by the separate ordering of the individual CH_2 units found in the alkyl chains. This notion works extremely well and has aided in the early structure determination of many different molecules using single-crystal X-ray diffraction (Vand and Bell, 1951). An excellent overview of the different subcells can be found in the article by Abrahamsson et al. (1978).

The diffraction pattern can be thought to arise from two separate structures: the subcell giving a simple description of a set of diffraction peaks and the larger crystal structure describing them all. This can easily be seen by considering our previously derived expression for the structure factor of a crystalline solid, which consists of a sum over all the atoms in a unit cell. This sum could be broken into parts, with one (or more) of the parts only containing atoms contributing to the formation of the subcell.

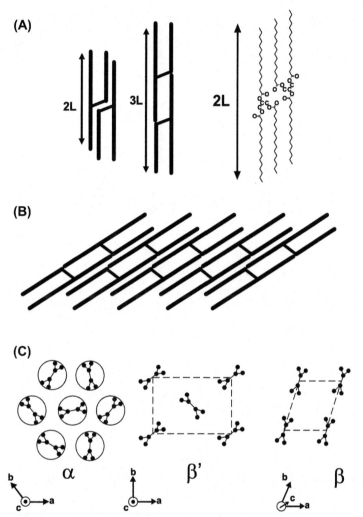

Figure 3.6 (A) Sketch of the vertical packing of triglyceride (TAG) molecules in different conformations. (B) Tilted TAG molecules in a plane showing the greatly reduced width of the resulting plane. (C) Schematic of the three common subcells found in TAG polymorphs (α, β', and β). *(Derived from Smith, K.W., 2001. Crystallization of palm oil and its fractions. In: Garti, N., Sato, K. (Eds.), Crystallization Processes in Fats and Lipid Systems. Marcel Dekker, Inc., New York.)*

This is easy to do in the structure factor but gets a bit more complicated when considering the X-ray diffraction intensities. Of course, any subcell description has to be fully consistent with the full crystal structure of the material being considered, and diffraction peaks that are described by the subcell must also be fully identifiable by the full crystal description as well.

Some of the typical subcells found in long-chain molecule crystals (Abrahamsson et al., 1978; Bunn, 1939; Smith, 2001) are seen in Fig. 3.6C. In the first subcell on

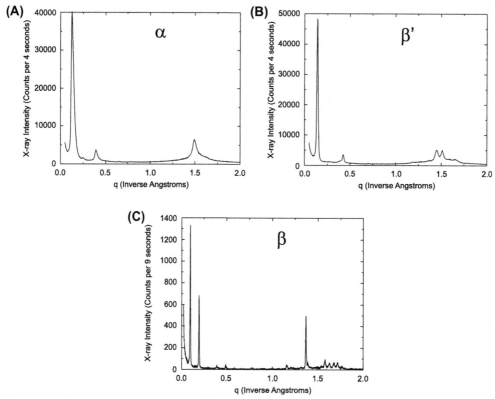

Figure 3.7 Characteristic X-ray powder diffraction patterns showing X-ray intensity as a function of scattering vector (**q**) for cocoa butter in three different forms. (A) α, (B) β′, and (C) β.

the left, the hydrocarbon chains are arranged perpendicular to the sketch, with the chains arranged in a hexagonal lattice. The relative orientation of the zigzag of the chains is not fixed for all molecules; therefore, the chain packing is shown as having cylindrical symmetry.

The middle sketch in Fig. 3.6C represents the orthorhombic perpendicular packing of the chains. Again, the chains are parallel with each other and are perpendicular to the sketch. The chains are arranged in a rectangular fashion with the relative orientation of the zigzag of adjacent chains being perpendicular with respect to each other as shown in the sketch (i.e., the zigzags are in different directions but are fixed). A complete unit cell is shown in the sketch, with two carbon atoms from each chain, and their associated four hydrogen atoms, depicted.

The sketch on the right represents a triclinic unit subcell with chains still parallel with each other but with the CH$_2$ groups, no longer on an orthorhombic lattice but a triclinic one.

The Three Polymorphic Forms (α, β', and β)

Many reviews describing the different polymorphic forms have been written, which can provide additional details beyond the scope of this chapter (Abrahamsson et al., 1978; Chapman, 1962; Clarkson and Malkin, 1934; Ferguson and Lutton, 1941; Larsson 1966, 1972; Lutton, 1950, 1972; Sato 2001a,b; Garti and Sato, 2001). Fig. 3.6 graphically describes the different forms. In the α form, it is thought that the alkyl chains are arranged parallel with each other in a hexagonal lattice, with molecules stacked on top of each other to form planes. These chains are not locked into any particular rotational orientation and are generally shown as having cylindrical symmetry. Free rotation of the chains is unlikely due to the steric constraint of having three alkyl chains on each TAG molecule, which should provide rotational hindrance with each other. This structure is typically characterized by a single-diffraction peak in the short spacing regime ($d = 4.15$ Å), which is said to correspond to the hexagonal packing of the chains. Please note that no definitive crystallographic analysis of the α form has ever been published, and it is very unreliable to base a structure on a single-diffraction peak. This model depends solely on how other long-chained molecular systems crystallize and not on an actual measurement of a TAG.

The subcell is probably not a useful concept for understanding the structure of the α form, if it is truly hexagonal as stated. Any disorder in the chains or free or random orientational order of the alkyl chains will inherently destroy any subcell. Again, if the α form structure is as traditionally described, there is no subcell that can be used in its characterization in 3D. Single-crystal diffraction measurements are sorely needed to verify this crystalline structure. Unfortunately, the inherent instability of the form may make such a measurement extremely difficult.

The β' form is more stable than the α but less than the β and is generally characterized by two different diffraction peaks corresponding to short d-spacings of 4.2 and 3.8 Å. The common description of the β' form states that the CH_2 groups of the alkyl chains are arranged in an orthorhombic fashion, with adjacent chain "zigzag" planes being fixed and oriented at different angles with respect to each other as shown in the middle sketch of Fig. 3.6C, where the chains are coming straight out of the page.

Several different types of possible β and β' structures have been reported in the literature and can be referred to by different notations. A good summary of the different forms and characteristics can be found in D'souza et al. (1990). In addition, other types of forms can be found when fats are exposed to external shear (Mazzanti et al., 2004). For this reason, it is difficult to say that the existence of d-spacings at 4.2 and 3.8 Å is sufficient for determining whether a form is a β' or not.

Two single-crystal studies have been performed on β' materials. In the first by van Langevelde et al. (2000), the β' form of CCC was determined. The lattice parameters for this structure, and for all the other single-crystal work reported here, can be found in Table 3.5. The orthorhombic unit cell for this β' structure is quite large, containing eight

Table 3.5 Unit cell parameters (in angstroms) for single-crystal diffraction studies of triglycerides

Material	Form	Crystal type	a	b	c	α	β	γ	Molecules per unit cell	Reference
LLL	β	Triclinic	12.31	5.40	31.77	94.27	96.87	99.20	2	Vand and Bell (1951)
CCC	β	Triclinic	5.50	12.10	27.03	85.25	87.30	79.64	2	Jensen and Mabis (1963)
CCC	β	Triclinic	5.488	12.176	26.93	85.35	87.27	79.28	2	Jensen and Mabis (1966)
SSS	β	Triclinic	11.97	5.45	45.88	87.2	103.0	101.0	2	Skoda et al. (1967)
CL^{Br}C	β	Triclinic	12.35	5.51	29.3	96	102	96	2	Doyne and Gordon (1968)
LLL	β	Triclinic	12.084	31.617	5.468	94.82	100.45	96.41	2	Gibon et al. (1984)
PP2	β	Monoclinic	5.375	8.286	42.96	90	93.30	90	2	Goto et al. (1992)
PPP	β	Triclinic	5.4514	11.945	40.482	84.662	86.97	79.77	2	van Langevelde et al. (1999)
Trielaiden	β	Triclinic	11.665	44.933	5.432	87.034	100.17	89.09	2	Culot et al. (2000)
CLC	β'	Orthorhombic	57.368	22.783	5.6945	90	90	90	8	van Langevelde et al. (2000)
PPM	β'	Monoclinic	16.534	7.537	81.626	90	90.28	90	8	Sato et al. (2001)

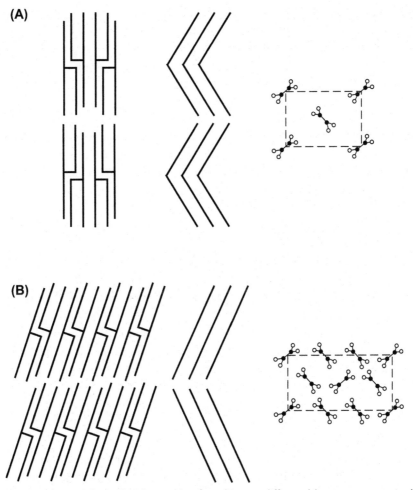

Figure 3.8 Schematics of the molecular packing found in two different b′ structures seen in different materials. (A) CLC (From van Langevelde et al., 2000). Reproduced with permission of the International Union of Crystallography. (B) PPM (From Jensen and Mabis, 1966). Reproduced with permission of the International Union of Crystallography.

full TAG molecules. The alkyl chains take on the conventional orthorhombic perpendicular subcell of the β′ form as seen in the right of Fig. 3.8A. The molecules stack on top of each other like chairs arranged seat on seat within layers as seen in the left part of the figure. The repeat distance is two molecular lengths, or roughly four times the average length of the alkyl chains. Looking along the edge of the molecules yields the arrangement seen in the middle of the figure clearly showing that the molecules are bent in the middle, resulting in an interesting chevron-like structure. Another way of looking at this is imagining the middle of the molecules in the left part of the figure is coming out of the page, while the ends of the chains go into the page.

The second single-crystal study by Sato et al. (2001) of PPM yielded a completely different type of β′ form. Here the TAG molecules still stack like chairs, but the molecules are tilted in their plane, as seen in the left part of Fig. 3.8B. The monoclinic unit cell also contains eight molecules, with the repeat distance corresponding to the length of two molecules (i.e., a 4L conformation). There is still a chevron-like arrangement of the molecules when looked on edgewise as seen in the middle of the figure, but in this case the individual molecules remain fairly straight and alternate layers of molecules tilt in different directions, yielding a longer period for the chevron. The subcell is also quite different as seen in the right side of Fig. 3.8B. It is clearly not the orthorhombic perpendicular subcell expected for a β′ form but a unique type of hybrid subcell.

Clearly, much work remains to be done characterizing the β′ form with one being orthorhombic and the other monoclinic. The two single-crystal works show that there are large differences in structure, both in stacking of the molecules and in the alkyl chain subcell. Yet both describe β′ forms.

The structure of the β form, on the other hand, is fairly well characterized. Traditionally, this form is identified in powder diffraction experiments by a large diffraction peak corresponding to a d-spacing of 4.6 Å. Several single-crystal studies of different molecules (Doyne and Gordon, 1968; Jensen and Mabis, 1963, 1966; Skoda et al., 1967; van Langevelde et al., 1999; Vand and Bell, 1951; Gibon et al., 1984) are summarized in Table 3.5 and have confirmed that the structure does indeed look like that shown in Fig. 3.6C with a triclinic unit cell and a triclinic subcell with the exception that the individual molecules are generally tilted in their planes.

SOME OPINIONS

Powder X-ray diffraction is an indispensable tool for studying the polymorphism of TAGs. It gives a clear signature of the different forms and provides definitive evidence of structural transformations from one form to another. With the increased use of synchrotron based diffraction experiments in the study of TAG polymorphism, diffraction patterns can be obtained at relatively high rates, with full patterns readily obtained every few seconds. This is essential for studying the dynamics of transitions between the forms and how these transitions are affected by external constraints, such as shear.

Care needs to be applied when interpreting these time-resolved experiments when attempting to calculate relative mass fractions of the different components. It is clear from the expression for the structure factor that any intensity measured depends on the actual position (and type) of all the atoms in the unit cell. Raw intensity measurements have traditionally been used as a measure of mass fraction, but care must be taken to ensure that the comparison of the intensities is valid. The calculated structure factor for the (001) peak in a β′ crystal is going to be different from that of a β crystal. Because of this difference, the relative mass fraction of the β to β′ found in the material is not simply

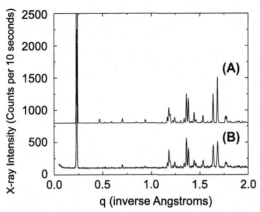

Figure 3.9 (A) Calculated diffraction pattern for CCC using a published single-crystal structure (Jensen and Mabis, 1966) and correcting for the Lorentz polarization factor. (B) Measured X-ray diffraction pattern for CCC.

equal to the relative fraction of the intensities of the two measured (001) diffraction peaks. Sometimes, when the structures are similar enough, there may not be much difference between the calculated peak intensities for the two different structures, but this must be determined in a rigorous manner before attempting this type of analysis. This characterization can be done either from calculations of the structure factor (if the structure is known) or calibration of the relative peak intensities against some kind of known standard.

Although it is straightforward to calculate a simulated powder diffraction pattern for a known structure, with an example of CCC given in Fig. 3.9A being a good illustration (this was calculated exactly the same way as the NaCl calculation performed earlier, with the atom positions given by Jensen and Mabis (1966) and remembering to consult the International Tables for X-ray Crystallography (Hahn, 2006) for the positions of the atoms in the second molecule). The measured diffraction pattern, taken with our in-house diffractometer is shown in Fig. 3.9B. It is clear that the agreement is excellent, with peak positions exactly in the right spot and peak intensities for the short spacings matching extremely well (there is some disagreement in the intensity of the low-angle peaks).

Unfortunately, the reverse is not true. It is very hard to calculate a structure from a powder diffraction pattern, particularly if there are many atoms in the unit cell. Of course, the availability of similar solved structures can be helpful, but care must always be taken to not overstate the models generated.

It is reasonably clear that the β form made from different molecular species have similar diffraction patterns when considering their short spacings. This is due to the similar subcells between the different crystals. The subcell arises from the packing of the alkyl chains, which will be fairly similar regardless of reasonable chain length (although different chains lengths can provide different signatures (Bunn, 1939)). Similar subcells would be expected to generate similar diffraction patterns.

The subcell has been extremely useful in solving early TAG single-crystal structures (Vand and Bell, 1951), as it has in the determination of other long-chain molecular systems. If the subcell is not known, however, or cannot be readily determined from the powder diffraction pattern, then indexing of the pattern can be extremely difficult for TAGs. This is due to the large asymmetry of the unit cell, with the c axis being much longer than the a and b axes. In this case, an (hk1) peak and an (hk2) peak can be fairly close together in diffraction angle and perhaps are indistinguishable if their separation is smaller than the instrumental resolution of the diffractometer. This would then manifest itself as a peak broadening in the measured diffraction pattern and not as two distinct peaks. Fairly high instrumental resolution diffractometers must be used to avoid this complication.

More work needs to be done toward understanding the relationship between the different β′ forms, particularly, when one considers the two published crystal structures, which clearly have different subcells. Perhaps, subcell characterization is not enough to describe structures as being in similar forms and that the β and β′ forms are actually many different types of significantly different forms that should not be grouped together. It is clear that a significant amount of single-crystal diffraction studies need to be done before a fairly compatible description of the different polymorphs of TAGs can be obtained.

REFERENCES

Abrahamsson, M., Groningsson, K., Castensson, S., 1978. Separation of pth-amino acids by isocratic high-performance liquid-chromatography. J. Chromatogr. 154 (2), 313–317.

Als-Nielsen, J., McMorrow, D., 2001. Elements of Modern X-ray Physics. John Wiley & Sons, New York.

Baym, G., 1973. Lectures on Quantum Mechanics. Benjamin Cummings, Menlo Park, CA.

Bunn, C.W., 1939. The crystal structure of long-chain normal paraffin hydrocarbons. The "shape" of the > CH_2 group. Trans. Faraday Soc. 35 (1), 0482–0490.

Chapman, D., 1962. Polymorphism of glycerides. Chem. Rev. 62 (5), 433–456.

Clarkson, C.E., Malkin, T., 1934. Alternation in long-chain compounds Part II an x-ray and thermal investigation of the triglycerides. J. Chem. Soc. 666–671.

Cullity, B.D., Stock, S.R., 2001. Elements of X-Ray Diffraction, 3 ed. Prentice-Hall, Inc., Upper Saddle River, NJ.

Culot, C., Norberg, B., Evrard, G., Durant, F., 2000. Molecular analysis of the beta-polymorphic form of trielaidin: crystal structure at low temperature. Acta Crystallogr. Sect. B Struct. Sci. 56, 317–321.

D'souza, V., Deman, J.M., Deman, L., 1990. Short spacings and polymorphic forms of natural and commercial solid fats - a review. J. Am. Oil Chem. Soc. 67 (11), 835–843.

Dauter, Z., Jaskolski, M., Wlodawer, A., 2010. Impact of synchrotron radiation on macromolecular crystallography: a personal view. J. Synchrotron Radiat. 17, 433–444.

Doyne, T.H., Gordon, J.T., 1968. Crystal structure of a diacid triglyceride. J. Am. Oil Chem. Soc. 45 (5), 333–334.

Duffy, P., 1853. XVIII.—on certain isomeric transformations of fats. Q. J. Chem. Soc. 5, 197–210.

Ferguson, R.H., Lutton, E.S., 1941. The polymorphic forms or phases of triglyceride fats. Chem. Rev. 29 (2), 355−384.

Filer, L.J., Sidhu, S.S., Daubert, B.F., Longenecker, H.E., 1946. X-ray investigation of glycerides .3. Diffraction analyses of symmetrical monooleyl-disaturated triglycerides. J. Am. Chem. Soc. 68 (2), 167−171.

Garti, N., Sato, K. (Eds.), 2001. Crystallization Processes in Fats and Lipid Systems. Marcel Dekker, Inc., New York.

Gibon, V., Blanpain, P., Norberg, B., Durant, F., 1984. New data about molecular-structure of beta-trilaurin. Bull. Soc. Chim. Belg. 93 (1), 27−34.

Goto, M., Kodali, D.R., Small, D.M., Honda, K., Kozawa, K., Uchida, T., 1992. Single-crystal structure of a mixed-chain triacylglycerol - 1,2-dipalmitoyl-3-acetyl-Sn-glycerol. Proc. Natl. Acad Sci USA 89 (17), 8083−8086.

Hahn, Th. (Ed.), 2006. International Tables for X-ray Crystallography Volume A: Space-group Symmetry. International Union of Crystallography.

Jensen, L.H., Mabis, A.J., 1963. Crystal structure of beta-tricaprin. Nature 197 (486), 681−682.

Jensen, L.H., Mabis, A.J., 1966. Refinement of structure of beta-tricaprin. Acta Crystallogr. 21, 770−781.

Larsson, K., 1966. Classification of glyceride crystal forms. Acta Chem. Scand. 20 (8), 2255−2260.

Larsson, K., 1972. Molecular arrangement in glycerides. Fette, Seiffen. Anstrichmittel 74, 136−142.

Lutton, E.S., 1950. Review of the polymorphism of saturated even glycerides. J. Am. Oil Chem. Soc. 27 (7), 276−281.

Lutton, E.S., 1972. Lipid structures. J. Am. Oil Chem. Soc. 49 (1), 1−9.

MacMillan, S.D., Roberts, K.J., Rossi, A., Wells, M.A., Polgreen, M.C., Smith, I.H., 2002. In situ small angle X-ray scattering (SAXS) studies of polymorphism with the associated crystallization of cocoa butter fat using shearing conditions. Cryst. Growth Des. 2 (3), 221−226.

Mazzanti, G., Guthrie, S.E., Sirota, E.B., Marangoni, A.G., Idziak, S.H.J., 2003. Orientation and phase transitions of fat crystals under shear. Cryst. Growth Des. 3 (5), 721−725.

Mazzanti, G., Guthrie, S.E., Sirota, E.B., Marangoni, A.G., Idziak, S.H.J., 2004. Novel shear-induced phases in cocoa butter. Cryst. Growth Des. 4 (3), 409−411.

Prince, E. (Ed.), 2006. International Tables for X-ray Crystallography Volume C: Mathematical, Physical and Chemical Tables. International Union of Crystallography.

Sato, K., 2001a. Crystallization behaviour of fats and lipids - a review. Chem. Eng. Sci. 56 (7), 2255−2265.

Sato, K., 2001b. Uncovering the structures of beta' fat crystals: what do the molecules tell us? Lipid Technol. 36−40.

Sato, K., Goto, M., Yano, J., Honda, K., Kodali, D.R., Small, D.M., 2001. Atomic resolution structure analysis of beta ' polymorph crystal of a triacylglycerol: 1,2-dipalmitoyl-3-myristoyl-sn-glycerol. JLR (J. Lipid Res.) 42 (3), 338−345.

Skoda, W., Hoekstra, L.L., Vansoest, T.C., Bennema, P., Vandente, M., 1967. Structure and morphology of beta-crystals of glyceryl tristearate. Kolloid-Zeitschrift and Zeitschrift Fur Polymere 219 (2), 149−156.

Smith, K.W., 2001. Crystallization of palm oil and its fractions. In: Garti, N., Sato, K. (Eds.), Crystallization Processes in Fats and Lipid Systems. Marcel Dekker, Inc., New York.

van Langevelde, A., van Malssen, K., Driessen, R., Goubitz, K., Hollander, F., Peschar, R., Zwart, P., Schenk, H., 2000. Structure of CnCn+2Cn-type (n = even) beta '-triacylglycerols. Acta Crystallogr. Sect. B Struct. Sci. 56, 1103−1111.

van Langevelde, A., van Malssen, K., Hollander, F., Peschar, R., Schenk, H., 1999. Structure of mono-acid even-numbered beta-triacylglycerols. Acta Crystallogr. Sect. B Struct. Sci. 55, 114−122.

Vand, V., Bell, I.P., 1951. A direct determination of the crystal structure of the beta-form of Trilaurin. Acta Crystallogr. 4 (5), 465−469.

CHAPTER 4

Melting and Solidification of Fats

Arun S. Moorthy[1]
University of Guelph, Guelph, ON, Canada

Naturally occurring fats are multicomponent mixtures of triacylglycerols (TAGs). A single TAG is a triester of glycerol with three fatty acid molecules. Schematic representations of TAGs can be seen throughout this book; however, for consistency with chapter conventions, we include Fig. 4.1 for reference. Similarly, Table 4.1 contains chapter-labeling conventions for describing fatty acids.

At practically relevant pressures and temperatures, fats will exist as liquids and solids. The molecular structure of TAGs, particularly the long hydrocarbon chain tails, allows for various packing structures during solidification (crystallization). This *polymorphism* contributes to the functional properties of the TAG and the subsequent fat system. There

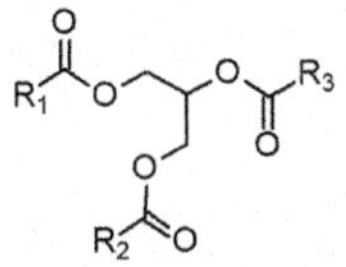

Figure 4.1 Schematic representation of triacylglycerol structure referred to throughout this chapter.

Table 4.1 Fatty acid descriptors used throughout the chapter. Number of carbons in hydrocarbon chain is indicated by "nC." Number of double bonds in hydrocarbon chain is indicated by "nU"

Symbol	Fatty acid	nC	nU	Symbol	Fatty acid	nC	nU
U	Butyric	4	0	G	Lignoceric	24	0
K	Caproic	6	0	F	Ceric	26	0
R	Caprylic	8	0	T	Palmitoleic	16	1
C	Capric	10	0	O	Oleic	18	1
L	Lauric	12	0	J	Linoleic	18	2
M	Myristic	14	0	N	Linolenic	18	3
P	Palmitic	16	0	D	Elaidic	18	1
S	Stearic	18	0	E	Erucic	22	1
A	Arachidic	20	0	H	Arachidonic	20	4
B	Behenic	22	0	#	# (odd chain)	#	0

[1] Present Address: Mass Spectrometry Data Center, National Institute of Standards and Technology, Gaithersburg, Maryland 20899, USA.

Structure-Function Analysis of Edible Fats, Second Edition
ISBN 978-0-12-814041-3, https://doi.org/10.1016/B978-0-12-814041-3.00004-6

are three basic polymorphic forms of TAGs: α, β', β. Basic polymorphism and additional *submodifications* have been observed in many fat systems, and are discussed at length in literature; a few references are (Himawan et al., 2006; Marangoni et al., 2012; Sato, 2001; Sato et al., 1989). The term *phase behavior* in the context of fats describes (1) how a fat material transitions from liquid to solid and vice versa, (2) the solid fraction of the material during the transitions between complete phases, (3) the polymorphism of any solid phase in the overall materials, and (4) the transition between solid phase polymorphic forms. These phase *properties* contribute to the overall function of the material, for example, spreadability, stability, hardness, etc.

Quantifying fat phase behavior is of significant interest across numerous application areas. Being able to correlate phase properties from a molecular structure (system composition), which can be extrapolated into function, is the first step toward *solving* for structure that create desired properties/functions. That is to say, by creating mathematical expressions that describe the relationship between structure and properties, we move the edible oil research paradigm from *discovering* new material compositions with desirable properties to *designing* desirable material compositions. In this chapter, we discuss known correlations between molecular structure (including polymorphism) and melting properties of fat systems. Our focus is primarily on pure TAG systems; however, notes about analyzing phase diagrams for mixed TAG systems are provided for consideration.

PURE TRIACYLGLYCEROL SYSTEMS

A **pure TAG system**, which we may interchangeably refer to as a single-component TAG system, can be characterized by the molecular composition of its component TAGs and its phase (and polymorphic form if solid). Examples of pure TAG systems include liquid triolein or solid β-tripalmitin. In this section, we investigate empirical correlations between pure TAG systems and their physical measurements.

Data

The methods described in the chapter can be broadly described as empirical model building. As with any data-driven method, strict scrutiny must be placed on the quality of data used. In this chapter, melting data for "pure" TAGs is adapted from Tables 9.A.1 through 9.A.12 in Wesdorp et al. (2012). The level of purity was not included in the original reference and so we work under the assumption that the reported values are for TAGs nearing ideal purity. The methods described should be reevaluated as more data are made available. In Table 4.2, we have revised the presentation of the aforementioned data set to be consistent with our nomenclature and for improved clarity as a resource for readers. Included are measurements for 196 TAGs, 13 of which have complete melting temperature and enthalpy measurements for α, β' and β polymorphic forms (boldfaced and italicized).

Table 4.2 Melting properties for pure triacylglycerols in various polymorphic forms

TAG/polymorph	Melting temperature (°C)			Melting enthalpy (kJ/mol)		
	α	β'	β	α	β'	β
RRR	−51	−19.5	9.1	18.4		69.2
GGG			86	160.9		
KKB	31		34			
KKP	−7.4	12				
KKS	6.8	17	22.6			
KSK	0		32			
RRB	26		38			
RRC		5.5	11.5	26		59
RRD	3					
RRS	5	25	31			
RCR		18.5	20.5	51	62	
RCC		12.8	19			
RSR			41			
AAA	62.9	69.5	77.8			
BBB	69.1	74.8	81.7	143.2		
BSB	61.1	71.8				
CRC			20			
CCC	−11.5	16.8	31.6	57.3		95
CCD	15					
CCL	0	26	30			
CCJ			−0.5			
CCM	3	31	34.5			
CCO		4.4				
CCP	2		35			
CCS	32	38	42.5			
CDD		25				
CLC	6	37.7	34	67	87.5	
CLL	5	31	34.1			
CLM		36.7				
CLP				57		95
CLS		40	44			
CMC	3	30	34			
CMM	15	38	43.5			
CMO			13.9			
CMS		42	45			
COC	−16.4		−4.8			
COO			−0.3			
CPC	6	36	40			
CPP	23	41	45.5	53		89
CSC	34	40	44.5			
CSS	42.5	46	48.3			

Continued

Table 4.2 Melting properties for pure triacylglycerols in various polymorphic forms—cont'd

Property	Melting temperature (°C)			Melting enthalpy (kJ/mol)		
TAG/polymorph	α	β'	β	α	β'	β
DDD	15.8	37	42.2	78		148
DDS	28.8		49.7			
DPD	26		44.5	79		130
DSD	34	43.2	49.7	92		155
DSS	43	56.7	61.1			
LCL	5		37.4			
LCS			41.8			
LDD			35.5			
NNJ		−16.5				
NNN	−44.6		−24.2			
NNS			−0.5			
NJN		−15.5				
NJO	−22.4					
NJS	−8.3					
NOJ		−28.5				
NON		−11.1				
NOO			−13.1			
NOS	−2.5					
NPJ		−4				
CSS	42.5	46	48.3			
NSJ	−9.2					
NSO	5.2					
NSS		27.8				
LLD			27			
JNO	−24.7					
LLL	***15.6***	***35.1***	***45.7***	***69.8***	***86***	***122.2***
LLJ	15.5					
LJJ	−11.5					
JJJ		−25.3	−12.3			84
LLM	19	37.8	42.3			116
LLO		18				
JJO	−16.4					
LLP	***20***	***43***	***45.6***	***67***	***90***	***117***
LLS	20.5	39.8	45.1			
JJS		2.5				
LML	24	49.8		83	112	
LMM	22	42	47.5			118
LMP			48.5	74	94	125
LMS		45.5	49			
LOL			16.5			
JOJ		−39				
LOO			5.1			

Table 4.2 Melting properties for pure triacylglycerols in various polymorphic forms—cont'd

Property	Melting temperature (°C)			Melting enthalpy (kJ/mol)		
TAG/polymorph	α	β'	β	α	β'	β
JOO	−2.2					
JOS		−3.5				
LPL	19	46.7				
LPL		42.5				
JPJ		−3				
LPO			29.5			
JPO	11.7					
LPP	*32*	*49.5*	*54.4*	*83*	*110*	*146*
LPS		47	52			
JPS	34.2					
LSL	21	43	49.8	66	87	131
JSJ	−3					
JSO	16.5					
LSP				76	107	124
LSS	36	51.4		70	104	123
JSS			35.8			
MAM			59	88	117	157
MCM		40	43.5			
MCS			51.7	67		88
MDD			40			
MJJ	−8.5					
MLM	24		49.8			
MLS			54.5			
MMD			39.5			
MMJ	20.5					
MMM	*32.6*	*45.9*	*57.1*	*81.9*	*106*	*146.8*
MMO			23.9			
MMP	*34.5*	*48.5*	*53.3*	*82*	*100*	*131*
MMS	*35.6*	*49.3*	*56.6*	*87*	*93*	*145*
MOM	11.7	26.4	28			
MOO			12.8			
MOP			27			
MOS			27			
MPA				72		122
MPM	36.2	59.5	55	93	127	
MPP	36	52	55.8	89		140
MPS	41.9		58.5	86	111	137
MSM	33	55.2		99	148	
NSJ	−9.2					
MSP				91	143.5	
MSS	*44*	*58.3*	*60.9*	*92*	*128*	*139*
ONO	−15					

Continued

Table 4.2 Melting properties for pure triacylglycerols in various polymorphic forms—cont'd

TAG/polymorph	Melting temperature (°C)			Melting enthalpy (kJ/mol)		
	α	β'	β	α	β'	β
ONS		14.2				
OJO			−9.5			
OJS		−10.4				
OOG		36.1				
OOA		29.2				
OOB		33.3				
OOO	**−33.7**	**−10**	**4.8**	**37**	**79**	**100**
OOS			23.5		110	
OPO			19.6			126
OPS	17.9		40.2		126	126
OSO	1	20.5	23.9			
OSS	30.3	41.9		71	125	
PKP	14.6		45.4			
PRP			48.5			122
PBB	55.9	66.1				
PBP	47.4	61.5	65.5			
PCP	20	44.5	51.8			122.5
PCS	16		54.2	80		125
PDD	22.8		44.2	81		134
PDP	39.4		54	122	135	150
PDS		48.5				
PNJ		−7.5				
PNN		−10.5				
PJJ		−4.2				
PJO	13.2					
PLP	32.6	49.6	53.9	65		121.5
PJP			27.1			100
PLS			57	70		123
PJS			24.5			
PMP	39.1		59.9	79		137
PMS	40.9	56.1	59.6	93		152
POJ	13.3					
POO			18.5		95	
POP	**16.6**	**33.2**	**37.2**	**70**	**104**	**140**
POS	19.6		31	78	114	150
PPD			50.2	118		157
PPJ		26.5				
PPO	18.4	34.6		53	111	
PPP	**44.7**	**55.7**	**65.9**	**95.8**	**126.5**	**171.3**
PPS	**46.4**	**58.7**	**62.6**	**100**	**124**	**166.3**
PSJ	36.5					
PSO	25.9	40			111	
PSP	**47.2**	**67.7**	**65.3**	**112.2**	**165.5**	**166**

Table 4.2 Melting properties for pure triacylglycerols in various polymorphic forms—cont'd

Property	Melting temperature (°C)			Melting enthalpy (kJ/mol)		
TAG/polymorph	α	β'	β	α	β'	β
PSS	50.1	61.8	64.4	106		175
SKS	27.8		53.1			
SRS	30		54			141
SBB	61.3	71.5	73.5			
SBS	56	64	69.9	128	189	
SCS		53	57.4			143
SDD				89		155
SDS	41.5		60.5			163
SNS			35.8			
SLS	36	57.3	60.3	70		132
SJS	37.9					
SMS	43.8	58.8	63.3	90		146.5
SOA			41.5			158
SOS	22.9	37		73	111	154
SPS	50.7		68	103		170.3
SSB	56.7	69.7	70.7			
SSS	*54.7*	*64.3*	*72.5*	*108.5*	*156.5*	*194.2*
9.9.9	−26	4				
11.11.11	2	27.9				
11.11.L	7.5	28.5	29.2			
11.13.11		42.6				
13.13.13	24.5	41.8	44.1			
15.15.15	39.5	50.8	54.6			
15.15.17			54			
17.17.17	49.9	61.5	63.9			
19.19.19	59.1	65.6	70.6			
21.21.21	65	71	77			
S.17.S	53.1		65.7			
S.19.S	55.5	69.8				
P.15.P	43.4		56.5			
P.17.P	48.2	61.2				

"Continuous" Empirical Models

Fatty acids are often distinguished by the "length," measured in number of carbons, and saturation properties of their hydrocarbon chains. For example, a palmitic acid contains a fully saturated hydrocarbon chain of 16 carbons (see Table 4.1 for fatty acid details). Historically, a similar approach has been used in describing a complete TAG—using the total number of carbons in its three constituent fatty acid hydrocarbon chains. This is often referred to as the "carbon number" of the TAG. For example, a 1-myristoyl-2-palmitoyl-3-stearoyl-glycerol (MPS) would have a carbon number of 48. In this chapter, we refer to empirical models that are developed around a "carbon number" description of TAGs as "continuous" models.

Defining TAGs, and other lipid materials (alkanes, fatty acids, methyl esters), by carbon number has proven valuable in developing correlations between TAG length and thermodynamic properties. In particular, many have shown that the enthalpy (ΔH_m) and entropy (ΔS_m) of fusion for **monoacid saturated TAGs** can be modeled as linear functions of carbon number:

$$\Delta H_m(n) = hn + h_o, \tag{4.1}$$

$$\Delta S_m(n) = sn + s_o, \tag{4.2}$$

where n is the total number of carbons, h and s are the hydrocarbon chain contributions to the measured enthalpy and entropy, respectively, and h_o and s_o are contributions due to head group (i.e., glycerol) to the measured enthalpy and entropy, respectively (see Wesdorp, 1990 and references therein). This linear correlation between heat (enthalpy) of fusion and carbon number for saturated monoacid TAGs as β-solids is visualized in Fig. 4.2.

The melting point (T_m) is simply the ratio of enthalpy and entropy of fusion:

$$T_m(n) = \frac{\Delta H_m(n)}{\Delta S_m(n)} = \frac{hn + h_o}{sn + s_o}. \tag{4.3}$$

Expanding the denominator in this equation as a power series of $1/n$ gives

$$T_m(n) = \frac{h}{s}\left(1 + \left(\frac{h_o}{h} - \frac{s_o}{s}\right)\cdot\frac{1}{n} - \frac{s_o}{s}\left(\frac{h_o}{h} - \frac{s_o}{s}\right)\cdot\frac{1}{n^2} + \dots\right). \tag{4.4}$$

Figure 4.2 Enthalpy of fusion as a function of total carbon number for β polymorphs of pure saturated monoacid TAGs.

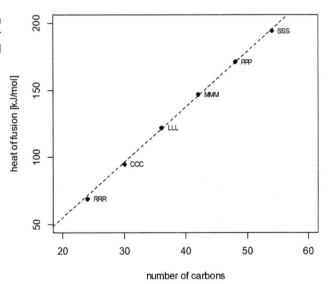

Truncating, we can approximate the melting point as

$$T_m(n) \approx T_\infty \left(1 + \frac{A}{n} - \frac{AB}{n^2} \right),$$

(4.5)

where the constants T_∞, A and B are given by

$$T_\infty = \frac{h}{s}, A = \frac{h_o}{h} - \frac{s_o}{s}, \quad and \quad B = \frac{s_o}{s}.$$

Multiple efforts have been made in extending (4.1) and (4.5) to model enthalpy (and melting temperatures) for more than simply monoacid TAGs. In particular, many authors have suggested adding additional terms to (4.1) such that the factors that differentiate TAGs can be accounted for when estimating thermodynamic properties, for example, differences in compositional fatty acids (chain length differences at R_1 through R_3 locations in Fig. 4.1), asymmetry, odd–number length chains, level of unsaturation, polymorphic form, etc.

To date, the semiempirical model of Wesdorp (Wesdorp, 1990; Wesdorp et al., 2012) is the most comprehensive in its attempt to capture properties for all combinations of fatty acids and all solid polymorphic forms. The published model parameters have been revised (Moorthy et al., 2017b), and a numerical implementation of the model has been made available by the American Oil Chemists' Society (AOCS) Lipids Library as the "triglyceride property calculator (TPC)." The model, although published elsewhere and implemented as a freely available software tool, has enough subtleties that we can benefit from an additional, revised description.

Wesdorp's method for predicting enthalpies of fusion for pure TAGs is presented as

$$\Delta H_m = \overbrace{hn + h_0}^{1} + \underbrace{h_{xy} f_{xy}(x, y)}_{2} + \overbrace{h_{odd} f_{odd} f_\beta}^{3} + \underbrace{\hat{h}_O n_O + \hat{h}_E n_E + \hat{h}_J n_J}_{4}.$$

(4.6)

The first contributor to the predicted enthalpy is the same as previously described for fully saturated monoacid TAGs. The second term in (4.6) "corrects" the enthalpy for differences in chain length between the three fatty acid moieties. In particular, the effect of fatty acid residue difference is modeled by

$$f_{xy} = 2 - \exp\left(-\left(\frac{x - x_o}{k_x} \right)^2 \right) - \exp\left(-\left(\frac{y}{k_y} \right)^2 \right),$$

(4.7)

where x is the difference in carbon number between the shorter terminal fatty acid (R_1 or R_3) and the medial fatty acid, and y is the absolute difference in carbon number between the two terminal fatty acids. The parameter h_{xy} in (4.6) and parameters x_o, k_x, and k_y in (4.7) are found empirically. A visual representation of the evaluation of (4.7) across a range of x and y values is provided as Fig. 4.3.

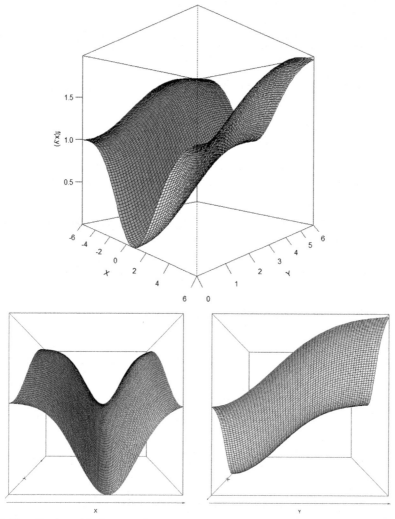

Figure 4.3 Visualization of f_{xy} function (Eq. 4.7) used to estimate the change in head group contribution to enthalpy of fusion for various differences in fatty acid chain length.

The third term in (4.6) corrects the enthalpy for the inclusion of odd length fatty acid chains in the TAG. The effect of odd chain fatty acids is only observed in β-solid polymorphs, and so switch functions (4.8) and (4.9) are used in term 3 of (4.6).

$$f_{odd} = \begin{cases} 1, & \text{if any of the fatty acids has an odd number chain length} \\ 0, & \text{if all fatty acids have even number chain length} \end{cases} \tag{4.8}$$

$$f_{\beta} = \begin{cases} 1, & \text{if TAG is } \beta \text{ polymorph} \\ 0, & \text{otherwise} \end{cases} \tag{4.9}$$

The parameter h_{odd} in term 3 of (4.6) is determined empirically. The fourth term in (4.6) captures the effect of unsaturated hydrocarbon chains on the predicted enthalpy of fusion. In particular, n_O, n_E, and n_J are the number of oleic, elaidic, and linoleic acid moieties, respectively, in the TAG. Note that the symbol used for elaidic acid in Tables 4.1 and 4.2 is "D"; however, "n_E" is used to represent the number of elaidic acids in a given TAG. This is to be consistent with other descriptions of the Wesdorp method found in literature (Moorthy et al., 2017b; Wesdorp, 1990; Wesdorp et al., 2012). The parameters $\widehat{h}_O$, $\widehat{h}_E$, and $\widehat{h}_J$ describe the increase (or decrease) in enthalpy due to the presence of oleic, elaidic, and linoleic acid moieties, respectively. In total, (4.6) through (4.9) contain 10 parameters that are tuned empirically, and these parameters are assigned differently for α, β', and β polymorphs. Accordingly, given any molecular composition and polymorphic form, and correct values for the 30 aforementioned parameters (10 parameters × 3 polymorphs), the enthalpy of fusion of the pure TAG can be predicted. Parameters are available in (Moorthy et al., 2017b).

Wesdorp's method predicts melting temperatures in two ways; both natural extensions of (4.3) through (4.5) using (4.6) to describe enthalpy of fusion. The general equation for melting temperature is

$$T_m = T_\infty \left(1 + \frac{A_u}{n} - \frac{A_u B_u}{n^2} \right), \tag{4.10}$$

where T_∞, A_u, and B_u are approximated in two ways. In **Case I**, the parameters are defined as follows:

$$T_\infty = \frac{h}{s}, \tag{4.11}$$

$$\begin{aligned} A_u = {}& \frac{h'_o}{h} - \frac{s'_o}{s} + \widehat{A}_O n_O + \widehat{A}_E n_E + \widehat{A}_J n_J + \widehat{A}_N n_N \\ & + \widehat{A}_{OO} n_{OO} + \widehat{A}_{EE} n_{EE} + \widehat{A}_{JJ} n_{JJ} + \widehat{A}_{NN} n_{NN} \\ & + \widehat{A}_{OJ} n_{OJ} + \widehat{A}_{ON} n_{ON} + \widehat{A}_{JN} n_{JN}, \end{aligned} \tag{4.12}$$

$$B_u = \frac{s'_o}{s} + \widehat{B}_O n_O + \widehat{B}_J n_J + \widehat{B}_N n_N, \tag{4.13}$$

where variables of the form n_{ij} indicate the number of pairs of fatty acids i and j. For example, n_{OJ} is the number of oleic acid and linoleic acid pairs contained in the molecule. Values for obtained fitting parameters are listed in (Moorthy et al., 2017b).

In **Case II**, the values of (4.10) are defined as follows:

$$T_\infty = T_{\infty,e}, \tag{4.14}$$

$$A_u = A_0 + A_{odd}f_{odd} + A_x x + A_{x^2}x^2 + A_{xy}xy + A_y y + A_{y^2}y^2$$
$$+ \hat{A}_O n_O + \hat{A}_E n_E + \hat{A}_J n_J + \hat{A}_N n_N$$
$$+ \hat{A}_{OO}n_{OO} + \hat{A}_{EE}n_{EE} + \hat{A}_{JJ}n_{JJ} + \hat{A}_{NN}n_{NN}$$
$$+ \hat{A}_{OJ}n_{OJ} + \hat{A}_{ON}n_{ON} + \hat{A}_{JN}n_{JN},$$
(4.15)

$$B_u = B_O + B_{odd}f_{odd} + B_x x + B_{x^2}x^2 + B_{xy}xy + B_y y + B_{y^2}y^2 + \hat{B}_O n_O + \hat{B}_J n_J$$
$$+ \hat{B}_N n_N,$$
(4.16)

where parameters are obtained empirically using only melting point data. The parameter values are detailed in full in Moorthy et al., 2017b. Usage of these models is made convenient through a software implementation as discussed previously and has been shown to fit 91% of the known TAG melting temperatures to within 10% of their actual values. A comparison of predicted melting temperatures and experimentally collected Mettler dropping points, for several example TAGs, is described in Moorthy et al. (2017a).

"Discrete" Empirical Models

In this section, we discuss approaches to predicting thermodynamic properties of TAGs while not reducing their description to carbon numbers, referring to these approaches as "discrete" empirical models. We begin with discussion of a fairly well-established method for fully saturated TAGs and then some preliminary thoughts on using *machine learning* approaches to predicting properties of pure TAGs.

Group Contribution Methods

Group Contribution (GC) is a family of methods that assume physical properties of a molecule can be efficiently described as the sum of physical properties of its "constituents." In fact, models (4.1) and (4.2) are GC methods where the constituent group is a CH_2 molecule, and h and s are the assumed enthalpy and entropy of fusion of CH_2, respectively. GC methods have proven to be very effective in approximating a variety of physical properties, including but not limited to boiling points (Stein and Brown, 1994), solubility (Nordström and Rasmuson, 2009), and other molecular properties (Benson, 1976; Irikura and Frurip, 1998). In this section, we describe one such GC method that is used to approximate the melting properties of pure saturated TAGs with an even number of carbons (10—22) in each acyl group. This method is detailed in Zeberg-Mikkelsen & Stenby, 1999. A software implementation of this method is also incorporated in the TPC available with the AOCS Lipid Library.

In this method, the "constituents" are the two pairs of terminal—medial acyl groups. For example, in MPS, one pair would be myristic—palmitic, and the second pair would be stearic—palmitic. The method sums the contribution of each pair and corrects for "class." We define four "classes" of TAGs discriminated by the type and position of fatty acids contained in the TAG. These can be generalized as follows:

Type 1: III, which we had previously referred to as a monoacid TAG that includes one unique acyl group,

Type 2: IIJ, an asymmetric TAG with two unique acyl groups,

Type 3: IJI, a symmetric TAG with two unique acyl groups, and

Type 4: IJK, a TAG with three unique acyl groups.

The approximate heat of fusion for a given saturated TAG is computed as

$$\Delta H_m = K_{H,IJK} \cdot \sum H_{ij}, \qquad (4.17)$$

where H_{ij} are the melting enthalpy contributions of the two terminal–medial acyl group pairs and $K_{H,IJK}$ is the TAG class parameter that adjusts the sum based on the four previously described classes. Parameters H_{IJ} and $K_{H,IJK}$ are obtained empirically for a variety of terminal–medial acyl group pairs and for all polymorphic forms and are available in Table 3 of Zeberg-Mikkelsen and Stenby (1999). Similarly, the approximate melting point of a saturated TAG is computed as

$$T_m = K_{T,IJK} \cdot \sum T_{ij}, \qquad (4.18)$$

where T_{ij} are the melting temperature contributions of the two terminal–medial acyl group pairs and $K_{T,IJK}$ is the TAG class parameter. Parameters T_{IJ} and $K_{T,IJK}$ are obtained empirically for a variety of terminal–medial acyl group pairs and for all polymorphic forms and are available in Table 4 of Zeberg-Mikkelsen and Stenby (1999).

As noted, both the "continuous" Wesdorp method (with revised parameters by Moorthy et al. (2017b)) and the "discrete" GC method (Zeberg-Mikkelsen and Stenby, 1999) are implemented in the TPC (available through the AOCS Lipid Library or through personal correspondence). Using the aforementioned software implementation, we can assess the "quality" of either approximation by computing the difference between the approximate measure (melting temperature or enthalpy of fusion) and the known measure. More precisely,

$$\delta = X_m - X_a, \qquad (4.19)$$

where X_m is the known measure and X_a is the approximate. Fig. 4.4 summarizes the quality of enthalpy approximations for β polymorph saturated TAGs using both the Wesdorp and GC methods categorized by the class of TAG.

It is clear that the approximations generated using the GC method are closer to real values, but this is easily justified by the number of parameters involved in these estimates. There are seven saturated fatty acids with an even length between 10 and 22 carbons: capric, lauric, myristic, palmitic, stearic, arachidic, and behenic acids. Using these fatty acids, we can create a total of 291 unique TAGs. Wesdorp's method allows us to

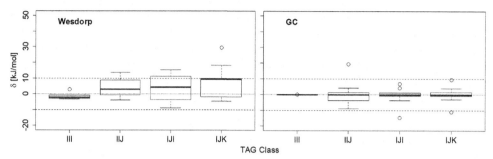

Figure 4.4 Comparison of error in heat of fusion prediction for various classes of beta-solid saturated TAGs using the Wesdorp and Group Contribution (GC) methods described in this chapter.

approximate melting temperatures and heats of fusions for all 291 TAGs in all polymorphic forms (1746 total property values) given 30 parameter values (or 37 if melting temperature is approximated using Case II). In comparison, the GC method needs an estimated contribution for all pairs of fatty acids, in both medial and terminal positions, in all polymorphic forms. In other words, the same 1746 approximations would require 318 parameters (294 contribution estimates, 24 class parameters). At present, such a data set does not exist. However, in practice, many of the 291 unique TAGs may not exist in any polymorphic form, and so the described GC method may be adequate for most applications.

Machine Learning

There has been significant interest in the application of *machine learning* methods to the approximation of melting properties for pure or naturally occurring fat systems. This class of methods is an extension of traditional pattern recognition and can be summarized as computational techniques that construct complex functional relationships between input variables and output data. A *training set* of data is used to construct the functional relationships (i.e., for the computer to learn how the variables might be related to the data) through an iterative procedure, and the identified functional construct (or pattern) is tested on a *validation set*. A few excellent introductions to pattern recognition and machine learning are cited (Alpaydin, 2014; Bishop, 2006; Mohri et al., 2012). There has been some success using machine learning approaches to predict molecular properties. For example, the application of artificial neural networks to predict enthalpy of fusions of pure compounds is described in Gharagheizi and Reza (2011).

As of the writing of this chapter, and to the best of our knowledge, there has not been any successful attempt at predicting melting properties of pure TAGs using a machine learning—based approach. The major issue is that the learning and subsequent model predictions are directly dependent on the quality and comprehensiveness of the training data set. Table 4.2 provides the most comprehensive set of melting data currently available, and yet only 13 of the included TAGs have complete information for all polymorphic forms. Furthermore, there is limited information about the uncertainty of the reported measurements in Table 4.2. To best make use of the data that are available, a clever "encoding" methodology must be developed to describe TAGs. In the two methods discussed earlier in this chapter, we saw both an abstract-encoding system (carbon numbers) and a very minimal abstraction (pairs of terminal—medial fatty acids). The strength of the abstraction was that we were able to generate predictions using fairly few parameters; however, these predictions were not always accurate. The strength of the minimal abstraction is that the predictions generated were very accurate, yet required assignment of many parameters resulting in limited applicability. Thus, one of the open challenges in this area will be determining an efficient TAG-encoding system that allows us to achieve good predictions for a variety of TAGs while requiring as few parameters as possible.

BINARY TRIACYLGLYCEROL SYSTEMS

This chapter has primarily focused on the correlation of melting properties for pure (single-component) TAGs. Although there have been substantial attempts in the area of multicomponent multiphase modeling of TAGs (see Rocha and Guirardello, 2009; Rocha et al., 2014; Wesdorp et al., 2012), the novelty of these methods makes comprehensive discussion difficult. Work toward numerical implementation and verification of these methods (similar to the work done with the TPC) is being undertaken and will be included in future editions of this book. For the current monograph, we focus our attention on developing a fundamental qualitative understanding of binary systems and the construction of binary phase diagrams.

A binary fat system is composed of two component TAGs. These systems may exist as liquids or solids, as ideal or nonideal solutions. In practice, binary TAG systems may not reach equilibrium, resulting in the system existing as multiple solid polymorphs. Additionally, the fatty acid chains of a mixed TAG may interact further contributing to its nonlinear behavior. This level of *freedom* is the basic challenge of quantifying melting and solidification behavior—the multicomponent system can exist in numerous states even at a fixed composition, temperature, and pressure. If we assume that the solid phase is limited to a single polymorphic form, as is most often the case when the system has reached equilibrium, general phase principles can be used to define the system.

A phase diagram of a binary fat system (at constant pressure) is a temperature—composition map detailing the equilibrium phases present at a given temperature, the fraction of these phases in the system, and the composition of each phase. Traditionally, the overall system composition is detailed along the horizontal axis and temperature along the vertical axis. Consider an idealized binary system with phase diagram shown in Fig. 4.5. This system is composed of components that display ideal solubility in

Figure 4.5 Binary phase diagram of a mono-
tectic system displaying ideal solid solubility.
*(Adapted from Bailey A.E., 1950. Melting and
Solidification of Fats. Interscience Publishers,
New York.)*

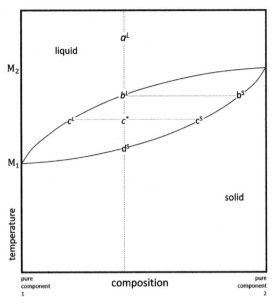

both the solid and liquid phase. Although such an ideal system is unlikely to be found in
practice, we can use this system to explain how a temperature–composition map can be
read. This approach is adapted from the work of Bailey (Bailey, 1950). There are two
curved lines, which split the temperature–composition map into three distinct regions
in Fig. 4.5. The topmost line is the *liquidus*, and any choice of temperature and composition mapping to a point above this line will result in a liquid system at equilibrium. The
bottommost line is the *solidus*, and any choice of temperature and composition mapping
to a point below this line will result in a solid system at equilibrium.

At point a^L in Fig. 4.5, the system is a completely liquid mixture of components 1 and
2. If the composition of the system remains unchanged but the temperature is reduced,
onset of solidification will begin at the temperature coordinate of b^L. The composition of
the initially formed solid can be determined by drawing a horizontal *tie* line from b^L to the
solidus line—marked as b^S on Fig. 4.5. If the system continues to be cooled until complete solidification (d^S), the composition of the final solid will be the same as the initial
liquid a^L. If the cooling process from state a^L to d^L is halted before full solidification,
e.g., at point c^*, the system will contain liquids with composition indicated by point c^L
and solids with composition indicated by point c^s, both of which were determined by
drawing a horizontal tie line from c^* to the liquidus and solidus lines, respectively, at equilibrium. The fractional amount of each phase (liquid and solid) can be determined using
the tie line information according to the "lever principle": the liquid fraction of the system at c^* can be computed as the difference in composition from the solidus line to the

system composition, over the difference in composition from the liquidus line to the solidus line. More precisely, the fraction of liquid in the system can be computed as

$$liq.\,frac = \frac{c^* - c^S}{c^L - c^S},$$

and the fraction of solid material in the system can be computed as

$$sol.\,frac = \frac{c^* - c^L}{c^S - c^L}.$$

Additional resources regarding the construction and analysis of phase diagrams with scope beyond fat systems are cited (Gordon, 1968; Kattner, 1997; Prausnitz et al., 1998).

As demonstrated, significant information can be gained from a binary phase diagram. If a phase diagram exists for a given binary or multicomponent system, one can determine the solid fraction of a mixture at a given temperature and initial composition. Ideally, and of incredible practical interest, one should be able to use the phase diagram to determine the initial composition that provides a desired solid fraction at a desired temperature. The limitations, which were alluded to earlier in the chapter but not expanded on, are that the majority of existing theory regarding phase behavior in multicomponent systems assumes that the system is near, or rapidly approaching, equilibrium. Developing mathematical methods that can lessen the burden of equilibrium assumptions and account for the practical kinetics experienced in the system (e.g., specialized heating/cooling treatment) is the future of this research area.

CONCLUSIONS

In this chapter, we discussed some of the fundamental data and correlations that govern our understanding of fat melting and solidification. In particular, we discussed the melting properties (i.e., melting point, heat of fusion) of pure TAGs as a function of a "continuous" variable, the total number of carbons in the component fatty acid hydrocarbon chains, and as a function of "discrete" combinations of fatty acids. We discussed the future of data correlations for pure TAGs (machine learning) and the value of generating quantitative models of melting behavior for multicomponent systems (phase diagrams). Continuing to add to our understanding of melting behavior, and building quantitative models relating measurable properties to molecular structure, will greatly advance the process of engineering fats with desirable functionality.

ACKNOWLEDGMENTS

It is my pleasure to thank Gary List, Gianfranco Mazzanti, and Leendert Wesdorp for sharing their notes and resources throughout the past several years. I would like to specifically acknowledge Alejandro Marangoni for introducing me to this rich research area and providing invaluable support.

REFERENCES

Alpaydin, E., 2014. Introduction to Machine Learning. MIT press.

Bailey, A.E., 1950. Melting and Solidification of Fats. Interscience Publishers, New York.

Benson, S.W., 1976. Thermochemical Kinetics. Wiley.

Bishop, C.M., 2006. Pattern Recognition and Machine Learning. springer.

Gharagheizi, F., Reza, G., 2011. Thermochimica acta prediction of enthalpy of fusion of pure compounds using an artificial neural network-group contribution method. Thermochim. Acta 521, 37–40.

Gordon, P., 1968. Principles of Phase Diagrams in Materials Systems. McGraw Hill, New York, p. 232.

Himawan, C., Starov, V.M., Stapley, A.G.F., 2006. Thermodynamic and kinetic aspects of fat crystallization. Adv. Colloid Interface Sci. 122, 3–33.

Irikura, K.K., Frurip, D.J., 1998. Computational Thermochemistry. ACS Publications.

Kattner, U.R., 1997. The thermodynamic modeling of multicomponent phase equilibria. JOM 49, 14–19.

Marangoni, A.G., Acevedo, N., Maleky, F., Peyronel, F., Mazzanti, G., Quinn, B., Pink, D., others, 2012. Structure and functionality of edible fats. Soft Matter 8, 1275–1300.

Mohri, M., Rostamizadeh, A., Talwalkar, A., 2012. Foundations of Machine Learning. MIT press.

Moorthy, A.S., List, G.R., Adlof, R.O., Steidley, K.R., Marangoni, A.G., 2017a. Using mettler dropping point data from dilute soybean oil-triglyceride mixtures to estimate thermodynamic properties for corresponding pure triglyceride. J. Am. Oil Chem. Soc. 94, 519–526.

Moorthy, A.S., Liu, R., Mazzanti, G., Wesdorp, L.H., Marangoni, A.G., 2017b. Estimating thermodynamic properties of pure triglyceride systems using the Triglyceride Property Calculator. J. Am. Oil Chem. Soc. 94, 187–199.

Nordström, F.L., Rasmuson, Å.C., 2009. Prediction of solubility curves and melting properties of organic and pharmaceutical compounds. Eur. J. Pharm. Sci. 36, 330–344.

Prausnitz, J.M., Lichtenthaler, R.N., de Azevedo, E.G., 1998. Molecular Thermodynamics of Fluid-phase Equilibria. Pearson Education.

Rocha, S.A., Guirardello, R., 2009. An approach to calculate solid-liquid phase equilibrium for binary mixtures. Fluid Phase Equilib. 281, 12–21.

Rocha, S.A., Lincoln, K., Boros, L.A.D., Krahenbuhl, M.A., Guirardello, R., 2014. Solid-liquid equilibrium calculation and parameters determination in thermodynamic models for binary and ternary fatty mixtures. Chem. Eng. Trans. 37, 535–540.

Sato, K., Arishima, T., Wang, Z.H., Ojima, K., Sagi, N., Mori, H., 1989. Polymorphism of POP and SOS. I. Occurrence and polymorphic transformation. J. Am. Oil Chem. Soc. 66, 664–674.

Sato, K., 2001. Crystallization behaviour of fats and lipids — a review. Chem. Eng. Sci. 56, 2255–2265.

Stein, S.E., Brown, R.L., 1994. Estimation of normal boiling points from group contributions. J. Chem. Inf. Comput. Sci. 34, 581–587.

Wesdorp, L., Van Meeteren, J., De Jon, S., Giessen, R., Overbosch, P., Grootscholten, P., Struik, M., Royers, E., Don, A., de Loos, T., 2012. Liquid-multiple solid phase equilibria in fats: theory and experiments. In: Marangoni, A.G., Wesdorp, L.H. (Eds.), Struct. Prop. Fat Cryst. Networks, second ed. CRC Press, Boca Raton.

Wesdorp, L., 1990. Liquid-multiple Solid Phase Equilibria in Fats: Theory and Experiments. TU Delft, Delft University of Technology.

Zeberg-Mikkelsen, C.K., Stenby, E.H., 1999. Predicting the melting points and the enthalpies of fusion of saturated triglycerides by a group contribution method. Fluid Phase Equil. 7–17.

CHAPTER 5

Rheology and Mechanical Properties of Fats

Joamin Gonzalez-Gutierrez[a], Martin G. Scanlon
University of Manitoba, Winnipeg, MB, Canada

IMPORTANCE OF RHEOLOGY TO FAT SYSTEMS

Rheology is a part of mechanics that studies the deformation and flow of matter in response to an applied stress. The capacity to undergo extensive flow, but yet exhibit a substantial solid character, is a desirable material property for essentially all industrially useful fat systems (Chrysam, 1985; Devi and Khatkar, 2016). The rheological and mechanical properties of fat materials arise directly from their composition and how these components are structurally organized (see Chapter 1 by Acevedo). In turn, the structure is markedly dependent on the various heat and momentum transfer processes that take place in the various unit operations used in the manufacture of the fat material (Ghotra et al., 2002). It is also important to recognize that fat materials are dynamically evolving systems; therefore, the duration of storage and the storage temperature can profoundly alter the rheology of lipid-rich food products.

The rheological properties displayed by a given fat system depend strongly on the physical state of the material. It will be appreciated from other chapters in this book that the biphasic nature (liquid and solid) of a fat system in large measure governs product structure, and as a consequence it governs fat rheology. For fat materials, melting point temperatures extend over a wide range, e.g., those of butter are from $-40°C$ to $+40°C$ (Wright et al., 2001), and thus in the technologically relevant temperature range for non-frying fat usage in food systems ($-20°C$ to $120°C$), the degree of solidity of the fat can be strongly temperature-dependent; therefore, temperature has a pronounced effect on the rheological response of fat. Further complicating the formulation of an easy definition of fat rheology are the effects of an additional fluid, such as when air is dispersed as bubbles in shortening (O'Brien, 1998) or water is dispersed as droplets in butter (Kalab, 1985).

Mechanical performance during processing is an essential quality feature when fats are used as ingredients in food, e.g., chocolate (Chen and Mackley, 2006; Yang et al., 2017) or dough (Campbell and Martin, 2012; Mert and Demirkesen, 2016; Macias-Rodriguez

---

[a] Current address: Institute of Polymer Processing, Department of Polymer Engineering and Science, Montanuniversitaet Leoben, Leoben, Austria.

Structure-Function Analysis of Edible Fats, Second Edition
ISBN 978-0-12-814041-3, https://doi.org/10.1016/B978-0-12-814041-3.00005-8

and Marangoni, 2017). In addition, fats confer desirable mouthfeel characteristics, such as smoothness and creaminess (van Vliet et al., 2009), and these sensory properties are largely determined by flow of the material to form thin layers in the oral cavity (Giasson et al., 1997; Chojnicka-Paszun et al., 2012). Accordingly, an understanding of rheological properties and how they are measured is an important consideration for improving the quality of fat products and improving the efficiency of their manufacturing process (Afoakwa et al., 2008; Ren and Barringer, 2016).

Two classical extremes in terms of mechanical behavior are the elastic solid and the Newtonian liquid. In the former, assuming testing rates are subsonic (Christensen, 1982), no rate dependency in properties is observed (i.e., no matter how fast or how slow the fat is tested, its properties remain constant). In the Newtonian liquid, the liquidity insinuates an inability to bear shear stresses (therefore, no matter how viscous the material is, it will continue to flow as long as a stress is applied). If the fat system is viewed as a solid then mechanical properties such as elastic modulus, Poisson's ratio, and yield stress (hence plasticity) are used to describe its mechanical behavior. If the material is viewed as a fluid, properties such as viscosity are apt descriptors of the fat's rheological behavior. However, fat systems, like all food materials, exhibit liquid and solid characteristics concurrently. Therefore, additional viscoelastic parameters, such as creep compliance, relaxation modulus, and relaxation time, are used to describe the rheological response. In addition, even at small strains, the elastic modulus of viscoelastic materials may be rate-dependent and will need to be defined in real and imaginary parts (often referred to as the storage and loss modulus (Ferry, 1970; Christensen, 1982; Barnes et al., 1989; Steffe, 1996; van Vliet, 2014)), and the magnitude of the real and imaginary parts may depend markedly on the rate at which the material is tested.

In attempting to depict simple rheological descriptions for fat systems, we will examine various constitutive models that appear to be valid. We will start from the solid "end" of the mechanics continuum because it is the simplest path to defining rheological properties that are most widely applicable to the great diversity of fat systems. However, it should be emphasized that many liquid fat systems are even more rheologically simple, requiring only one parameter, the viscosity, to fully define their rheological behavior over a wide range of rates of testing (Coupland and McClements, 1997). On the other hand, other researchers have reported that oils such as canola oil are mildly shear thinning, so that additional rheological parameters are required to adequately define their rheology (Mirzayi et al., 2011).

FAT AS A PERFECT ELASTOPLASTIC MATERIAL

In Fig. 5.1, the stress—strain curve of a specimen of an experimental milk fat blend subject to compressive deformation is shown as discrete points. The line that approximates the behavior of the fat blend is the stress—strain curve of an elastic perfectly plastic material

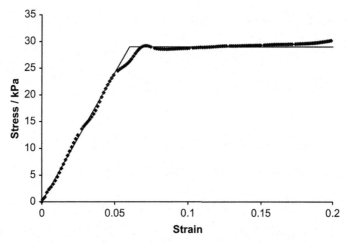

Figure 5.1 Stress–strain curve for uniaxial monotonic compression of specimen made from a blend of the high melting point fraction from butterfat and canola oil (25:75) compressed at 40 mm/min and an elastic perfectly plastic material with $E = 480$ kPa and yield strain of 0.06 (*solid line*).

whose elastic modulus (E) is 480 kPa and whose yield strain (ε_y) is 0.06. Although real fat systems (as will be seen later) do display rate dependency, elastic perfectly plastic materials do not. The test was performed at a specific rate of crosshead displacement, but if the same results were obtained at other rates of testing, we would have to conclude that the two-parameter constitutive model (E and ε_y) does a good job of defining the mechanical properties of the milk fat blend to a strain of 0.2. Therefore, both parameters that define this constitutive model will be examined.

Elasticity

Elastic Modulus

The elastic nature of a solid fat can be quantified using any one of three elastic moduli, the specific type of modulus being defined according to how the material deforms when subject to a particular simple state of stress. The most common means of measuring elasticity is to conduct a compression test, where a cylinder or prism of the solid fat is placed between two well-lubricated compression platens. From the measured force and the displacement of the platens toward each other, the stress and strain can be calculated. Because solid fats only display elastic behavior over a small strain region (Kloek et al., 2005; Macias-Rodriguez and Marangoni, 2016a; Renzetti and Jurgens, 2016), the stress and strain are accurately approximated by engineering stress (σ_E) and strain (ε_E):

$$\sigma_E = \frac{F}{A_0} \tag{5.1}$$

$$\varepsilon_E = \frac{h}{H}, \tag{5.2}$$

where F is the reaction force exerted by the specimen on the load cell, A_0 is the initial cross-sectional area of the specimen, H is the initial specimen height, and h is the current height, which for compression is the original height minus the current deformation of the specimen.

Once the stress and strain are calculated, they can be used to determine the elastic modulus (E) of the fat (Steffe, 1996):

$$E = \frac{\sigma_E}{\varepsilon_E}. \tag{5.3}$$

For a fat specimen that has been subject to compression (or indeed, in the unusual situation of being tested in tension (Svanberg et al., 2012)), the elastic modulus is the Young's modulus. Because of the common use of compression tests for determinations of elasticity, Young's modulus frequently defines the elastic character of fats (Rohm, 1993; Kloek et al., 2005; Goh and Scanlon, 2007; Gonzalez-Gutierrez and Scanlon, 2013; Yoshikawa et al., 2017). However, despite its common usage, a more comprehensive definition of the elasticity of a fat system requires measurement of the shear elastic modulus, G, and the bulk elastic modulus, K. For any elastic material, measurement of any two of the four common elastic constants fully defines the elastic behavior of the material (Edelglass, 1966). As will be seen in the sections on shear-based tests and bulk compression tests, when the shear modulus, G, is substituted for E in Eq. (5.3), shear stress and shear strain are the defining parameters, whereas for the bulk modulus (K), hydrostatic pressure and voluminal strain are the parameters in Eq. (5.3). Using G and K to define elasticity has the advantage that the former is a measure of the elastic resistance to a change in shape of the material without a change in volume, whereas in the latter case, volume changes but shape does not (Ferry, 1970).

Young's modulus is usually calculated from the initial linear part of the stress–strain curve. In the case of Fig. 5.1, this linear portion appears to extend all the way to the apparent yield strain at $\varepsilon = 0.06$. Not all elastic regions are this linear: where a nonlinear stress–strain relationship is evident, Young's modulus can be expressed as the secant modulus, either evaluated at a given stress or at a specific strain, e.g., 5% strain as done by Charalambides et al. (1995) for cheese (Fig. 5.2).

The modulus of elasticity is an important parameter that describes the deformability of a food material: it is related to the sensory attributes of such a material when one deforms it in the mouth (Narine and Marangoni, 2000; Renzetti and Jurgens, 2016), and it is related to other parameters that are used to define fat quality (Ronn et al., 1998). The elastic moduli, regardless of the type of stress state imposed are also a strong function of structure (Vreeker et al., 1992; Narine and Marangoni, 1999a,b; Ramirez-Gomez et al., 2016), including known quality parameters such as solid fat content (Kloek et al., 2005;

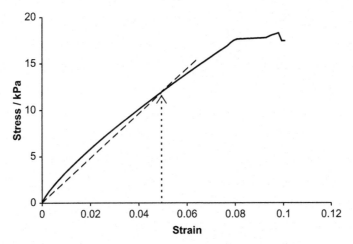

Figure 5.2 Schematic stress—strain curve showing how to calculate the secant modulus at 5% strain.

Liu et al., 2015). Indeed, over the past 25 years, significant progress has been made in understanding how the shear modulus of solid fat systems is related to the volume fraction of solid fat crystallites and their structural organization (Vreeker et al., 1992; Narine and Marangoni, 1999a; Schaink and van Malssen, 2007; Gregersen et al., 2016). Some of the impetus toward a better understanding of structure—shear modulus relations has been driven by the goal of eliminating trans fats (De Graef et al., 2007; Macias-Rodriguez et al., 2017) while still retaining the functional performance that governs fat quality. In contrast, few studies have been conducted where the bulk modulus of a solid fat system is related to structure or quality, and this is likely due to lower availability of equipment for measuring the bulk properties of materials.

Poisson's Ratio

When a fat material is compressed in one direction by a force (F) from original height (H) to a new height (h), it usually extends outward in the other two directions. In the case of the cylinder shown in Fig. 5.3, compression causes the cylinder to expand from an original diameter (D) to a new diameter (d). The elastic characterization of the extent of this lateral expansion relative to the axial contraction is through definition of the fat's Poisson's ratio (v). Poisson's ratio is defined (Eq. 5.4) as the negative ratio of the strain normal to the applied load (lateral strain) with respect to the strain in the direction of the applied load (axial strain) (Hjelmstad, 2005). The negative sign indicates that the lateral dimensions increase as the axial dimension decreases under compression and vice versa during tension.

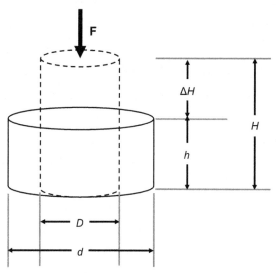

Figure 5.3 Geometric parameters used to calculate Poisson's ratio of a cylindrical specimen under compression. *(Adapted from Gunasekaran, S., Ak, M.M., 2003. Cheese Rheology and Texture. CRC Press LLC, Boca Raton, FL.)*

$$\nu = -\frac{\varepsilon_{lateral}}{\varepsilon_{axial}} = -\frac{\left[\dfrac{d-D}{D}\right]}{\left[\dfrac{h-H}{H}\right]}. \tag{5.4}$$

Most materials have a Poisson's ratio between 0.0 and 0.5, but some materials have negative values, i.e., materials that contract laterally as they are compressed (Lakes, 1987). A completely incompressible material will have a Poisson's ratio of 0.5, and in many studies of fats, it is assumed that $\nu = 0.5$ (Marangoni and Rogers, 2003; Vithanage et al., 2009). However, the presence of highly compressible regions, such as the air bubbles in commercial shortening (O'Brien, 2005), belies the applicability of an incompressibility assumption to all fat systems.

Interrelations Between Elastic Moduli

The three elastic moduli and the Poisson's ratio are the four common elastic constants relating stress to strain for isotropically elastic materials. Interrelationships between them are given by the following equations (Ferry, 1970):

$$E = \frac{9KG}{3K+G} \tag{5.5}$$

$$\nu = \frac{3K-2G}{3K+G} \tag{5.6}$$

$$E = 2G(1+\nu). \tag{5.7}$$

From Eq. (5.7), it can be seen that for an incompressible solid, where $v = 0.5$ (Edelglass, 1966; Ferry, 1970):

$$E = 3G. \tag{5.8}$$

Yield Stress and Plasticity

A linear proportionality between stress and strain remains only if the stress is less than the yield stress (Narine and Marangoni, 2000) and nonlinear elasticity is absent. Once the yield stress (σ_y) is exceeded, plastic flow occurs in the fat because the yield stress represents the stress required to induce permanent deformation; in other words, the yield stress marks the end of the elastic region of a material, and continued application of the stress causes the material to deform (at that yield stress for a perfectly plastic material). This appears to be the case for the elastic perfectly plastic model in Fig. 5.1 that mimics the milk fat blend's mechanical response. In such a material, when the stress is removed, the plastic strain, ε_p, is irrecoverable and thus the material remains deformed (White, 1999).

If the specimen is subject to large deformation, such as in Fig. 5.1, then true or Hencky stress (σ) and strain (ε) should be used to calculate the modulus of elasticity rather than engineering stress and strain (Eqs. 5.1 and 5.2). Eqs. (5.9) and (5.10) define Hencky stress and strain, respectively (Charalambides et al., 2001):

$$\sigma = \frac{F}{A} \tag{5.9}$$

$$\varepsilon = -\ln\frac{h}{H}, \tag{5.10}$$

where A is the cross-sectional area of the specimen after a force F has been applied to it. From Fig. 5.3, it can be appreciated that A ($\pi d^2/4$) is changing, whereas A_0 ($\pi D^2/4$) is not. The apparent yield strain of the milk fat blend in Fig. 5.1 is 0.058 when calculated using engineering strain, compared with its true strain counterpart of 0.06 (a difference of only 3%).

Measuring the actual value of the cylinder's area, A, is not easy; therefore, evaluation of true stress is experimentally challenging. True stress can be approximated if it is assumed that the material does not undergo a change in volume when deformed, i.e., the material is incompressible ($v = 0.5$); this assumption has been used in rheological testing of fat-based materials such as cocoa butter alternatives (Gregersen et al., 2015b), shortening, butter, and margarine (Goh and Scanlon, 2007), as well as foods that can have substantial amounts of fat, such as cheese (Thionnet et al., 2017) and some dough systems (Charalambides et al., 1995, 2001, 2006). Assuming incompressibility, the true stress can be calculated from the displacement, which for a prismatic specimen is given by

$$\sigma = \frac{Fh}{L^2H}, \tag{Eq. 5.11}$$

where L is the original side dimension of the specimen and other parameters are as previously defined.

It can be appreciated from Eq. (5.3), that for the elastic perfectly plastic material, yield stress, and yield strain are interrelated through the elastic modulus. Therefore, for the milk fat blend of Fig. 5.1, the apparent yield strain of 0.06 corresponds to an apparent yield stress of 29 kPa (because the modulus is 480 kPa). In real fat systems, the yield strain is much lower than this (Kloek et al., 2005; Macias-Rodriguez and Marangoni, 2016a), and for this reason, the yield stress and strain are denoted as "apparent," because, as in the case of the milk fat blend, the true yield stress may not be detected or even *be* detectable with many measuring techniques.

The yield stress is one of the most important macroscopic properties of economically important food products such as chocolate, butter, and shortening because it strongly correlates to the sensory attributes of hardness and spreadability, as well as to the fat system's stability (Marangoni and Rogers, 2003). Because the elastic modulus and the yield stress are both elastic descriptors of the mechanical performance of the fat system, strong correlations exist between them as a function of solid fat content (Vreeker et al., 1992; Marangoni and Rogers, 2003; Kloek et al., 2005; Reyes-Hernandez et al., 2007; Renzetti and Jurgens, 2016) and crystallite particle size (Afoakwa et al., 2008). Because of the importance of the correct fat crystal habit and microstructure to the processability and quality of chocolate, considerable research effort has been expended to derive reliable values for the yield stress (Afoakwa et al., 2009; De Graef et al., 2011).

FAT AS A RATE-INDEPENDENT STRAIN-HARDENING MATERIAL

Although the simplistic approach of the previous section appears attractive in adequately defining the properties of solid fat systems, it is not mechanistically tenable. The apparent yield strain observed for the milk fat blend in Fig. 5.1 is not a true yield strain, as it occurs at too large a strain. From linear shear strain/stress sweeps in milk fat/oil blends, yield strains of the order of 0.0005—0.002 (depending on solid fat content) have been reported (Awad et al., 2004). Using the same technique, a yield strain of 0.002 was reported for chocolate (Nigo et al., 2009). Likewise, compression tests can be used to show that a value of 0.06 strain is not feasible for the true yield strain of a fat system, with Kloek et al. (2005) reporting yield strains to be less than 0.001 in hydrogenated palm oil blends. This is illustrated in Fig. 5.4, where a specimen of vegetable shortening was subject to cyclic compression to a strain of 0.08 at a slow rate of crosshead displacement (so that rate-dependent effects are minimized). It can be seen that permanent deformation occurs in the shortening, even though the applied stress is well below the apparent yield stress of 10 kPa that is observed for this material (Gonzalez-Gutierrez and Scanlon, 2013). Therefore, the shortening is no longer responding in a solely elastic manner at a very low strain, and the continued rise in stress in the milk fat blend at larger strain is due to nonlinear elastic behavior associated with postyielding structural changes in the fat crystallites

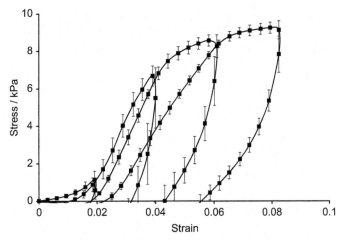

Figure 5.4 Average cyclic uniaxial compression of vegetable shortening at 4 mm/min; four cycles up to 8% strain (error bars are one standard deviation).

(emphasizing again at this stage, that discussion of a constitutive model for solid fats excludes rate-dependent effects).

Many materials, including fats, that have been subjected to stresses beyond their yield point undergo a phenomenon known as strain-hardening on subsequent deformation (Dasgupta and Hu, 1992). In this situation, plastic deformation does not occur at a constant stress, but the stress needs to increase incrementally for additional strain to occur (Bulatov et al., 2006). Based on a model for solid fat structures (Vreeker et al., 1992; Narine and Marangoni, 1999a), it can be conceived that initial yielding is associated with deformation occurring at the boundaries between flocs, but more strain energy is required to induce slip of crystal planes across one another (Chen and Mackley, 2006).

To define the rheological properties of a rate-independent strain-hardening material, additional mechanical parameters are required. For example, Goh and Scanlon (2007) used two parameters—a strain-hardening strain limit (ε_{y2}), where perfect plasticity was attained, and a power law strain-hardening constant (n) to characterize the increase in stress associated with the static elastic character (σ_s) beyond the yield stress. It is worth emphasizing here that up to this point all stresses are elastic static stresses and should have had the subscript "s" associated with them. The distinction is reiterated here because in the next section, the rate-dependency contribution to the overall stress–strain curve will be introduced that adds to the elastic contribution. The constitutive equations used to define the elastic nature of these materials were thus a function of strain (Goh and Scanlon, 2007).

$$\sigma_s = E\varepsilon \quad \text{for } \varepsilon \leq \varepsilon_y \tag{5.12}$$

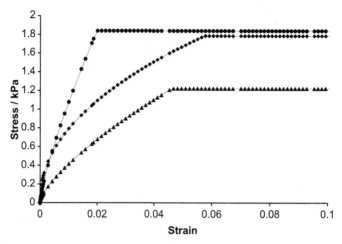

Figure 5.5 Static elastic contribution to stress–strain curves for vegetable shortening (diamond), margarine (circle) and butter (triangle), based on material constants obtained from the compressive stress–strain data of Goh and Scanlon (2007).

$$\sigma_s = E\varepsilon_\gamma \left(\frac{\varepsilon}{\varepsilon_\gamma}\right)^n \quad \text{for } \varepsilon_\gamma \leq \varepsilon \leq \varepsilon_{\gamma2} \tag{5.13}$$

$$\sigma_s = E\varepsilon_\gamma \left(\frac{\varepsilon_{\gamma2}}{\varepsilon_\gamma}\right)^n \quad \text{for } \varepsilon \geq \varepsilon_{\gamma2}. \tag{5.14}$$

The differences in the static elastic contribution to the stress–strain curves is shown for three different fat-rich materials (shortening, butter, and margarine) in Fig. 5.5 using the material constants determined by Goh and Scanlon (2007).

FAT AS A RATE-DEPENDENT MATERIAL

Solid fats, like most food materials, are not ideal elastic materials; their combination of both liquid and solid characteristics confers rate dependency, making them viscoelastic. This is evident in Fig. 5.6, where pronounced differences in the stress–strain curves are observed when the shortening is compressed at various loading rates. Therefore, in addition to the elastic parameters that were required to define rheological properties in the previous sections "Fat as a Perfect Elastoplastic Material" and "Fat as a Rate-Independent Strain-Hardening Material," rate-dependent parameters must also be incorporated into the constitutive models used to define the properties of fat systems (Macias-Rodriguez and Marangoni, 2017). Before defining parameters that are appropriate for rate-dependent descriptions of fat systems, a brief introduction to viscoelastic behavior is given.

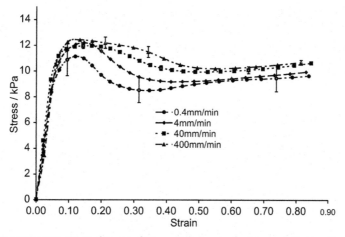

Figure 5.6 True stress—true strain diagram from uniaxial monotonic compression of vegetable shortening at four different loading rates. Each curve is the average of nine specimens with three different contact areas and error bars are 95% confidence limits from nine specimens at each data point.

Viscoelastic Solids and Liquids

In defining rate dependency for fats, we need models that add liquidity. Viscoelastic behavior is often modeled by a system of dashpots (visualized as pistons moving in oil) connected to springs. The spring represents the elastic component of the response (instantaneous deformation on application of a load and immediate relaxation on load release). The spring's elasticity here will be defined by G, its elastic resistance to a simple shear stress. The dashpot represents the viscous component whose deformation increases in proportion to time as long as a load is applied and the piston continues to be drawn out. The rate at which the piston is withdrawn is defined by the Newtonian viscosity, η, associated with the viscosity of the oil in the dashpot. When the spring is connected in series to the dashpot, a Maxwell material is formed (Fig. 5.7A). This material exhibits steady

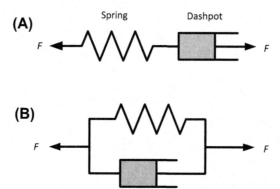

Figure 5.7 Basic (A) Maxwell and (B) Voigt elements consisting of a spring and a dashpot.

state flow a long time after a fixed shear stress has been applied; therefore, the model is characterized as a viscoelastic liquid. When the spring is connected in parallel to the dashpot, the material is known as the Kelvin or Voigt model (Fig. 5.7B). In this material, the full-scale deformation of the spring prevents unrestricted flow of the material; therefore, the Voigt model represents the simplest viscoelastic solid.

In evaluating the mechanical response of a viscoelastic material, an important consideration is how fast the material is being deformed relative to a characteristic time defining the rate at which the material flows. This characteristic time can be defined by the ratio of the viscosity of the dashpot to the elastic modulus of the spring:

$$\tau = \frac{\eta}{G}. \tag{5.15}$$

Although descriptions of rheological test procedures come in the next section, "Measuring Rheological Properties," in the context of time-dependent phenomena in viscoelastic liquids and solids, it is worthwhile examining two extremes in test protocols.

The ability of a material to alleviate stress under conditions of constant strain as a function of time is called stress relaxation (Hassan et al., 2005). In a stress relaxation test, the material is deformed to a specific strain (ε_0), and the stress in the material is measured as a function of time ($\sigma(t)$). Thus, in a complementary manner to how the modulus of elasticity is defined (Eq. 5.3), a relaxation modulus as a function of time is described:

$$G(t) = \frac{\sigma(t)}{\varepsilon_0}. \tag{5.16}$$

The relaxation modulus is a measure of how the stress decays with time when a constant strain is applied. The simplest way of characterizing the decay in stress uses the relaxation time of Eq. (5.15) in an exponential decay function, where the relaxation time is the time needed for the stress to fall to $1/e$ of its initial value under the constant strain, both of which can be measured during stress relaxation tests (Roylance, 1996).

The other extreme in test protocol is the creep test (Shimizu et al., 2012; Macias-Rodriguez and Marangoni, 2017). Here, the fat system is loaded with a specific applied stress (σ_0), and the strain ($\varepsilon(t)$) is monitored as a function of time. In an analogous manner to stress relaxation, the strain in this test is related to the stress by a function ($J(t)$) that defines the compliance of the material as a function of time (Ferry, 1970):

$$J(t) = \frac{\varepsilon(t)}{\sigma_0}. \tag{5.17}$$

Stress relaxation and creep tests can be performed under shear, bulk, or uniaxial compression ("Measuring Rheological Properties" section). In the case of stress relaxation and creep testing performed with shear strains and stresses, the distinction between a viscoelastic liquid and solid is apparent: in a relaxation test, $G(t)$ at long time is zero for a viscoelastic liquid, while it has a finite value for a viscoelastic solid; in a creep test, a viscoelastic solid displays a constant value at long time for $J(t)$, while $J(t)$ for a viscoelastic liquid will increase indefinitely, flowing in a steady state manner. Macias-Rodriguez and Marangoni (2017) reported steady state flow for a bakery shortening that was subject to shear creep testing with a 900 Pa stress. This viscoelastic liquid definition contrasts with the viscoelastic solid characterization of a different bakery shortening in the section "Measuring Rheological Properties" (Fig. 5.16).

Fat as an Elastoviscoplastic Material

Because of the viscous character of fat systems (Rohm, 1993; Vithanage et al., 2009), Eqs. (5.12)–(5.14) are inadequate for a full definition of a fat's mechanical response because they only define the static elastic response of the fat. An elastoviscoplastic model has been shown to be a suitable model of the mechanical behavior of materials, such as plasticine (Adams et al., 1996). The obvious analogy between the structures of plasticine, where solid clay particles are dispersed in a liquid hydrocarbon matrix (Adams et al., 1993), and fat, where crystal agglomerates interdigitate with a liquid oil matrix, was the motive for using an elastoviscoplastic model such as the overstress power law to characterize both the elastic and the rate-dependent response of fat systems (Goh and Scanlon, 2007).

In Eq. (5.13), where the yield stress of the fat has been exceeded, the strain is comprised of two parts, elastic strain (ε_s) and plastic strain (ε_p). In an elastoviscoplastic material, the viscous nature of the material modifies the stress values of Eqs. (5.13) and (5.14), and two additional material parameters, the fluidity ($\dot{\varepsilon}_0$) and the overstress power law constant (m) are required to model the stress–strain curve:

$$\sigma = \sigma_s(\varepsilon_p)\left[1 + \left(\frac{\dot{\varepsilon}_p}{\dot{\varepsilon}_0}\right)^m\right],\tag{5.18}$$

where $\dot{\varepsilon}_p$ is the plastic strain rate. Using this model, Goh and Scanlon (2007) showed that the compressive stress of three diverse lipid systems could be predicted with reasonable accuracy at various rates of testing (as seen in Fig. 5.8 for margarine).

MEASURING RHEOLOGICAL PROPERTIES

Because lipid-based particle gels, and food materials in general, exhibit a broad range of rheological characteristics, a variety of measuring techniques have been developed to characterize their rheological properties. These measuring techniques can be classified according to the mode of stress that is applied to the sample, either as bulk compression, shear, uniaxial compression (tension) or a combination (McClements, 2003). In turn, each stress mode can be subclassified according to the magnitude of the stress, so that

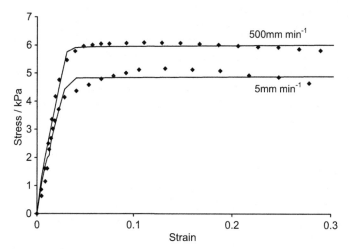

Figure 5.8 Comparison of compressive stress–strain curve of margarine with that predicted by elastoviscoplastic model (Eq. 5.18).

the fat system is subjected to small or large deformation. The nature of the information required from the test dictates this stress magnitude. For example, in examining the relationship between the native structure of fat blends and their elasticity (Narine and Marangoni, 1999a,b), it is essential that the yield stress is not exceeded and thus small stresses should be applied during testing. In contrast, when examining how well a fat system such as butter will spread at a given temperature (Wright et al., 2001; Vithanage et al., 2009), testing at large strains is appropriate. A further subclassification of both large and small strain stress applications for each stress mode is whether monotonic loading or cyclic load–unload actions are conducted. Because of their rate-dependent properties, the mechanical and rheological properties of fats may differ according to the rate at which the cycling is performed.

Shear-Based Tests

Many of the techniques used to measure the rheological properties of particle gels have been shear-based (Shukla and Rizvi, 1995). Simple shear deformation occurs when two opposite faces of a specimen are displaced by a certain distance after forces are applied on opposite faces (Fig. 5.9A). The angle φ defines the strain magnitude. In pure shear displacements, opposite faces are rotated relative to each other, and the angle of twist φ defines the strain magnitude (Fig. 5.9B). During shear tests, a change in shape of the specimen occurs without a change in volume, and this can be useful when interpreting the mechanical behavior of materials in molecular terms (Ferry, 1970).

Shear displacements within a specimen can also be induced by pressure-driven flow. In pressure-driven flow, pressure forces a sample to flow through a straight channel,

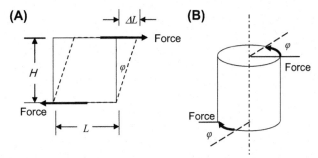

Figure 5.9 Schematic representation of simple (A) and pure (B) shear deformations in a specimen.

which may be a capillary or a slit causing shearing within the sample (Ouriev and Windhab, 2002; Gunasekaran and Ak, 2003). Pressure-driven rheometers are primarily used for the measurement of viscosity at high shear rates (Hatzikiriakos and Migler, 2004), a requirement for understanding how fats will perform in processes such as lamination and sheeting (Renzetti et al., 2016; Macias-Rodriguez and Marangoni, 2017).

Small Strain Shear Tests

Because fat-based particle gels have a fragile viscoelastic nature, many shear-based rheological tests are conducted as load—unload cycles (at a specific frequency) with applied shear stresses being well below those of the yield stress (the test is conducted in the linear viscoelastic regime). A typical application would be for measurements of the mechanical properties of the native material and relating those measurements to its structure (Vreeker et al., 1992; Narine and Marangoni, 1999a,b) or to its process history (Lupi et al., 2013). A typical test consists of applying a sinusoidal pure shear, sometimes referred to as an oscillatory or dynamic test. A sample is placed between two parallel plates; one is fixed, whereas the other one oscillates back and forth, either to a specified maximum stress or strain amplitude (in rheometers that are stress-controlled or strain-controlled, respectively). For the example of a strain-controlled experiment (Fig. 5.10A), by measuring the amplitude ratio (stress amplitude divided by strain amplitude) and the phase shift between the strain and the stress during the harmonic deformation (Fig. 5.10B), the shear storage (G') and loss (G'') moduli can be obtained (Ferry, 1970; Higaki et al., 2004; Macias-Rodriguez et al., 2017). The storage modulus is a parameter that directly relates to the elasticity of a material under shear, whereas the loss modulus relates to its viscous character during shearing. At a given frequency, G' and G'' must have constant values regardless of the stress or strain amplitude, i.e., the modulus is not a function of strain because testing occurs within the linear viscoelastic regime where the yield stress has not been transgressed.

A further means of evaluating the storage and loss moduli of fat materials, but generally at higher frequency, is by propagating shear waves. It is analogous to simple shear

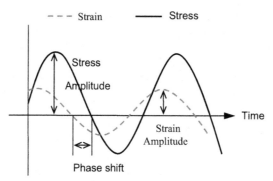

Figure 5.10 Schematic representation of dynamic shear test with sinusoidally varying shear strain. *(Adapted from Ferry, J.D., 1970. Viscoelastic Properties of Polymers. John Wiley and Sons, Inc., Toronto, ON.)*

(Fig. 5.9A), except that a rapid disturbance of the top face induces transversely polarized displacements that propagate to the bottom face of the sample at a particular velocity (Scanlon and Page, 2015). From the velocity and the attenuation coefficient of the recorded shear pulse (in addition to a separate determination of the fat's density), the shear storage and loss moduli of the fat can be determined. Difficulties in reliably transmitting shear wave pulses through soft solids such as fats mean that ultrasonic reflectance techniques can be a better means of acquiring a direct measurement of the shear modulus (Saggin and Coupland, 2004; Leroy et al., 2010). Using a reflectance technique, Saggin and Coupland (2004) used shear pulses reflected from fat crystal dispersions to measure their rheology. The technique has also been used to monitor the crystallization of cocoa butter (Rigolle et al., 2015).

Shear Tests to Measure Large Strain Viscoelastic Properties

As mentioned above, there are two common experiments for studying the viscoelastic behavior of materials: stress relaxation and creep tests, and both can be performed under shear. In recent years, large amplitude oscillatory shear (LAOS) has also been used to study the nonlinear properties of materials (Ewoldt et al., 2008), including fat systems (Macias-Rodriguez et al., 2017). As with the oscillatory tests, stress relaxation and creep tests can be performed at small and large strains or stresses, i.e., within the linear and the nonlinear viscoelastic regions. Here we address results acquired in the nonlinear region.

During stress relaxation tests, a constant shear strain is imposed on a specimen in a short period of time, and the shear stress is monitored over time. The material response is defined by the shear relaxation modulus ($G(t)$) according to Eq. (5.16). The test is continued for as long as it is necessary to obtain a good characterization of the relaxation modulus (Ferry, 1970).

During a shear creep test, the shear stress that is applied within a brief period of time is maintained constant, whereas the shear strain is monitored for a specific amount of time. The shear strain and the shear stress are related through the shear creep compliance ($J(t)$) using a rearrangement of Eq. (5.17) (Ferry, 1970).

In many situations where fats are used, and in particular during processing, flow of the fat is required. The flow properties can be acquired from the long-time response of a shear creep test. This approach was used by Vithanage et al. (2009) in assessing the temperature sensitivity of different butters to large deformation and by Macias-Rodriguez and Marangoni (2017) in determining the functionality of bakery shortenings. Alternatively, a steady state shearing experiment can be conducted on the fat using a rotational viscometer. A sample of the fat is placed between two symmetrical rotating fixtures (plates, concentric cylinders, or cones); some researchers achieve this by solidification of a fat melt in the fixtures, thus avoiding damage during sample preparation. The fat exerts a resistance to the applied torque of the instrument that is driving movement of the fixture, and this is related to the fat's viscosity. The viscosity is calculated as the ratio of the shear stress (σ) to the shear rate ($\dot{\varepsilon}$):

$$\eta = \frac{\sigma}{\dot{\varepsilon}} \tag{5.19}$$

More often than not, a viscosity that is constant over all shear rates is not evident. Instead, an apparent viscosity, valid at a specific shear rate, is cited (Nigo et al., 2009).

In an LAOS test, a standard linear viscoelastic test is extended to investigate nonlinear properties by gradually increasing the shear strain or the shear rate. A strain-controlled LAOS analysis of bakery shortenings was conducted at a fixed angular frequency by Macias-Rodriguez et al. (2017). They discovered that shortenings used for lamination applications (e.g., puff pastry) differed from standard cake shortenings in their nonlinear viscous dissipation behavior, an outcome that was compatible with the need for these shortenings to undergo massive strains during their normal use. The authors linked differences in the nonlinear shear properties of the fats to their microstructure determined by X-ray scattering and cryo—scanning electron microscopy analyses.

Bulk Compression Tests

The other desired elastic modulus that permits a full characterization of the elasticity of fat systems is the bulk modulus. There are two main means of measuring the bulk modulus: statically and dynamically.

Static Tests

In bulk or volumetric tests, a specimen is subjected to pressure that is uniform on all axes. In doing this, the shape of the specimen is conserved, and the volume decreases (assuming the pressure is larger than atmospheric pressure). The pressure imposed on the material is generally achieved by submerging the sample in a surrounding fluid, which is pressurized. Control of the fluid pressure is often achieved by an electrohydraulic loading piston and a pressure cell (Ferry, 1970). The relative change in volume of the specimen undergoing a bulk measurement test is called the voluminal strain ($\delta V/V$). In an exact analogy of shear stress relaxation and creep tests, a specific voluminal strain can be induced in the specimen, or it can be subjected to a specific hydrostatic pressure.

Assuming that the hydrostatic pressure (P_H) surrounding the specimen is the same on all axes and that voluminal strain is accomplished in a very short time interval, the measured hydrostatic pressure (with appropriate corrections for the finite compressibility of the intermediary fluid) changes as the specimen relaxes, and the degree of relaxation is regulated by the bulk relaxation modulus:

$$P_H(t) = -\frac{\Delta V}{V} K(t), \tag{5.20}$$

where $K(t)$ is the bulk relaxation modulus, which is analogous to the shear relaxation modulus $G(t)$ from shear-based relaxation tests.

As with shear creep measurements, if the pressure, P_H, is held constant after being applied "instantaneously" to the specimen and the volume change as a function of time is followed, a bulk creep experiment can be performed. During bulk creep tests the relative volume change is related to the applied hydrostatic pressure by the creep bulk compliance function $B(t)$ (Ferry, 1970):

$$\frac{\Delta V}{V}(t) = -P_H B(t). \tag{5.21}$$

Dynamic Bulk Tests

Ferry (1970) describes a number of techniques that have been used to evaluate the complex bulk modulus of polymers as a function of frequency. However, none of them appear to have been used to evaluate the bulk modulus of fat materials. The same information, but at higher frequencies, can be acquired from propagation of longitudinal pulses if the complex shear modulus has been acquired at the same frequencies and the density is determined. If the attenuation coefficient in the fat is small, then the bulk modulus can be determined from the following equation:

$$K(\omega) = \rho[v(\omega)]^2 - \frac{4}{3} G(\omega), \tag{5.22}$$

where ρ is the density and $v(\omega)$ is the ultrasonic velocity as a function of angular frequency. Maleky and Marangoni (2011) used an ultrasonic transmission measurement using longitudinal pulses and Eq. (5.22) to derive values for the bulk modulus of fats crystallized under different shear fields. Biofuel use of liquid fats has led to increased interest in acoustic methods for this purpose because bulk modulus measurements allow prediction of fuel performance (Kiełczynski et al., 2017).

Uniaxial Compression Tests

Although shear and bulk measurement tests have special significance in determining elastic and viscoelastic moduli, the ubiquity of universal testing machines (UTMs) and the ease of sample loading makes uniaxial compression tests the methods of choice. These tests have been used extensively to determine mechanical properties that are related to various properties of interest in a number of different fat systems; for example, how mechanical performance is affected by composition (Gregersen et al., 2015b; Macias-Rodriguez and Marangoni, 2016b), storage temperature (Rohm, 1993), or process conditions (Herrera and Hartel, 2000; Kloek et al., 2005). Compressive tests can also provide valuable information that relates mechanical characteristics to data obtained from sensory analysis (Di Monaco et al., 2008; Gregersen et al., 2015b). For this reason, compression tests are used routinely in the shortening industry to measure quality parameters, such as hardness and plasticity (Metzroth, 2005).

The UTM is a versatile instrument that provides precise control of displacement while accurately measuring force. A universal testing machine is so named because it can perform tension, bending, and shear tests through the use of different attachments (Gunasekaran and Ak, 2003). The development of closed-loop servohydraulic technology means that whereas UTMs were only able to evaluate the properties of materials at constant rates of displacement (so that strain rates were not controllable), now tests can be conducted as a function of true strain rates.

For a compression test, a specimen is typically prepared as a cylinder or a prism and is placed between two parallel rigid lubricated flat platens on the universal testing machine. Using this setup allows a uniform stress distribution within the specimen if frictionless conditions are achieved and the specimen contact surfaces are completely parallel to the compression platens (Charalambides et al., 2001).

A variety of compression-based tests can be performed including monotonic compression, cyclic compression, creep, and stress relaxation. During monotonic compression tests, the top platen moves down on the specimen at a constant rate of displacement until a target displacement is reached (Wright et al., 2001). Cyclic compression is similar to monotonic compression except that the specimen is subject to several cycles of loading and unloading. Creep and stress relaxation tests are performed with the top platen moving to compress the specimen with a specified load or to a fixed displacement, and the displacement or the force are recorded as a function of time.

Uniaxial Monotonic Compression

Uniaxial monotonic compression tests have been widely used to study the mechanical properties of materials and are one of the most popular tests for determining rheological properties of fat systems, including butter (Rohm, 1993), hydrogenated palm oil and sun-flower oil blends (Kloek et al., 2005), shortening (Gonzalez-Gutierrez and Scanlon, 2013), and margarine (Goh and Scanlon, 2007). The main reason for their popularity is ease of specimen preparation, certainly in comparison with tensile specimens, and there is no concern about invalidity of test results because of improper gripping (Gunasekaran and Ak, 2003). For soft fat materials, the latter consideration is likely paramount because it is very hard to prepare and grip specimens and obtain meaningful results. However, the results of compression tests can be affected by friction between compression fixtures and the sample (Charalambides et al., 1995). Friction leads to inhomogeneous deformation; if the region next to the compression platens is restrained from moving radially, then the material appears to be stiffer than it truly is (Charalambides et al., 2006). The use of lubricants, such as low viscosity oils, can reduce or eliminate frictional effects (Macias-Rodriguez and Marangoni, 2016a). An alternative is to bond the samples to the compression platens using adhesives such as cyanoacrylate (Casiraghi et al., 1985), but for soft food materials, such as shortening, it is very difficult to bond specimens without damaging them.

Uniaxial Cyclic Compression Tests

Cyclic compression consists of loading and unloading a single specimen more than once. For materials with considerable elastic character, loading is conducted to the same displacement at each cycle. This type of compression test is used to study the fatigue of materials, especially metals for which this type of loading is very common to assess their useful life (Hakamada et al., 2007). For soft materials such as food, increasing increments of displacement are frequently used with cyclic loading tests to study the onset of plastic behavior and the true elasticity of the material from its unloading modulus (Goh and Scanlon, 2007). Alternatively, cyclic loading tests using dynamic mechanical tests conducted in the linear viscoelastic region can be used to determine the storage and loss moduli in compression (Herrera and Hartel, 2000), and these moduli are then reported as a function of frequency.

It will be appreciated from above that determination of a valid elastic modulus in a fat system is confounded by plastic deformation, even at very low strains. Thus, although it appears that linear elasticity is maintained up to a strain of 0.06 (Fig. 5.1), any modulus calculated within this strain region will be lower than the fat system's true elasticity due to plastic compliance. To ascertain the true elasticity of the material, the unloading modulus (E_U) can be determined (Goh and Scanlon, 2007; Gonzalez-Gutierrez and Scanlon, 2013). The unloading modulus is the tangent taken from the stress–strain diagram at the point where the load is removed (Fig. 5.11). At this point, only the elastic character of the fat is "pushing back" on the platen because any ongoing plastic compliance ceases as soon as the stress is lowered below the fat's true yield stress. In an ideal situation, if Young's modulus during loading (E) was determined within the elastic region

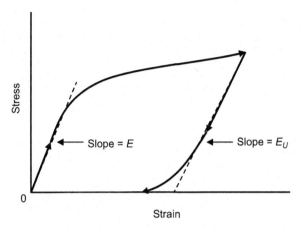

Figure 5.11 Schematic loading–unloading curve showing the modulus of elasticity (E) and the unloading modulus (E_U).

of the material, the unloading modulus (E_U) and the Young's modulus (E) would be identical. However, fats are complex structures and their degree of elastic character is influenced by structural rearrangements that occur during the permanent deformation of the compression process (Gonzalez-Gutierrez and Scanlon, 2013). Therefore, even a true elastic modulus determined during loading can differ from the unloading modulus because the overall structure of the solid is modified by the onset of plasticity through mechanisms such as slip and twinning, which tend to occur in crystalline solids (de With, 2006).

Compressive Creep Tests

As well as providing quantitative information on rheological properties through the creep compliance function (Eq. 5.17), creep tests can classify materials according to their creep behavior (Gunasekaran and Ak, 2003; Rao, 2007). For example, an ideal elastic solid material (Fig. 5.12A) will exhibit a constant strain with time due to its lack of flow, and a complete recovery of the strain will occur after the load is removed. On the other hand, an ideal viscous liquid material (Fig. 5.12B) exhibits a linear change of the strain as time passes due to steady flow, and zero recovery occurs after unloading (Steffe, 1996). For compositional and structural reasons, fat systems show viscous and elastic behavior simultaneously. Therefore, the creep test provides valuable information on their viscoelastic behavior. In a viscoelastic liquid (Fig. 5.12C), strain increases with time until it approaches a steady state where the rate of strain is constant; in other words, there is a linear increase in deformation with time. In contrast, a viscoelastic solid (Fig. 5.12D) eventually reaches an equilibrium strain (Ferry, 1970).

Performing compressive creep tests on soft solids such as lipid-based particle gels is a simple and convenient procedure. The compressive creep response can be related to

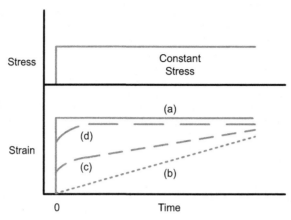

Figure 5.12 Creep curve for (a) elastic solid, (b) viscous liquid, (c) viscoelastic liquid, and (d) viscoelastic solid. *(Adapted from Gunasekaran, S., Ak, M.M., 2003. Cheese Rheology and Texture. CRC Press LLC, Boca Raton, FL.)*

shear and bulk creep tests, and this has merit because most of the theory of creep was developed from shear and bulk test studies on polymers and other nonfood materials (Ferry, 1970; Christensen, 1982). During a compression-based creep test, a suddenly applied compressive constant stress (σ_o) produces time-dependent strains ($\varepsilon(t)$), which are related in the following manner (Ferry, 1970):

$$\varepsilon(t) = \sigma_o S(t), \qquad (5.23)$$

where $S(t)$ is the compression creep compliance function. During uniaxial compression of materials, a simultaneous change in both volume and shape occurs, and for this reason, $S(t)$ is related to the bulk creep compliance (B), which is related to the change in volume, and to the shear creep compliance (J), which is related to the fat specimen's change in shape. The following equation summarizes this relationship (Ferry, 1970):

$$S(t) = \frac{J(t)}{3} + \frac{B(t)}{9}. \qquad (5.24)$$

For fats over a broad range of time scales, the bulk creep compliance, $B(t)$, is small compared with the shear creep compliance, $J(t)$, so that Eq. (5.24) can be simplified to

$$S(t) = J(t)/3. \qquad (5.25)$$

Therefore, shear and compression creep tests give results that are interconvertible by virtue of Eq. (5.25). It is important to recognize that Eq. (5.25) is only valid for materials in which the change in volume during compression is negligible compared with the change in shape; in other words, it is applicable to essentially incompressible materials (Ferry, 1970). Because plastic flow is assumed to take place isovolumetrically (Johnson, 1987), the stipulation is not overly restrictive.

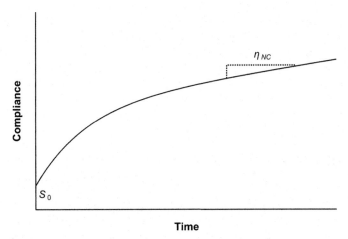

Figure 5.13 Schematic compliance–time curve for a viscoelastic material. *(Adapted from Rao, M.A., 2007. Rheology of Fluid and Semisolid Foods: Principles & Applications. Springer, New York, NY.)*

Typically, the results of a creep test are presented as a compliance–time curve (Fig. 5.13). Material parameters such as the creep compliance function ($S(t)$) and the retardation time (τ) can potentially be derived from creep tests, and these mechanical parameters are useful for describing the viscoelastic character of fat materials (Rohm, 1993; Vithanage et al., 2009). The instantaneous compliance, S_0, which represents the region where the bonds between the different structural units are stretched elastically, and is the inverse of the modulus of elasticity ($S_0 = 1/E$), can also be obtained from such a test. Another parameter that can be obtained from the compliance–time curve is the Newtonian compliance (S_N); in this region, the bonds between structural components break and flow past one another. The Newtonian compliance is related to Newtonian viscosity (η_{NC}) derived from the creep test and time of steady flow (t) through (Rao, 2007):

$$S_N = \frac{t}{\eta_{NC}}. \qquad (5.26)$$

Usually it is not possible to model the creep behavior of fats through a single retardation time. A generalized Kelvin–Voigt model with multiple dashpots connected in parallel to springs is then used (Fig. 5.14). Individual compliances (S_i) are assigned to each of the springs to describe their elasticities, and viscosities (η_i) are assigned to each of the dashpots. The Kelvin–Voigt model is summarized by the expression:

$$S(t) = S_0 + \frac{t}{\eta_{NC}} + \sum_{i=1}^{p} S_i\left(1 - e^{(-t/\tau_i)}\right), \qquad (5.27)$$

where τ_i is the retardation time of each Kelvin–Voigt element (Gunasekaran and Ak, 2003).

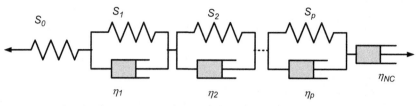

Figure 5.14 Generalized Kelvin–Voigt model for creep behavior. *(Adapted from Gunasekaran, S., Ak, M.M., 2003. Cheese Rheology and Texture. CRC Press LLC, Boca Raton, FL.)*

Compressive Stress Relaxation Tests

Stress relaxation is one of the most important tests for determining the viscoelastic properties of biological materials (Cenkowski et al., 1992). Depending on parameters such as solid fat content and the structural organization of the solid fat, different behaviors can be observed during stress relaxation. An ideal elastic material (Fig. 5.15A) will (essentially instantaneously) reach a finite constant stress with no stress relaxation over time (Del Nobile et al., 2007); all the energy used during straining will be stored, and this energy used to return the specimen to its original shape and size after the strain is removed (Gunasekaran and Ak, 2003). An ideal liquid material (Fig. 5.15B) instantaneously shows stress decay to zero because liquids do not store strain energy nor retain a memory of their initial state (Del Nobile et al., 2007; Gunasekaran and Ak, 2003). Viscoelastic solid materials (Fig. 5.15C) under stress relaxation tests gradually relax and reach an equilibrium stress (σ_{eq}) that is greater than zero, a behavior that can be observed in permanent gels with covalent cross-links (Steffe, 1996). In contrast, viscoelastic liquids (Fig. 5.15D) show a residual stress vanishing to zero, but this may be hard to observe because the relaxation to zero stress can occur at times far beyond the experimental time scale (Steffe, 1996).

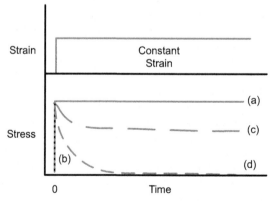

Figure 5.15 Stress relaxation curve for (a) elastic solid, (b) viscous liquid, (c) viscoelastic solid, and (d) viscoelastic liquid. *(Adapted from Gunasekaran, S., Ak, M.M., 2003. Cheese Rheology and Texture. CRC Press LLC, Boca Raton, FL.)*

The stress relaxation experiment can be viewed as consisting of two parts: the straining stage when the material is being squeezed up to a set displacement or strain, and the relaxation stage. Ideally, the straining stage should be instantaneous, but in reality it takes time, and because the stress relaxation behavior of materials is affected by the history of deformation, the time it takes to reach the desired strain will affect the stress relaxation results (Christensen, 1982; Gunasekaran and Ak, 2003).

During compressive stress relaxation tests, the stress ($\sigma(t)$) and the strain (ε_0) are related in the following manner:

$$\sigma(t) = E(t)\varepsilon_0, \tag{5.28}$$

where $E(t)$ is the compressive relaxation modulus. For elastic solids, $E(t)$ has a constant value—the Young's modulus (Ferry, 1970).

Stress relaxation behavior is generally described using equations that were derived from the generalized or discrete Maxwell model, using a combination of the basic Maxwell element shown in Fig. 5.7A (Peleg and Pollak, 1982; Hassan et al., 2005):

$$\sigma(t) = \sigma_{eq} + \sum_{i=1}^{p} E_i e^{-(t/\tau_i)}, \tag{5.29}$$

where σ_{eq} is the equilibrium stress, p is the number of Maxwell elements, E_i is the stress relaxation constant for each Maxwell element, and τ_i is the relaxation time for each Maxwell element. The stress relaxation constant, E_i, represents the contribution of each Maxwell element to the overall elastic character of the fat, and the relaxation time, τ_i, is related to the viscous contribution of each Maxwell element to the overall viscous behavior of the material. In Fig. 5.16, a stress relaxation test on vegetable shortening is compared with a generalized Maxwell model that uses three Maxwell elements. It can be seen that the fit is excellent and that for a loading strain of ~ 0.04, the shortening behaves as a viscoelastic solid.

Indentation Tests

Indentation, or penetrometry, tests are a convenient means of measuring the mechanical properties of solid materials; they are generally inexpensive and not totally destructive to the specimen (Huang et al., 2002; Ma et al., 2003). No standard specimen preparation is required as with other tests, and indeed, as long as specimen geometry and indenter size requirements are not transgressed (Anand and Scanlon, 2002), fat specimens can be tested in situ in molds or pans in which they have been solidified. In the test, a probe accesses and evaluates the mechanical properties from a localized region of the specimen, rather than the properties being derived from the specimen as a whole. For this reason, indentation is an ideal method for the evaluation of rheological behavior in localized areas of a given specimen (Lin and Horkay, 2008).

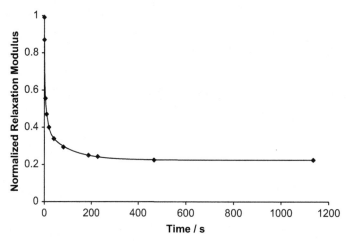

Figure 5.16 Comparison between experimental stress relaxation test data (points) and prediction by a generalized Maxwell model with 3 Maxwell elements; imposed strain (loaded at 4 mm/min) is 0.04.

In a somewhat analogous manner to the different principles of testing of fats by creep and stress relaxation, indentation testing can be load-controlled or displacement-controlled. The long-established means of indentation testing is load control, where a specific load is applied to the probe that indents the specimen, and a visual quantification of the plastic deformation on the specimen surface is used to define a value of hardness (Johnson, 1987). A more common approach, due to the ubiquity of universal testing machines, is the displacement-controlled approach, such as depth-sensing indentation tests, where the load is measured as a function of indentation depth during a load−unload cycle of a probe inserted and withdrawn from the specimen.

Indentation has proven to be a useful technique for assessing the mechanical properties of soft food materials (Anand and Scanlon, 2002; Liu and Scanlon, 2003; Goh et al., 2004a), including lipid particle gels (Goh and Scanlon, 2007; Yoshikawa et al., 2017). Although indentation tests are easy to perform, deriving fundamental mechanical parameters from them is not straightforward because of the complex strain field induced in the specimen during the indentation process (Huang et al., 2002). For this reason, many results are interpreted empirically (Goh et al., 2004a). Depending on the material to be tested, indenters can be made of diamond, hard carbon steel, or tungsten carbide, but given the low relative stiffness of fat systems, indenters made from hard polymers will furnish accurate results. The geometry of indenters is also diverse; the most common ones include conical (Vandamme and Ulm, 2006; Gregersen et al., 2015b), cylindrical (Liu and Scanlon, 2003; Gregersen et al., 2015a), prismatic (Bae et al., 2006), pyramidal (Menčik, 2007), and spherical (Beghini et al., 2006; Yoshikawa et al., 2017).

Hardness Tests

Penetrometry is the most common test of the mechanical properties of fat systems (deMan and Beers, 1987). Because of its simplicity, and the fact that results correlate well with sensory evaluation (Dixon, 1974; Staniewski et al., 2006; Gregersen et al., 2015b), it is still widely used to evaluate the mechanical properties of fats (Chen and Mackley, 2006). In penetrometry, a penetrating body (a cone, needle, or sphere) of a specific load is allowed to indent the fat system. The depth of penetration at a specified length of time, or the rate at which the penetrating body indents the system, is measured (Sherman, 1976). Standardized tests and commercial standards of design are available; the most widely used being that of the American Oil Chemists' Society (American Oil Chemists' Society, 1960, 2017), where a mechanical parameter denoted fat consistency is defined (deMan, 1983). The relationship between consistency and other material parameters of a fat system depends not only on the properties of the fat but also on the geometry and the load of the penetrometer. A common means of delineating instrument attributes and material properties is to define the hardness of the fat system.

Classically (Tabor, 1948; Vandamme and Ulm, 2006), hardness values, such as the Brinell hardness (P), take account of the projected contact area (A_c) measured after removal of the penetrometer after indentation with a specific load (F_{max}):

$$P = \frac{F_{max}}{A_c}. \tag{5.30}$$

This is close to the definition of penetrometer test hardness, the ratio of load to the area of the impression made by the penetration, on the grounds that "the cone will sink into the fat until the stress exerted by the increasing contact surface of the cone is balanced by the hardness of the fat" (deMan, 1983). But, the definition of hardness of fats differs from the Brinell hardness because it includes the elastic strain and any residual viscoelastic strain that has not relaxed at the time of measurement of hardness. For simple indenter geometries, unique geometric relationships exist between the depth of penetration (thus defined by fat consistency) and the penetration area (thus defined by classical hardness). For example, for a sharp-ended cone defined by the angle, 2α, where smaller angles represent sharper cones, the relationship between the applied load, the hardness, the penetration impression area (A_{imp}), and penetration depth (z) are given by

$$P = \frac{F_{max}}{A_{imp}} = \frac{F_{max} \cos \alpha}{\pi z^2 \tan \alpha}. \tag{5.31}$$

The rationale for elimination of the effect of indentation load in metallurgical hardness testing is that the contact pressure defined by hardness in Eq. (5.31) is used to deduce the yield stress of the material (Tabor, 1996). However, the work of plastic deformation is not the only resistance to cone penetration because the elastic properties of the fat and the coefficient of friction between the cone and the fat also impede cone penetration (Tabor, 1948).

Numerous examples of the use of hardness measurements could be cited, for example, de Man (1983). Because conventional penetrometers tend not to be used in research nowadays, researchers cite hardness values that have been acquired on displacement-controlled devices. Hardness values will then have different units associated with them, e.g., penetration depth (Campos et al., 2002; Dassanayake et al., 2012), load per unit penetration depth (Marangoni and Rousseau, 1998), and load at a specific depth (Foubert et al., 2006; Gregersen et al., 2015a) have been cited. Nevertheless, measurement of indentation area has also been used to characterize the hardness of chocolate, with Chen and Mackley (2006) using indent area to quantify hardening as a function of postextrusion time.

Depth-Sensing Indentation Tests

Although the original penetrometers were designed to indent a specimen with a constant load, Tanaka et al. (1971) advocated the use of constant speed penetrometers. Here, a penetrating body is mechanically driven into a fat specimen at a constant speed while the force is continuously recorded. Constant speed penetrometry permits control over penetration depth and allows hardness values as a function of penetration depth to be evaluated from the load—displacement curve as the fat specimen is indented (Boodhoo et al., 2009). An additional advantage of constant speed penetrometers, where the load—displacement response is monitored on unloading, is that the elasticity of the sample can be derived from the penetrometer test (Page, 1996). Load—unloading cycles of a cone penetrating a simulated vegetable shortening to two different depths are shown in Fig. 5.17.

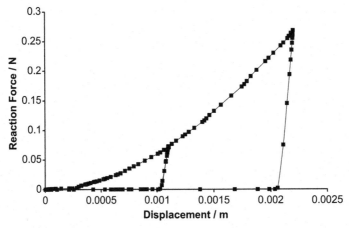

Figure 5.17 Simulated load—unload cycles for conical indentation of vegetable shortening (defined by material constants of Goh and Scanlon, 2007) to two depths with indenter moving at 5 mm/min.

Just as many equations have been proposed to relate indentation force and indentation depth in constant load tests, numerous constant speed equations have been proposed, but they all depend on indenter geometry and the type of material being tested. For indentation using self-similar indenters, such as cones, the relationship between indentation load (F_i) at a given indentation (or penetration) depth (z_i) is usually expressed as a parabolic relation known as Kick's law:

$$F_i = \beta z_i^2, \tag{5.32}$$

where β is a constant depending on the geometry of the indenter and the indented material properties (Ma et al., 2003). For rate-independent materials, research has shown that the indentation response follows Eq. (5.32) (Ma et al., 2003). However, for rate-dependent materials, the load—displacement response has been better described by an equation that includes both a rate-dependent and a rate-independent contribution of the following form (Goh and Scanlon, 2007):

$$F = F_0 \left[1 + C(m) \left(\frac{\dot{z}}{\dot{\varepsilon}_0 z} \right)^m \right], \tag{5.33}$$

where F_0 is the static indentation response, and the second term takes account of the fat's rate-dependent material parameters, fluidity and overstress power law constant (both defined in Eq. 5.18), as well as the nominal indentation strain rate. The function $C(m)$ is an empirical function of the overstress power law constant (Goh and Scanlon, 2007).

Because a valid elastic modulus should only be acquired from the unloading portion of a compression test, an indentation modulus should only be acquired from the unloading portion of an indentation test. The indentation modulus (M) relates the slope of the initial part of the unloading curve (Q) to the projected contact area (A_C). For a conical indenter, M can be obtained with the following equation (Vandamme and Ulm, 2006):

$$M = \frac{\sqrt{\pi} Q}{2\sqrt{A_C}}. \tag{5.34}$$

The force—depth curve produced during an indentation test is related to the stress—strain curve of the material being tested, just like in simple compression tests. However, this relation is not as straightforward as with compression due to the complex deformation process in the indentation region. The indentation region of the material is subject to multiaxial stresses with high gradients and large strains (Beghini et al., 2006).

Given that fat systems have complex rheological behavior and that a commonly used test for determination of their rheology has to be carefully interpreted, it is useful if computational tools are available to assist in separating true constitutive properties from artifacts attributable to specimen geometry or test protocols. One technique that has been applied to help understand both the structure (Janssen, 2004) and the rheology of fat systems (Goh and Scanlon, 2007; Gonzalez-Gutierrez, 2008), and their interrelations (Engmann and Mackley, 2006), as well as how rheological properties can be back-extracted from indentation testing (Goh and Scanlon, 2007; Gonzalez-Gutierrez, 2008), is the finite element method (FEM).

FINITE ELEMENT METHOD

The FEM is used to find approximate solutions to complex problems by visualizing the solution region as being composed of many small, interconnected subregions referred to as finite, or discrete, elements (Rao, 1982). FEM was developed as a means of performing structural analysis in nonrectangular geometries and solving complex elasticity problems in civil and aeronautical engineering. Nowadays, the FEM has been generalized into a branch of applied mathematics for numerical modeling of physical systems and can be used to solve complex problems of soft solid mechanics, fluid dynamics, electromagnetism, and heat and mass transfer (Liu and Scanlon, 2003; Goh et al., 2004b; Roduit et al., 2005; Bermudez et al., 2007; Farhloul and Zine, 2008; Le Reverend et al., 2011; Fryer and Bakalis, 2012).

A continuum of matter can be subdivided into small elements with properties equal to those of the entire body (Fung, 1969); if the continuum has known boundaries then it is called a domain. A domain can be an entire physical object or a portion of it depending on the known boundaries. Within a domain, there are an infinite number of solutions to the field variables because they change from point to point within the domain. The finite element method relies on the decomposition of the domain into a finite number of elements for which an approximating function can be used to solve for a finite number of unknown field variables. The approximating functions are defined in terms of the values at specific points along the boundaries of elements, which are called nodes. Nodes also connect adjacent elements as seen in Fig. 5.18 (Madenci and Guven, 2006).

Steps in Finite Element Method

The FEM can be divided into six major steps (Madenci and Guven, 2006):
1. Discretization of the domain into a finite number of elements
2. Selection of interpolation functions
3. Development of the element matrices or element equations for individual elements
4. Assembly of the element matrices for each element to obtain the global equilibrium matrix of the entire domain, also known as overall equilibrium equations
5. Determination of the boundary conditions of the domain
6. Solution of the equations in the global matrix

A simple mechanical problem consisting of calculating the stress in a fat crystallite will be used to illustrate these six steps. For ease of illustration, the crystallite can be viewed simplistically as a stepped bar that is axially loaded (Fig. 5.19). The crystallite (bar) is made

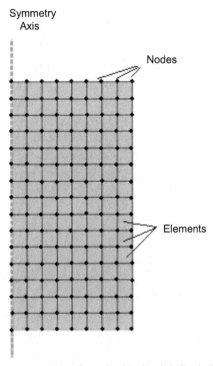

Figure 5.18 Axisymmetric representation of a cylindrical solid divided into square elements. Each square element has four nodes.

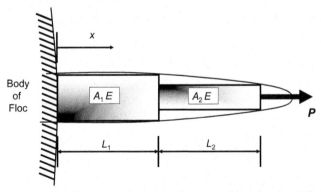

Figure 5.19 Fat crystallite viewed as a simple stepped bar that is axially loaded.

of a material with a modulus of elasticity, E, has cross-sectional areas of $A1$ and $A2$ over the lengths $L1$ and $L2$ and is subjected to a load P. $A1 = 2\,\mu m^2$, $A2 = 1\,\mu m^2$, $L1 = L2 = 10\,\mu m$, $E = 6000\,MPa$, and $P = .001\,N$. The description relevant to this problem follows (Rao, 1982).

Discretization

The first step in the FEM is the discretization of the domain, i.e., its subdivision into elements. During this step, the domain is replaced by a system that has a finite number of degrees of freedom. The basic elements that constitute the domain are selected with characteristics that simulate the original body as closely as possible without excessively increasing the computational efforts required to obtain the solution.

In the example shown in Fig. 5.19, the domain is the crystallite that is divided into two elements (element 1 and 2) that have two nodes each with node 2 being common to both. Because the load (P) is acting axially along the crystallite length, the displacement (d) is also in the axial direction; therefore, each element has only one degree of freedom, which is displacement in the x-direction. Fig. 5.20 shows the discretization for the fat crystallite loading problem.

Selection of Interpolation Functions

Once the domain has been discretized, simple functions for the solution of each element must be selected. The functions simulating the behavior of the solution within each element are called interpolation functions; such functions are formulated to act at the nodes of each element. The most common types of interpolation function are polynomials because it is easier to perform differentiation and integration, and the accuracy of the results can be enhanced by increasing the order of the polynomial function. Several conditions must be met by the chosen interpolating function: the function must be expressed in terms of the nodal degrees of freedom (in mechanics, the number of displacements and rotations that a node can have), it should not change with a change in the local coordinate system, it should converge to the exact solution if element size is reduced successively, and the number of unknown coefficients in the polynomial equation should equal the number of nodal degrees of freedom.

Each of the elements shown in Fig. 5.20 can be generalized as in Fig. 5.21 and are referred to as element "e" to assign the interpolation functions at each of the nodes. The displacement (d) at each node is deemed to vary in a linear fashion, such that a polynomial of first degree can be used to describe this behavior:

$$d(x) = a + cx, \tag{5.35}$$

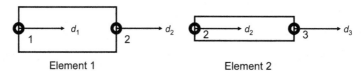

Element 1   Element 2

Figure 5.20 Discretization of stepped bar representing fat crystallite into two one-dimensional elements; dots 1, 2, and 3 are nodes while d_1, d_2, and d_3 are the displacements of these nodes. *(Adapted from Rao, S.S., 1982. The Finite Element Method in Engineering. Pergamon Press Canada Ltd., Willowdale, ON.)*

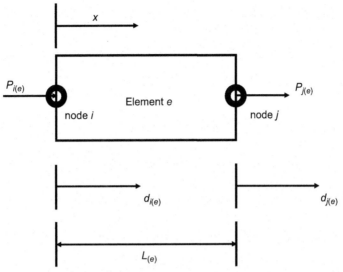

Figure 5.21 Displacements (*d*) and loads (*P*) for generalized element "*e*" (Rao, 1982).

where a and c are constants. The displacement at the end of each element ($x = 0$) is $d1^{(e)}$, and the displacement at the far end of each element ($x = L^{(e)}$) is $d2^{(e)}$, so that the interpolation function (Eq. 5.35) can be expressed in terms of the nodal displacements at each element as

$$d(x) = d1^{(e)} + \frac{d2^{(e)} - d1^{(e)}}{L^{(e)}} x. \tag{5.36}$$

Only one unknown is in Eq. (5.36) because each node has only one degree of freedom.

Development of Matrix for Individual Elements

The third step in the FEM involves the formulation of matrices and vectors characteristic of each element. The nature of developing a matrix for each element depends on the complexity of the problem. If the problem is simple, direct physical reasoning suffices. If the problem can be formulated so that calculations associated with maxima and minima are required, a variational approach can be used (Mura and Koya, 1992). Alternatively, a weighted residual approach can be used to obtain approximate solutions to linear- and nonlinear-governing differential equations (Rao, 1982).

The crystallite loading problem is rather simple; therefore, element matrices can be derived directly from the principle of minimum potential energy. The potential energy of the fat crystallite (U) is given by the difference between the strain energy from each element, $w^{(e)}$, and the work done by external forces (W):

$$U = w^{(1)} + w^{(2)} - W. \tag{5.37}$$

The strain energy of each element "e," $w^{(e)}$, can be calculated by the following equation:

$$w^{(e)} = \frac{1}{2} A^{(e)} E^{(e)} \int_0^{L^{(e)}} \varepsilon^{(e)2} dx, \tag{5.38}$$

where $A^{(e)}$ is the cross-sectional area of each element, $L^{(e)}$ is the length of each element, $\varepsilon^{(e)}$ is the strain of each element, and $E^{(e)}$ is the modulus of elasticity of each element. The strain of each element can be derived from the interpolating function (Eq. 5.36) because

$$\varepsilon^{(e)} = \frac{\partial d}{\partial x} = \frac{d2^{(e)} - d1^{(e)}}{L^{(e)}}. \tag{5.39}$$

Therefore,

$$w^{(e)} = \frac{A^{(e)} E^{(e)}}{2L^{(e)}} \left(d1^{(e)2} + d2^{(e)2} - 2d1^{(e)} d2^{(e)} \right). \tag{5.40}$$

Eq. (5.40) can be written in matrix notation as

$$\pi^{(e)} = \frac{1}{2} \vec{d}^{(e)T} \left[K^{(e)} \right] \vec{d}^{(e)}, \tag{5.41}$$

where $\vec{d}^{(e)} = \left\{ \begin{array}{c} d1^{(e)} \\ d2^{(e)} \end{array} \right\}$ is the vector of nodal displacements of each element:

for element 1, $\vec{d}^{(e)} = \left\{ \begin{array}{c} d1 \\ d2 \end{array} \right\}$ and for element 2, $\vec{d}^{(e)} = \left\{ \begin{array}{c} d2 \\ d3 \end{array} \right\}$; while

$\vec{d}^{(e)T} = \left\{ d1^{(e)} \quad d2^{(e)} \right\}$ and $\left[K^{(e)} \right] = \frac{A^{(e)} E^{(e)}}{L^{(e)}} \begin{bmatrix} 1 & -1 \\ -1 & 1 \end{bmatrix}$, which is the stiffness

matrix of each element.

The work done by external forces (W) on the whole crystallite can be expressed as

$$W = d1P1 + d2P2 + d3P3 = \begin{bmatrix} d1 & d2 & d3 \end{bmatrix} \left\{ \begin{array}{c} P1 \\ P2 \\ P3 \end{array} \right\}, \tag{5.42}$$

where $P1$ is the reaction at the fixed node, $P2 = 0$ and $P3 = P = .001$ N. If the whole crystallite is in equilibrium when subjected to load P then the potential energy, U, is equal to zero, and this can be expressed in matrix notation as

$$\sum_{e=1}^{2} \left(\left[K^{(e)} \right] \vec{d}^{(e)} - \vec{P}^{(e)} \right) = \vec{0}. \tag{5.43}$$

Development of Global Matrix

After the matrices and vectors, characteristic of each element, have been defined in a common global coordinate system, the next step is the construction of the overall (or system) equations. This procedure is based on the requirement of compatibility at each of the element nodes, which means that the values of variables are the same for all elements joined at that node (Rao, 1982).

The global matrix ($[K]$) is the addition of the stiffness matrices of each element ($[K^{(e)}]$), and the global vectors include nodal displacement ($\overrightarrow{d}$) and load ($\overrightarrow{P}$) vectors. For the values assigned in this fat crystallite example, the element stiffness matrices are

$$\left[K^{(1)}\right] = \frac{A1E}{L1}\begin{bmatrix} 1 & -1 \\ -1 & 1 \end{bmatrix} = \begin{bmatrix} 1200 & -1200 \\ -1200 & 1200 \end{bmatrix} \tag{5.44}$$

$$\left[K^{(2)}\right] = \frac{A2E}{L2}\begin{bmatrix} 1 & -1 \\ -1 & 1 \end{bmatrix} = \begin{bmatrix} 600 & -600 \\ -600 & 600 \end{bmatrix}. \tag{5.45}$$

The stiffness matrix of element 1 overlaps with the stiffness matrix at node 2 and so the displacement at this node ($d2$) should have contributions from both elements; therefore the global stiffness matrix becomes

$$[K] = \begin{bmatrix} 1200 & -1200 & 0 \\ -1200 & 1200+600 & -600 \\ 0 & -600 & 600 \end{bmatrix} = 600\begin{bmatrix} 2 & -2 & 0 \\ -2 & 3 & -1 \\ 0 & -1 & 1 \end{bmatrix}. \tag{5.46}$$

The overall equilibrium equation $[K]\overrightarrow{d} = \overrightarrow{P}$ can be expanded as

$$600\begin{bmatrix} 2 & -2 & 0 \\ -2 & 3 & -1 \\ 0 & -1 & 1 \end{bmatrix}\begin{Bmatrix} d1 \\ d2 \\ d3 \end{Bmatrix} = \begin{Bmatrix} P1 \\ 0 \\ 0.001 \end{Bmatrix}. \tag{5.47}$$

Determination of Boundary Conditions

Before solving the equations on the global matrix, some restrictions are specified to prevent the system from having an infinite number of solutions. By specifying the value of at least one, and sometimes more than one, boundary condition, the motion or the degrees of freedom of certain nodes are restricted. The physics of the problem dictates the number of boundary conditions that should be specified (Rao, 1982).

The overall equilibrium equations (Eq. 5.47) would have an infinite number of solutions because there are four unknowns and only three equations. However, from

Fig. 5.19, one can see that the crystallite is rigidly attached to the floc at node 1; therefore, the displacement $d1 = 0$. This boundary condition reduces Eq. (5.47) to

$$600 \begin{bmatrix} 2 & -2 & 0 \\ -2 & 3 & -1 \\ 0 & -1 & 1 \end{bmatrix} \begin{Bmatrix} 0 \\ d2 \\ d3 \end{Bmatrix} = \begin{Bmatrix} P1 \\ 0 \\ 0.001 \end{Bmatrix}. \tag{5.48}$$

Solving the Global Matrix

Once the global matrix has been constructed and the boundary conditions have been established, the equations in the global matrix can be solved. Linear equations can be solved by variations of the Gaussian elimination method. If the problem is nonlinear then nonlinear equations of the global matrix must be solved by an iterative procedure, such as Newton–Raphson, continuation, minimization, or perturbation methods (Rao, 1982). These numerical methods are implemented in computer software packages, thereby simplifying the solution of complex problems.

The global equilibrium equations defined for the crystallite problem are linear; therefore, they can be solved by the Gaussian elimination method, which consists in eliminating unknowns by expressing them in terms of the remaining unknowns until only one unknown appears in the system of equations. The system of equations (Eq. 5.48) can be solved using the Gaussian elimination method in the following manner.

The first equation of the system is $600(-2(d_2)) = P_1$; solving for d_2, one obtains $d_2 = P_1/-1200$; substituting d_2 into the second equation of the system, $600(3(P_1/-1200)-d_3) = 0$; solving for d_3, the second equation becomes $d_3 = 3P_1/-1200$; substituting d_2 and d_3 into the last equation allows us to get $600((P_1/120) - 3(P_1/1200)) = 0.001$; therefore $P_1 = -0.001$ N, $d_2 = 8.3 \times 10^{-7}$ m, and $d_3 = 2.5 \times 10^{-6}$ m. With these values, the strains in each of the elements that form the stepped bar can be calculated (Eq. 5.39):

$$\varepsilon^{(1)} = \frac{d2^{(1)} - d1^{(1)}}{L^{(1)}} = \frac{d2 - d1}{L1} = \frac{8.3 \times 10^{-7} - 0}{10 \times 10^{-6}} = 0.083$$

$$\varepsilon^{(2)} = \frac{d2^{(2)} - d1^{(2)}}{L^{(2)}} = \frac{d3 - d2}{L2} = \frac{2.5 \times 10^{-6} - 8.3 \times 10^{-7}}{10 \times 10^{-6}} = 0.17.$$

And the stresses in each element are calculated with $\sigma_{(e)} = E\varepsilon_{(e)}$ to give the numerical results of $\sigma_{(1)} = 500$ MPa and $\sigma_{(2)} = 1000$ MPa.

Computer Implementation of the Finite Element Method

Greater utilization and popularity of FEMs has occurred because of widespread availability of powerful personal computers. Nowadays, complex finite element analyses (FEAs) can be performed with results delivered in minutes, instead of hours or even days. As computers have become more powerful and cheaper, finite element software has also become more capable and more accessible.

FEA software first became commercially available in the early 1970s, and it was primarily used in the nuclear and aerospace industries. Examples of commercially available finite element software available in the 1970s were ANSYS and MSC/NASTRAN (Baran, 1988). Some FEA software packages available today are listed in Table 5.1.

Nowadays, FEA software is very user-friendly with the provision of graphic interfaces (that facilitate the setup of the problem) and drop-down menus (where many choices can be selected to set up realistic situations, thereby eliminating the need to write lengthy codes, which was the norm in the past). Today the FEM is integrated with other applications such as computer aided design to give rise to what is called computer aided engineering (CAE).

Table 5.1 Examples of finite element analysis software packages currently available

Software name	Company	Website
Abaqus	Dassault Systemes S.A., Suresnes, France	www.3ds.com
ADINA	ADINA R&D, Inc., Watertown, MA, USA	www.adina.com
Agros2D	Open-source code started by University of West Bohemia, Pilzen, Czech Republic	www.agros2d.org
ANSYS	ANSYS Inc., Canonsburg, PA, USA	www.ansys.com
Autodesk Nastran In-CAD	Autodesk Inc., San Rafael, CA, USA	www.autodesk.com
CAEplex	Open-source code started by SEAMPLEX, Rafaela, Argentina	www.caeplex.com
CalculiX	Open-source code started and maintained by Guido Dhondt and Klaus Wittig	www.calculix.de
COMSOL Multiphysics	COMSOL AB, Stockholm, Sweden	www.comsol.com
DIANA FEA	DIANA FEA BV, Delf, The Netherlands	www.dianafea.com
Elmer FEM Solver	Open-source code started by CSC—IT Centre for Science Ltd., Espoo, Finland	www.csc.fi/web/elmer
Femap	Siemens PLM Software, Plano, TX, USA	www.plm.automation.siemens.com
LS-DYNA	Livermore Software Technology Corporation, Livermore, CA, USA	www.lstc.com
MSC Nastran	MSC Software Corporation, Newport Beach, CA, USA	www.mscsoftware.com
RFEM	Dlubal Software, Inc., Philadelphia, PA, USA	www.dlubal.com
Strand7	Strand7 Pty Limited, Sidney, Australia	www.strand7.com

One popular FEA software package is Abaqus. Abaqus has been preferred by many academic and research institutions due to its wide material capability and its ability to be customized, but it is also used in the automotive, aerospace, and product manufacturing industries. One version of Abaqus, Abaqus/CAE, provides a graphic interface that facilitates the visualization of the problem to be solved, as well as the results after analysis. Abaqus/CAE prepares an input file from the parameters entered into the graphic interface, which is submitted into analytical software packages where the solutions are calculated (Abaqus, 2017).

Finite Element Method Considering Foods as Homogeneous Materials

Food materials have a complex rheological response due to their complex composition and structure. Because of this complexity, the FEM is an appropriate tool in studies of the rheology of various food products. Early FEA conducted on food materials arose from the need to predict the mechanical damage done to agricultural products during harvesting and processing (Puri and Anatheswaran, 1993), while current research has focused on obtaining models that accurately describe the rheological response of processed food such as butter, cheese, bread, dough, margarine, and shortening.

Rumsey and Fridley (1977) used a viscoelastic finite element computer model to predict stresses resulting from contact loads on fruits and vegetables after harvesting. The fruits and vegetables were simulated as perfect spheres. It was concluded that the finite element model concurred with the analytical solutions formulated for viscoelastic materials as long as deformation was small because viscoelastic theory used during the experiment was developed for small strains only. As such, this early analysis emphasized that an FEM solution is only as good as the constitutive model that it uses, thereby emphasizing the need to have a comprehensive understanding of the microstructural features that influence the rheological properties of the food system (Goh and Scanlon, 2007).

Two finite element elastic deformation models, axysymmetrical and three-dimensional, were implemented by Cardenas-Weber et al. (1991) on the commercially available FEA software ANSYS to analyze the compression of a melon by a robot gripper. The two models predicted lower stresses than the ultimate strength of the melon tissue for a v-shaped robot gripper and higher stresses for a flat gripper. It was concluded that the model could be used to predict the maximum force, which could be applied to a melon to prevent bruising by a gripping device, but again, research to determine the force magnitude that produces bruising damage to the melons was pivotal to successful utilization of the FEM.

One very useful function of FEM is its ability to perform virtual experiments. In this way, researchers can evaluate numerically the effects that are difficult or tiresome to investigate experimentally. Charalambides et al. (2001) used the commercial finite element software package Abaqus to investigate frictional effects on the stress—strain data obtained during uniaxial compression of Gruyere and mozzarella cheeses. Even though sample—platen interfaces may be well lubricated, it is not always possible to completely eliminate frictional effects; therefore, a series of experiments on samples of different geometries is necessary to quantify frictional effects. To obviate these experiments, simulations of uniaxial compression tests on cylindrical cheese specimens of different heights were set up in Abaqus and an iterative method was used to develop a correction factor as a function of strain for the force measured during imperfectly lubricated compression tests. It was concluded that iterative FEA was a more accurate alternative to quadratic extrapolation in converting the results of unlubricated compression into frictionless compression and that the FEM yielded values for the coefficient of friction of cheeses in agreement with values obtained from analytical methods.

Liu and Scanlon (2003) used FEA to study the rheological properties of white bread crumb. Using the material model Abaqus HYPERFOAM, based on the Ogden strain energy function, they were able to correlate the modulus of elasticity and the critical stress measured experimentally to the values predicted by FEA. The use of an axisymmetric indentation finite element simulation allowed good prediction of the load—displacement curves produced by cylindrical indenters of various sizes but underpredicted the measurements produced by spherical indenters. Liu and Scanlon (2003) concluded that FEA was a robust tool useful in assisting researchers to study the role of various factors contributing to the textural quality of food materials.

By using the Abaqus finite element software package and Microsoft Excel Solver function, Goh et al. (2004b) were able to extract parameters that defined the viscoelastic response of materials subject to a stress relaxation test with a finite initial loading rate. As noted in the section, "Fat as a Rate-Dependent Material," the loading history affects the relaxation modulus of viscoelastic materials; the FEM facilitated the extraction of viscoelastic constitutive constants to generate a model that matched the experimental data under any arbitrary strain history (Goh et al., 2004b).

Using Abaqus FEA software, Goh and Scanlon (2007) were able to develop a viscoplastic model to simulate the compression and conical indentation response of three lipid-based particle gel systems (margarine, butter, and shortening). In general, the model predicted the indentation response reasonably well, but the load at a given displacement was underpredicted for shortening and butter while it was overpredicted for margarine. Using the viscoplastic model, Goh and Scanlon (2007) were able to back-predict the stress—strain properties from the conical indentation response. Some discrepancies between the results predicted from the finite element model and the experimental data were surmised to be due to a lack of a full knowledge of the large strain compression, and time-dependent behaviors of the lipid gels. Goh and Scanlon (2007) suggested that large strain, stress relaxation, and creep tests should be performed to better understand the rheological response of lipid-based particle gels and to develop a more accurate model.

The Goh–Scanlon model was later modified by Gonzalez-Gutierrez and Scanlon (2013) to include a function that provides for a stress overshoot response at the end of the strain-hardening region. The stress overshoot region was surmised to be due to the strain-hardening process caused by the movement of defects in crystalline solids after the onset of plasticity (Gottstein, 2004; de With, 2006), while the descending part of the stress overshoot was putatively ascribed to flow of the liquid component of the shortening from compressed regions to shear band zones (Gonzalez-Gutierrez and Scanlon, 2013). To take this into account, Eqs. (5.13) and (5.14) were modified as

$$\sigma = E\varepsilon_y\left(\frac{\varepsilon}{\varepsilon_y}\right)^m e^{-(\varepsilon-\varepsilon_y)/\zeta} \quad \text{for } \varepsilon_y \leq \varepsilon \leq \varepsilon_{y2} \tag{5.49}$$

$$\sigma = E\varepsilon_y\left(\frac{\varepsilon_{y2}}{\varepsilon_y}\right)^m e^{-(\varepsilon_{y2}-\varepsilon_y)/\zeta} \quad \text{for } \varepsilon \geq \varepsilon_{y2}, \tag{5.50}$$

where ζ is a stress decay factor (a material constant). As with the original model (Eqs. 5.12–5.14), the modified model has three regions. An exponential function was added to the original model in the postyield regions; the function is analogous to the exponential function of the Maxwell model to describe viscoelastic behavior of materials, and therefore ζ was a stress decay factor—a plastic strain-dependent analogue of the relaxation time in the Maxwell model (Gonzalez-Gutierrez and Scanlon, 2013).

The modified model (Eq. 5.12, 5.49, and 5.50) was calibrated using the material constants obtained during monotonic and cyclic compression test using the nonlinear fit provided by the solver function of Excel (Goh and Scanlon, 2007). To demonstrate the robustness of the modified model in predicting the mechanical response of vegetable shortening, conical indentation simulations were set up in Abaqus/CAE (Gonzalez-Gutierrez, 2008). Indentation was simulated with a two-dimensional axisymmetrical model to save computational time.

The visual representation provided by Abaqus is shown in Fig. 5.22. One can see that as indentation progresses the deformation under the indenter is extremely large and a severe concentration of stresses occurs around the indentation area.

After the virtual indentation at crosshead speeds of 0.4 and 4.0 mm min^{-1} was set up in Abaqus/CAE with the modified constitutive model for shortening, the forces and the displacements were plotted and compared with experimental data acquired at the same indentation speeds (Fig. 5.23). As seen in Fig. 5.23, the modified constitutive model predicts the indentation response of vegetable shortening reasonably well, and with better matching of model and experimental data than that was achieved by Goh and Scanlon (2007). Nevertheless, the general underprediction of experimental data of vegetable shortening by Goh and Scanlon (2007) and by Gonzalez-Gutierrez (2008) indicates that additional stiffening mechanisms appear to play a role in the rheology of vegetable shortening.

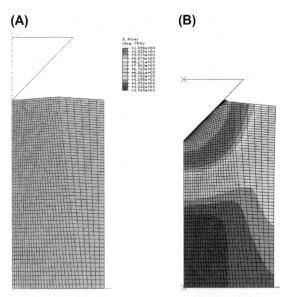

Figure 5.22 Visual representation of indentation response of vegetable shortening simulated using Abaqus with modified Goh–Scanlon model. (A) Undeformed specimen and (B) indentation to 2.9 mm.

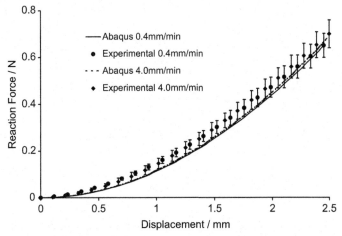

Figure 5.23 Average indentation response of vegetable shortening at two crosshead speeds and simulated results from Abaqus calibrated using the modified Goh–Scanlon model. Error bars on experimental data are one standard deviation.

Finite Element Method Considering Foods as Heterogeneous Materials

In the previous section, the physical properties and the structure of the food materials were considered to be homogeneous. However, food products, including fat systems, are heterogeneous materials with a complex morphology or microstructure. Fat systems have a biphasic nature (liquid–solid), which is spatially organized during processing and in large measure governs product morphology. This resulting morphology is clearly linked to the measured mechanical response, regardless of the type of test (Devi and Khatkar, 2016). FEM modeling based on explicit meshing of material microfeatures has been shown to bring new insights into the complex mechanical and rheological behavior observed in foods (Chen and Opara, 2013).

This microstructure meshing approach was pursued by Guessasma et al. (2008) and Guessasma and Nouri (2016) to model the compressive mechanical response of bread crumb and partially dried bread crumb. In both studies, structural models of the crumb were matched to nondestructive characterization of structure measured by X-ray micro-tomography. In the study by Guessasma and Nouri (2016), an innovative setup allowed simultaneous compressive loading and X-ray tomographic analysis of the bread crumb at various increments of strain. This permitted structure changes in models of this cellular material under compression to correspond to actual observations of structure changes acquired during compression. An Ogden strain energy function, as used by Liu and Scanlon (2003) for bread crumb, was used in the FEA of the mechanical response of the meshed bread crumb structure.

To simulate the heterogeneous microstructure of fat systems, Janssen (2004) generated three types of mesh networks found in the literature: a straight-through mesh, a regular cellular model, and a random needle model. The mesh represented the solid fraction of the fat system only. It was observed that the straight-through mesh and the regular cellular model underpredicted the mechanical response of fat systems. On the other hand, a random network of interlinked needles seemed to approach the actual mechanical response of fat systems. The limited computer capabilities of the time thwarted more realistic descriptions of the fat's microstructure, likely impairing predictions of the observed mechanical response (Janssen, 2004).

Kanit et al. (2006) took three-dimensional confocal images of two ice cream model materials processed under different conditions to obtain descriptions of their strongly contrasting morphology. From four-point bending tests conducted at −18°C, one material had a Young's modulus twice as high as the other. Four-point bending was also performed on each of the two principal components that make up the ice cream (i.e., polycrystal ice and cream). Using FEM simulations based on explicit meshing of the observed microstructure and different boundary conditions, the researchers estimated the effective elastic properties of the two ice cream samples. The voxels from digital images obtained by confocal microscopy were directly transformed into the mesh needed to do the FEA, i.e., the mesh was directly superimposed on the image. The material properties of either ice or cream were assigned at each integration point in the finite element because different material properties were found in different elements of the mesh. This approach solved the issue of matching element nodes to the boundaries between one phase and the other but increased the number of elements needed for convergence of local results close to the interfaces. After setting the mesh and assigning material properties, the boundary conditions were assigned. Contrasting results were obtained with different boundary conditions, with uniform stress boundary conditions generating predicted values of Young's modulus closest to the values measured by four-point bending.

Kanit et al. (2006) also investigated how big the sampling of the morphology has to be to be representative of the whole specimen. They concluded that for one of the materials, the images collected from confocal microscopy were representative, and hence FEM predictions of the ice cream's mechanical response were close to experimental values. However, for the second material, samples of twice the volume would be required to improve the accuracy of the FEM simulation.

As previously discussed, the voxel-to-mesh approach used by Kanit et al. (2006) required the use of a large number of elements for convergence to take place. Using a large number of elements leads to larger computing power requirements, which can be obtained by using supercomputers with large numbers of computing cores; however, this is not always available. In addition, voxel approaches can produce intrinsic jagged surfaces, which can cause high local stress oscillations that smoothing algorithms are not able to completely eliminate. Therefore, other FEM or meshing strategies have been developed. One such technique is referred to as extended finite element methods (X-FEM) (Lian et al., 2013).

X-FEM was developed to simulate systems with discontinuities, singularities, high gradients, or other nonsmooth properties. Earlier examples of its application include models to understand how cracks, shear bands, dislocations, inclusions, and voids affect mechanical properties (Belytschko and Black, 1999; Sukumar et al., 2001; Belytschko and Gracie, 2007). The appearance of shear bands in fats (Gonzalez-Gutierrez and Scanlon, 2013; Macias-Rodriguez et al., 2017) points to the utility of the X-FEM technique for studying the large strain response of fat systems. To model such phenomena, remeshing is required in traditional FEM solutions (Fries and Belytschko, 2010). Therefore, it can be expected that using X-FEM for materials with heterogeneous morphologies, such as fat systems, seems to be a step in the right direction. However, in our current overview of the literature not one example has been found where X-FEM has been applied to food materials, let alone fat systems. It has been used in materials with heterogeneous morphology such as highly filled composites (Legrain et al., 2011; Lian et al., 2013), making it suited to fat-rich composite materials such as cheese (Thionnet et al., 2017). Describing the details of X-FEM is beyond the scope of this chapter; the reader is referred to Fries and Belytschko (2010) and Khoei (2015) for more details about the method.

CONCLUSIONS

The richness of the microstructural and rheological details of fat systems ensures that simple models will only provide a limited quantitative description of their rheology. Although a very simple constitutive model, the elastic perfectly plastic model mimics the mechanical behavior of various fat systems reasonably well, but the fact that yield stresses in fat particle gels are very low negates its mechanistic legitimacy. Therefore, more complex models, such as elastoviscoplastic material models are required to furnish predictive capacities that will extend beyond one given set of rheological testing conditions. Given the rheological and structural complexity, FEMs have emerged to cope with demanding nonlinear problems that are intractable to analytical solutions and which can be used to perform virtual experiments that allow troubleshooting and elimination of experimental artifacts that confound elucidation of the real constitutive properties of fat systems. A further demand on the capabilities of the FEM is the highly heterogeneous distribution of stress and strain fields that are evident during indentation (penetrometry), a technique that is routinely used to define the hardness and quality of fat materials.

ACKNOWLEDGMENT

The authors thank NSERC, Canada, for Discovery Grant research support.

REFERENCES

Abaqus, 2017. Abaqus Unified FEA. https://www.3ds.com/products-services/simulia/products/abaqus/latest-release/.

Adams, M.J., Briscoe, B.J., Kamjab, M., 1993. The deformation and flow of highly concentrated dispersions. Adv. Colloid Interf. Sci. 44, 143−182.

Adams, M.J., Briscoe, B.J., Sinha, S.K., 1996. An indentation study of an elastoviscoplastic material. Phil. Mag. A 74, 1225−1233.

Afoakwa, E.O., Paterson, A., Fowler, M., 2008. Effect of particle size distribution and composition on rheological properties of dark chocolate. Europ. Food Res. Technol. 226, 1259−1268.

Afoakwa, E.O., Paterson, A., Fowler, M., Vieira, J., 2009. Comparison of rheological models for determining dark chocolate viscosity. Int. J. Food Sci. Technol. 44, 162−167.

American Oil Chemists' Society, 1960. Method Cc. 16−60. Official and Tentative Methods: Additions and Revisions. American Oil Chemists' Society (reapproved 1989), Champaign, IL.

American Oil Chemists' Society, 2017. AOCS Official Method 16−60. Official Methods and Recommended Practices, seventh ed. American Oil Chemists' Society, Champaign, IL.

Anand, A., Scanlon, M.G., 2002. Dimensional effects on the prediction of texture related mechanical properties of foods by indentation. Trans. Am. Soc. Agric. Eng. 45, 1045−1050.

Awad, T.S., Rogers, M.A., Marangoni, A.G., 2004. Scaling behavior of the elastic modulus in colloidal networks of fat crystals. J. Phys. Chem. B 108, 171−179.

Bae, W.C., Lewis, C.W., Levenston, M.E., Sah, R.L., 2006. Indentation testing of human articular cartilage: effect of probe tip geometry an indentation depth on intra-tissue strain. J. Biomech. 39, 1039−1047.

Baran, N.M., 1988. Finite Element Analysis on Microcomputers. McGraw-Hill Inc., Toronto, ON.

Barnes, H.A., Hutton, J.F., Walters, K., 1989. An Introduction to Rheology. Elsevier, San Diego, CA.

Beghini, M., Bertini, L., Fontanari, V., 2006. Evaluation of the stress-strain curve of metallic materials by spherical indentation. Int. J. Solids Struct. 43, 2441−2459.

Belytschko, T., Black, T., 1999. Elastic crack growth in finite elements with minimal remeshing. Int. J. Numer. Meth. Eng. 45, 601−620.

Belytschko, T., Gracie, R., 2007. On XFEM applications to dislocations and interfaces. Int. J. Plast. 23, 1721−1738.

Bermudez, A., Gomez, D., Muñiz, M.C., Salgado, P., 2007. A FEM/BEM for axisymmetric electromagnetic and thermal modelling and induction furnaces. Int. J. Numer. Meth. Eng. 71, 856–878.

Boodhoo, M.V., Humphrey, K.L., Narine, S.S., 2009. Relative hardness of fat crystal networks using force displacement curves. Int. J. Food Prop. 12, 129–144.

Bulatov, V.V., Hsiung, L.L., Tang, M., et al., 2006. Dislocation multi-junctions and strain hardening. Nature 440, 1174–1178.

Campbell, G.M., Martin, P.J., 2012. Bread aeration and dough rheology: an introduction. In: Cauvain, S.P. (Ed.), Breadmaking, second ed. Woodhead Press, Cambridge, pp. 299–336.

Campos, R., Narine, S.S., Marangoni, A.G., 2002. Effect of cooling rate on the structure and mechanical properties of milk fat and lard. Food Res. Int. 35, 971–981.

Cardenas-Weber, M., Stroshine, R.L., Haghighi, K., Edan, Y., 1991. Melon material properties and finite element analysis of melon compression with application to robot gripping. Trans. ASAE (Am. Soc. Agric. Eng.) 34, 920–929.

Casiraghi, E.M., Bagley, E.B., Christianson, D.D., 1985. Behaviour of Mozzarella, cheddar and processed cheese spread in lubricated and bonded uniaxial compression. J. Texture Stud. 16, 281–301.

Cenkowski, S., Zhang, Q., Bielewicz, J., Britton, M.G., 1992. Effect of maturity stage on mechanical properties of canola seeds. Trans. ASAE (Am. Soc. Agric. Eng.) 35, 1243–1248.

Charalambides, M.N., Williams, J.G., Chakrabarti, S., 1995. A study of the influence of ageing on the mechanical properties of Cheddar cheese. J. Mater. Sci. 30, 3959–3967.

Charalambides, M.N., Goh, S.M., Lim, S.L., Williams, J.G., 2001. The analysis of frictional effect on stress-strain data from uniaxial compression of cheese. J. Mater. Sci. 36, 2313–2321.

Charalambides, M.N., Wanigasooriya, L., Williams, J.G., Goh, S.M., Chakrabarti, S., 2006. Large deformation extensional rheology of bread dough. Rheol. Acta 46, 239–248.

Chen, Y.W., Mackley, M.R., 2006. Flexible chocolate. Soft Matter 2, 304–309.

Chen, L., Opara, U.L., 2013. Approaches to analysis and modeling texture in fresh and processed foods — a review. J. Food Eng. 119, 497–507.

Chojnicka-Paszun, A., de Jongh, H.H.J., de Kruif, C.G., 2012. Sensory perception and lubrication properties of milk: influence of fat content. Int. Dairy J. 26, 15–22.

Christensen, R.M., 1982. Theory of Viscoelasticity. Academic Press Inc., New York, NY.

Chrysam, M.M., 1985. In: Applewhite, T.H. (Ed.), Bailey's Industrial Oil and Fat Products, vol. 3. Wiley, New York, NY, p. 41.

Coupland, J.N., McClements, D.J., 1997. Physical properties of liquid edible oils. JAOCS (J. Am. Oil Chem. Soc.) 74, 1559–1564.

Dasgupta, A., Hu, J.M., 1992. Failure mechanism models for plastic deformation. IEEE Trans. Reliab. 41, 168–174.

Dassanayake, L.S.K., Kodali, D.R., Ueno, S., Sato, K., 2012. Crystallization kinetics of organogels prepared by rice bran wax and vegetable oils. J. Oleo Sci. 61, 1–9.

De Graef, V., Foubert, F.I., Smith, K.W., Cain, F.W., Dewettinck, K., 2007. Crystallization behavior and texture of trans-containing and trans-free palm oil based confectionery fats. J. Agric. Food Chem. 55, 10258–10265.

De Graef, V., Depypere, F., Minnaert, M., Dewettinck, K., 2011. Chocolate yield stress as measured by oscillatory rheology. Food Res. Int. 44, 2660–2665.

Del Nobile, M.A., Chillo, S., Mentana, A., Baiano, A., 2007. Use of the generalized Maxwell model for describing the stress relaxation behaviour of solid-like foods. J. Food Eng. 78, 978–983.

DeMan, J.M., 1983. Consistency of fats: a review. J. Am. Oil Chem. Soc. 60, 82–87.

DeMan, J.M., Beers, A.M., 1987. Review: fat crystal networks: structure and rheological properties. J. Texture Stud. 18, 303–318.

Devi, A., Khatkar, B.S., 2016. Physicochemical, rheological and functional properties of fats and oils in relation to cookie quality: a review. J. Food Sci. Technol. 53, 3633–3641.

de With, G., 2006. Structure, Deformation, and Integrity of Materials. Wiley-VCH Verlag GmbH & Co., Weinheim, Germany.

Di Monaco, R., Cavella, S., Masi, P., 2008. Predicting sensory cohesiveness, hardness and springiness of solid foods from instrumental measurements. J. Texture Stud. 39, 129–149.

Dixon, B.D., 1974. Spreadability of butter: determination. 1. Description and comparison of five methods of testing. Aust. J. Dairy Technol. 29, 15–22.

Edelglass, S.M., 1966. Engineering Materials Science. The Ronald Press Company, New York.

Engmann, J., Mackley, M.R., 2006. Semi-solid processing of chocolate and cocoa butter - modelling rheology and microstructure changes during extrusion. Food Bioprod. Process. 84, 102–108.

Ewoldt, R.H., Hosoi, A.E., McKinley, G.H., 2008. New measures for characterizing nonlinear viscoelasticity in large amplitude oscillatory shear. J. Rheol. 52, 1427–1458.

Farhloul, M., Zine, A.M., 2008. A posteriori error estimation for a dual mixed finite element approximation of non-Newtonian fluid flow problems. Int. J. Numer. Anal. Model. 5, 320–330.

Ferry, J.D., 1970. Viscoelastic Properties of Polymers. John Wiley and Sons, Inc., Toronto, ON.

Foubert, I., Vereecken, J., Smith, K.W., Dewettinck, K., 2006. Relationship between crystallization behavior, microstructure, and macroscopic properties in trans containing and trans free coating fats and coatings. J. Agric. Food Chem. 54, 7256–7262.

Fries, T.P., Belytschko, T., 2010. The extended/generalized finite element method: an overview of the method and its applications. J. Numer. Methods Eng. 84, 253–304.

Fryer, P.J., Bakalis, S., 2012. Heat transfer to foods: ensuring safety and creating microstructure. J. Heat Transfer – Trans. ASME 134, 031021.

Fung, Y.C., 1969. A First Course in Continuum Mechanics. Prentice-Hall, Inc, Englewood Cliffs, NJ.

Ghotra, B.S., Dyal, S.D., Narine, S.S., 2002. Lipid shortenings: a review. Food Res. Int. 35, 1015–1048.

Giasson, S., Israelachvili, J., Yoshizawa, H., 1997. Thin film morphology and tribology study of mayonnaise. J. Food Sci. 62, 640–652.

Goh, S.M., Charalambides, M.N., Williams, J.G., 2004a. Characterisation of non-linear viscoelastic foods by the indentation technique. Rheol. Acta 44, 47–54.

Goh, S.M., Charalambides, M.N., Williams, J.G., 2004b. Determination of the constitutive constants of non-linear viscoelastic materials. Mech. Time-Depend. Mater 8, 255–268.

Goh, S.M., Scanlon, M.G., 2007. Indentation of lipid-based particle gels: an experimental, theoretical and numerical study. Acta Mater. 55, 3857–3866.

Gonzalez-Gutierrez, J., 2008. An Investigation of the Rheology and Indentation Response of Vegetable Shortening Using Finite Element Analysis (MSc diss.). Univ. Manitoba.

Gonzalez-Gutierrez, J., Scanlon, M.G., 2013. The strain dependence of the uniaxial compression response of vegetable shortening. J. Am. Oil Chem. Soc. 90, 1319–1326.

Gottstein, G., 2004. Physical Foundations of Materials Science. Springer-Verlag Berlin, Berlin.

Gregersen, S.B., Miller, R.L., Andersen, M.D., Hammershoj, M., Wiking, L., 2015a. Inhomogeneous consistency of crystallized fat. Europ. J. Lipid Sci. Technol. 117, 1782–1791.

Gregersen, S.B., Povey, M.J.W., Kidmose, U., Andersen, M.D., Hammershoj, M., Wiking, L., 2015b. Identification of important mechanical and acoustic parameters for the sensory quality of cocoa butter alternatives. Food Res. Int. 76, 637–644.

Gregersen, S.B., Povey, M.J.W., Andersen, M.D., Hammershoj, M., Rappolt, M., Sadeghpour, A., Wiking, L., 2016. Acoustic properties of crystallized fat: relation between polymorphic form, microstructure, fracturing behavior, and sound intensity. Europ. J. Lipid Sci. Technol. 118, 1257–1270.

Guessasma, S., Nouri, H., 2016. Comprehensive study of biopolymer foam compression up to densification using X-ray micro-tomography and finite element computation. Europ. Polymer J. 84, 715–733.

Guessasma, S., Babin, P., Della Valle, G., Dendievel, R., 2008. Relating cellular structure of open solid food foams to their Young's modulus: finite element calculation. Int. J. Solids Struct. 45, 2881–2896.

Gunasekaran, S., Ak, M.M., 2003. Cheese Rheology and Texture. CRC Press LLC, Boca Raton, FL.

Hakamada, M., Kuromura, T., Chino, Y., Yamada, Y., Chen, Y., Kusuda, H., Mabuchi, M., 2007. Monotonic and cyclic compressive properties of porous aluminum fabricated by spacer method. Mater. Sci. Eng. A 459, 286–293.

Hassan, B.H., Alhamdan, A.M., Elansari, A.M., 2005. Stress relaxation of dates at khalal and rutab stages of maturity. J. Food Eng. 66, 439–445.

Hatzikiriakos, S.G., Migler, K.B., 2004. Polymer Processing Instabilities: Control and Understanding. Marcel Dekker, New York, NY.

Herrera, M.L., Hartel, R.W., 2000. Effect of processing conditions on physical properties of a milk fat model system: Rheology. JAOCS (J. Am. Oil Chem. Soc.) 77, 1189—1196.

Higaki, K., Koyano, T., Hachiya, I., Sato, K., Suzuki, K., 2004. Rheological properties of b-fat gel made of binary mixtures of high-melting and low-melting fats. Food Res. Int. 37, 799—804.

Hjelmstad, K.D., 2005. Fundamentals of Structural Mechanics, second ed. Springer, New York, NY.

Huang, Z., Lucas, M., Adams, M.J., 2002. A numerical and experimental study of the indentation mechanics of plasticine. J. Strain Anal. 37, 141—150.

Janssen, P.W.M., 2004. Modelling fat microstructure using finite element analysis. J. Food Eng. 61, 387—392.

Johnson, K.L., 1987. Contact Mechanics. Cambridge University Press, Cambridge.

Kalab, M., 1985. Microstructure of dairy foods. 2. Milk products based on fat. J. Dairy Sci. 68, 3234—3248.

Kanit, T., NGuyen, F., Forest, S., Jeulin, D., Reed, M., Singleton, S., 2006. Apparent and effective physical properties of heterogeneous materials: representativity of samples of two materials from food industry. Comput. Methods Appl. Mech. Eng. 195, 3960—3982.

Khoei, A., 2015. Extended Finite Element Method: Theory and Applications. John Wiley and Sons, Ltd., Chichester, West Sussex.

Kiełczynski, P., Ptasznik, S., Szalewski, M., Balcerzak, A., Wieja, K., Rostocki, A.J., 2017. Thermophysical properties of rapeseed oil methyl esters (RME) at high pressures and various temperatures evaluated by ultrasonic methods. Biomass Bioenerg. 107, 113—121.

Kloek, W., van Vliet, T., Walstra, P., 2005. Large deformation behaviour of fat crystal networks. J. Texture Stud. 36, 516—543.

Lakes, R.S., 1987. Foam structures with a negative Poisson's ratio. Science 235, 1038—1040.

Legrain, G., Cartraud, P., Perreard, I., Moës, N., 2011. An X-FEM and level set computational approach for image-based modelling: application to homogenization. Int. J. Numer. Meth. Eng. 86, 915—934.

Le Reverend, B.J.D., Smart, I., Fryer, P.J., Bakalis, S., 2011. Modelling the rapid cooling and casting of chocolate to predict phase behaviour. Chem. Eng. Sci. 66, 1077—1086.

Leroy, V., Pitura, K.M., Scanlon, M.G., Page, J.H., 2010. The complex shear modulus of dough over a wide frequency range. J. Non-newtonian Fluid Mech. 165, 475—478.

Lian, W.D., Legrain, G., Cartraud, P., 2013. Image-based computational homogenization and localization: comparison between X-FEM/level set and voxel-based approaches. Comput. Mech. 51, 279—293.

Lin, D.C., Horkay, F., 2008. Nanomechanics of polymer gels and biological tissues: a critical review of analytical approaches in the Hertzian regime and beyond. Soft Matter 4, 669—682.

Liu, Z., Scanlon, M.G., 2003. Modelling indentation of bread crumb by finite element analysis. Biosyst. Eng. 85, 477—484.

Liu, K., Stieger, M., van der Linden, E., De Velde, F.V., 2015. Fat droplet characteristics affect rheological, tribological and sensory properties of food gels. Food Hydrocoll. 44, 244—259.

Lupi, F.R., Gabriele, D., Greco, V., Baldino, N., Seta, L., de Cindio, B., 2013. A rheological characterisation of an olive oil/fatty alcohols organogel. Food Res. Int. 51, 510—517.

Ma, X., Yoshida, F., Shinbata, K., 2003. On the loading curve in microindentation of viscoplastic solder alloy. Mater. Sci. Eng. A 344, 296—299.

Macias-Rodriguez, B.A., Marangoni, A.G., 2016a. Rheological characterization of triglyceride shortenings. Rheol. Acta 55, 767—779.

Macias-Rodriguez, B.A., Marangoni, A.G., 2016b. Physicochemical and rheological characterization of roll-in shortenings. J. Am. Oil Chem. Soc. 93, 575—585.

Macias-Rodriguez, B.A., Marangoni, A.G., 2017. Bakery shortenings: structure and mechanical function. Appl. Rheol. 27, 10—17.

Macias-Rodriguez, B.A., Peyronel, F., Marangoni, A.G., 2017. The role of nonlinear viscoelasticity on the functionality of laminating shortenings. J. Food Eng. 212, 87—96.

Madenci, E., Guven, I., 2006. The Finite Element Method and Applications in Engineering Using ANSYS[r]. Springer Science + Business Media, (LLC), New York, NY.

Maleky, F., Marangoni, A., 2011. Ultrasonic technique for determination of the shear elastic modulus of polycrystalline soft materials. Cryst. Growth Des. 11, 941—944.

Marangoni, A.G., Rogers, M.A., 2003. Structural basis for the yield stress in plastic disperse systems. Appl. Phys. Lett. 82, 3239—3241.

Marangoni, A.G., Rousseau, D., 1998. Chemical and enzymatic modification of butterfat and butterfat-canola oil blends. Food Res. Int. 31, 595—599.

McClements, D.J., 2003. The rheology of emulsion-based food products. In: McKenna, B.M. (Ed.), Texture in Food, Semi-solid Foods, vol. 1. CRC Press LLC, Boca Raton, FL, pp. 3—32.

Menčik, J., 2007. Determination of mechanical properties by instrumented indentation. Meccanica 42, 19—29.

Mert, B., Demirkesen, I., 2016. Reducing saturated fat with oleogel/shortening blends in a baked product. Food Chem. 199, 809—816.

Metzroth, D.J., 2005. Shortenings: science and technology. In: Shahidi, F. (Ed.), Bailey's Industrial Oil and Fat Products, sixth ed. John Wiley & Sons, Inc., Hoboken, NJ, pp. 83—123.

Mirzayi, B., Heydari, A., Noori, L., Arjomand, R., 2011. Viscosity and rheological behavior of castor—canola mixture. Eur. J. Lipid Sci. Technol. 113, 1026—1030.

Mura, T., Koya, T., 1992. Variational Methods in Mechanics. Oxford University Press, Inc., New York, NY.

Narine, S.S., Marangoni, A.G., 1999a. Fractal nature of fat crystal networks. Phys. Rev. E 59, 1908—1920.

Narine, S.S., Marangoni, A.G., 1999b. Mechanical and structural model of fractal networks of fat crystals at low deformations. Phys. Rev. E 60, 6991—7000.

Narine, S.S., Marangoni, A.G., 2000. Elastic modulus as an indicator of macroscopic hardness of fat crystal networks. Lebensmitt. Wissen. Technol. 34, 33—40.

Nigo, R.Y., Chew, Y.M.J., Houghton, N.E., Paterson, W.R., Wilson, D.I., 2009. Experimental studies of freezing fouling of model food fat solutions using a novel spinning disc apparatus. Energy Fuels 23, 6131—6145.

O'Brien, R.D., 1998. Fats and Oils: Formulating and Processing for Applications. Technomic Publishing Company, Lancaster, PA, USA.

O'Brien, R.D., 2005. Shortenings: types and formulations. In: Shahidi, F. (Ed.), Bailey's Industrial Oil and Fat Products, sixth ed. John Wiley & Sons, Inc., Hoboken, NJ, pp. 125—157.

Ouriev, B., Windhab, E.J., 2002. Rheological study of concentrated suspensions in pressure-driven shear flow using a novel in-line ultrasound Doppler method. Experim. Fluids 32, 204—211.

Page, T.F., 1996. Nanoindentation testing. In: Adams, M.J., Biswas, B.K., Briscoe, B.J. (Eds.), Solid-solid Interactions. Imperial College Press, London, pp. 93—116.

Peleg, M., Pollak, N., 1982. The problem of equilibrium conditions in stress relaxation analyses of solid foods. J. Texture Stud. 13, 1—11.

Puri, V.M., Anantheswaran, R.C., 1993. The finite element method in food processing: a review. J. Food Eng. 19, 247—274.

Ramirez-Gomez, N.O., Acevedo, N.C., Toro-Vazquez, J.F., Ornelas-Paz, J., Dibildox-Alvarado, E., Perez-Martinez, J.D., 2016. Phase behavior, structure and rheology of candelilla wax/fully hydrogenated soybean oil mixtures with and without vegetable oil. Food Res. Int. 89, 828—837.

Rao, S.S., 1982. The Finite Element Method in Engineering. Pergamon Press Canada Ltd., Willowdale, ON.

Rao, M.A., 2007. Rheology of Fluid and Semisolid Foods: Principles & Applications. Springer, New York, NY.

Ren, S., Barringer, S., 2016. Electrohydrodynamic spraying quality of different chocolate formulations. J. Electrost. 84, 121—127.

Renzetti, S., Jurgens, A., 2016. Rheological and thermal behavior of food matrices during processing and storage: relevance for textural and nutritional quality of food. Curr. Opin. Food Sci. 9, 117—125.

Renzetti, S., de Harder, R., Jurgens, A., 2016. Puff pastry with low saturated fat contents: the role of fat and dough physical interactions in the development of a layered structure. J. Food Eng. 170, 24—32.

Reyes-Hernandez, J., Dibildox-Alvarado, E., Charo-Alonso, M.A., Toro-Vazquez, J.F., 2007. Physico-chemical and rheological properties of crystallized blends containing trans-free and partially hydrogenated soybean oil. JAOCS (J. Am. Oil Chem. Soc.) 84, 1081—1093.

Rigolle, A., Foubert, I., Hettler, J., Verboven, E., Demuynck, R., Van den Abeele, K., 2015. Development of an ultrasonic shear reflection technique to monitor the crystallization of cocoa butter. Food Res. Int. 75, 115–122.

Roduit, B., Borgeat, C.H., Cavin, S., Fragniere, C., Dudler, V., 2005. Application of finite element analysis (FEA) for the simulation of release of additives from multilayer polymeric packaging structures. Food Addit. Contam. 22, 945–955.

Rohm, H., 1993. Rheological behaviour of butter at large deformations. J. Texture Stud. 24, 139–155.

Ronn, B.B., Hyldig, G., Wienberg, L., Qvist, K.B., Laustsen, A.M., 1998. Predicting sensory properties from rheological measurements of low-fat spreads. Food Qual. Prefer. 9, 187–196.

Roylance, D., 1996. Mechanics of Materials. John Wiley & Sons, Inc., Toronto, ON.

Rumsey, T.R., Fridley, R.B., 1977. Analysis of viscoelastic contact stresses in agricultural products using a finite element method. Trans. ASAE (Am. Soc. Agric. Eng.) 20, 162–167, 171.

Saggin, R., Coupland, J.N., 2004. Shear and longitudinal ultrasonic measurements of solid fat dispersions. JAOCS (J. Am. Oil Chem. Soc.) 81, 27–32.

Scanlon, M.G., Page, J.H., 2015. Probing the properties of dough with low-intensity ultrasound. Cereal Chem. 92, 121–133.

Schaink, H.M., van Malssen, K.F., 2007. Shear modulus of sintered 'House of Cards'-like assemblies of crystals. Langmuir 23, 12682–12686.

Sherman, P., 1976. The textural characteristics of dairy products. In: De Man, J.M., Voisey, P.W., Rasper, V.F., Stanley, D.W. (Eds.), Rheology and Texture in Food Quality. AVI Publishing, Westport, CT, pp. 382–395.

Shimizu, Y., Ito, H., Sadakane, T., Yamaguchi, T., 2012. Multi-visco-elastic contact model in discrete element method - application to margarine's process engineering. J. Soc. Rheol. Japan 40, 257–266.

Shukla, A., Rizvi, S.S.H., 1995. Viscoelastic properties of butter. J. Food Sci. 60, 902–905.

Staniewski, B., Szpendowski, J., Panfil-Kuncewicz, H., Malkus, J., Bohdziewicz, K., 2006. The application of an AP4/2 conical penetrometer for the evaluation of spreadability of selected table fats. Milchwiss. Milk Sci. Int. 61, 292–296.

Steffe, J.F., 1996. Rheological Methods in Food Process Engineering, second ed. Freeman Press, East Lansing, MI.

Sukumar, N., Chopp, D.L., Moes, N., Belytschko, T., 2001. Modeling holes and inclusions by level sets in the extended finite element method. Comp. Methods Appl. Mech. Eng. 190, 6183–6200.

Svanberg, L., Ahrne, L., Loren, N., Windhab, E., 2012. A method to assess changes in mechanical properties of chocolate confectionery systems subjected to moisture and fat migration during storage. J. Texture Stud. 43, 106–114.

Tabor, D., 1948. A simple theory of static and dynamic hardness. Proc. Royal Soc. London A192, 247–274.

Tabor, D., 1996. Indentation hardness and material properties. In: Adams, M.J., Biswas, B.K., Briscoe, B.J. (Eds.), Solid-solid Interactions. Imperial College Press, London, pp. 6–15.

Tanaka, M., deMan, J.M., Voisey, P.W., 1971. Measurement of textural properties of foods with a constant speed cone penetrometer. J. Texture Stud. 2, 306–315.

Thionnet, O., Havea, P., Gillies, G., Lad, M., Golding, M., 2017. Influence of the volume fraction, size and surface coating of hard spheres on the microstructure and rheological properties of model mozzarella cheese. Food Biophys. 12, 33–44.

Vandamme, M., Ulm, F.J., 2006. Viscoelastic solutions for conical indentation. Int. J. Solids Struct. 43, 3142–3165.

van Vliet, T., 2014. Rheology and Fracture Mechanics of Foods. CRC Press, Boca Raton, FL.

van Vliet, T., van Aken, G.A., de Jongh, H.H.J., Hamer, R.J., 2009. Colloidal aspects of texture perception. Adv. Colloid Interf. Sci. 150, 27–40.

Vithanage, C.R., Grimson, M.J., Smith, B.G., 2009. The effect of temperature on the rheology of butter, a spreadable blend and spreads. J. Texture Stud. 40, 346–369.

Vreeker, R., Hoekstra, L., Denboer, D.C., Agterof, W.G.M., 1992. The fractal nature of fat crystal networks. Colloids Surf. 65, 185–189.

White, M.A., 1999. Properties of Materials. Oxford University Press Inc., New York, NY.

Wright, A.J., Scanlon, M.G., Hartel, R.W., Marangoni, A.G., 2001. Rheological properties of milkfat and butter. J. Food Sci. 66, 1056–1071.

Yang, F., Zhang, M., Bhandari, B., 2017. Recent development in 3D food printing. Crit. Rev. Food Sci. Nutr. 57, 3145–3153.

Yoshikawa, H.Y., Pink, D.A., Acevedo, N.C., Peyronel, F., Marangoni, A.G., Tanaka, M., 2017. Mechanical response of single triacylglycerol spherulites by using microcolloidal probes. Chem. Lett. 46, 599–601.

CHAPTER 6

Nonlinear Rheology of Fats Using Large Amplitude Oscillatory Shear Tests

Braulio A. Macias Rodriguez
University of Guelph, Guelph, ON, Canada

INTRODUCTION

As highlighted in the previous chapter, knowledge of rheological properties is essential for controlling the application of fat systems and their manufacturing processes. Rheology provides information on structure, interaction forces and associated elasticity, and plasticity of such structural elements. Understanding the relationship between structure and rheology is crucial for the design of fat products with tailored rheological properties and proper adjustment of formulation/crystallization conditions to meet this goal. The pursuit of this endeavor brings with it two major challenges: first, the rheology of fats depends on a hierarchical structure with multiple physical length scales spanning from the molecular to the micro, which largely dependent on crystallization (cooling and shearing) conditions and second, fats do not obey the principle of superposition and they "age," i.e., they undergo gradual recrystallization and crystal sintering. Despite these difficulties, great strides have been made toward understanding the relationship between linear viscoelasticity and the microstructure of fats. Although linear viscoelasticity determined at small amplitudes is important to probe the underlying microstructural network, it lacks practical significance (e.g., spreading of fats involves large amplitudes). On the other hand, nonlinear viscoelasticity determined at large amplitudes is of greater relevance for processing and use of fat-based materials (and soft materials in general). In this chapter, we briefly revisit conventional small amplitude oscillatory shear (SAOS) and present a succinct account of large amplitude oscillatory shear (LAOS). The information of this chapter is organized as follows: first, linear and nonlinear viscoelasticity theory is shortly recapped; second, quantitative methods for data analysis and interpretation are reviewed with emphasis on the Chebyshev stress decomposition method; third, experimental considerations for collection, processing, and analysis of data obtained from LAOS experiments are presented; and finally, results from two different samples (fat crystal networks and fat-filled gels) are presented and discussed to prove the value of the LAOS technique in understanding material functionality. For more traditional methods to measure large deformation properties (e.g., compression and indentation), the readers

Structure-Function Analysis of Edible Fats, Second Edition
ISBN 978-0-12-814041-3, https://doi.org/10.1016/B978-0-12-814041-3.00006-X

are referred to the preceding chapter (Chapter 5). Despite the simplicity of the latter methods, LAOS possesses several advantages given that oscillations (1) gradually lead up to yield, (2) can decompose energy storage and loss mechanisms, and (3) have a better signal-to-noise (SN) ratio for comparing between materials (because oscillations lock-in on each strain and frequency probed), etc. All these features are important for distinguishing the yielding behavior of the materials studied here.

OSCILLATORY SHEAR RHEOLOGY

Small Amplitude Oscillatory Shear

Most of the fundamental work on the rheology of fats concerns linear viscoelastic properties obtained from quite small amplitudes (deformation or load inputs) (Marangoni and Narine, 2002; Narine and Marangoni, 1999a; Rohm and Weidinger, 1993; Thareja, 2013; van den Tempel, 1961). In the linear regime, strain and stress linearly scale, and the viscoelastic moduli (independent of the input amplitude) are calculated based on well-developed theory (Bird et al., 1987; Dealy and Wissbrun, n.d.; Ferry, 1980; Tschoegl, 1989). For a sinusoidal strain excitation (i.e., strain-controlled test) $\gamma(t) = \gamma_0 \sin(\omega t)$, a sinusoidal stress response is obtained and expressed as

$$\sigma(t) = \gamma_0 [G'(\omega)\sin(\omega t) + G''(\omega)\cos(\omega t)], \qquad (6.1)$$

In which G' the in-phase elastic or stored energy and G'' the out-of-phase viscous or dissipated energy represent the real and imaginary components of the complex modulus G^* at frequency (ω), respectively (Ferry, 1980; Macosko, 1994; Tschoegl, 1989).

For fat crystal networks, linear deformations are typically in the order of $\gamma_0 \approx 0.01\% - 0.1\%$, a common feature of materials interacting via short-range van der Waals attractions. Because of the narrow linear regime, handling of the sample shall minimize disruption of the original network. This can be achieved by loading cylindrical preformed samples with controlled normal force (i.e., samples crystallized ex situ, cut, and shaped), loading with geometries that minimize sample disturbance (e.g., vane) or by crystallization in situ (Macias-Rodriguez and Marangoni, 2016; Thareja et al., 2011). Linear viscoelastic moduli of fats are in the order of G', $G'' \approx 10^4 - 10^6$ Pa. The storage modulus G' is larger than the loss modulus G'' by about an order of magnitude, and both moduli are nearly independent on frequency (a typical behavior of soft viscoelastic solids) (van den Tempel, 1961; Narine and Marangoni, 1999b; Macias-Rodriguez and Marangoni, 2016b; Rohm and Weidinger, 1993; Thareja et al., 2013). Linear viscoelastic properties are useful for developing the structure—property relations (e.g., interaction forces holding a network). The magnitude of the elastic modulus G' of fats depends in a complex manner on at least three factors: volume fraction (estimated as SFC/100), crystal microstructure, and crystalline

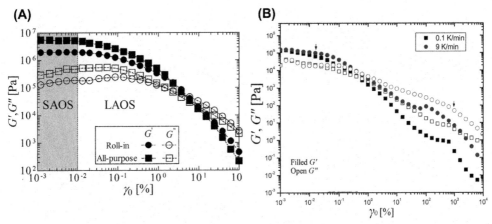

Figure 6.1 Linear viscoelastic moduli G', G'' as a function of strain input γ_0 of (A) fat shortenings and (B) 25% tribehenin in isopropyl myristate. Small amplitude oscillatory shear (SAOS) (linear) and large amplitude oscillatory shear (LAOS) (nonlinear) strain regions are approximately delineated in (A). Data in (A) and (B) were collected at $T = 16°C$, $\omega = 3.6$ rad/s and $T = 20°C$, $\omega = 10$ rad/s, respectively. *Arrows* indicate strain values nearly at the onset of yielding and deeply into the nonlinear regime in (B). ((B) *Reproduced from Ramamirtham, et al., 2017. Rheol. Acta 56 (12), 971–982. Copyright (2017), with permission from Springer-Verlag GmbH Germany.)*

interactions (Narine and Marangoni, 1999a). The elastic modulus G' provides a combined measure of primary crystallization, aggregation and network formation, as well as of material hardness or firmness (Graef et al., 2006; Narine and Marangoni, 1999b). Fig. 6.1 shows the viscoelastic moduli G' and G'' for fat shortenings formulated for various commercial "functionalities" and model fats crystallized at two cooling rates. As illustrated, linear viscoelastic moduli remain unremarkable across material classes irrespective of their functionalities (Fig. 6.1A) or their crystallization regimes (Fig. 6.1B). However, macroscopic and microscopic images of fats subject to large deformations (either in shear or compression) reveal a quite different story untold by SAOS (see Fig. 6.2). Observations of this nature motivate our investigation of nonlinear viscoelasticity of fats.

Large Amplitude Oscillatory Shear

During small amplitudes (SAOS), fats deform in an elastic or reversible manner, whereas at large amplitudes (LAOS) they undergo plastic flow or irreversible deformations. In other words, the material response shifts from linear to nonlinear, as deformation or load input increases. The point at which this transition occurs is known as the "yield point," associated with yield strain (γ_y) and yield stress (σ_y) values, respectively. The definition of a "yield point" is usually unclear (e.g., it can be defined at the onset of nonlinearity, at the "peak" maxima or at the "equilibrium" state, etc) because it changes according to the method of determination and it often cannot describe adequately the state of "flow" (Dinkgreve et al., 2016; Møller et al., 2006). Despite these ambiguities,

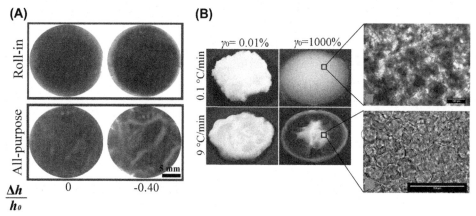

Figure 6.2 (A) Views from below a transparent bottom plate supporting different mechanical behavior of fat shortenings. A roll-in shortening behaves as a ductile-like solid, whereas an all-purpose shortening resembles a brittle-like solid under compression. (B) Macrographs and bright-field microscopic images of 25% tribehenin in isopropyl myristate before ($\gamma_0 = 0.01\%$) and after yielding ($\gamma_0 = 1000\%$). High strain inputs induce complete or partial "fluidization" depending on crystallization regime of the sample. Samples crystallized at slow cooling rates experience complete "fluidization" (i.e., viscous behavior) due to strong breakage of "interlinks" between bundles of needles. Samples crystallized at fast cooling rate retain some elasticity (i.e., they show more elastoviscoplastic behavior) due to spherical clusters that appear to hold together. *((B) Reproduced from Ramamirtham, et al., 2017. Rheol. Acta 56 (12), 971–982. Copyright (2017), with permission from Springer-Verlag GmbH Germany.)*

σ_y continues to be a valuable parameter influencing the processability and usability of materials (Coussot, 2007). From a practical standpoint, σ_y indicates the minimum pressure to initiate flow in a pipeline and material stiffness. For thixotropic yield stress materials (e.g., in fats where σ_y is dictated from the competition between buildup and breakdown of the microstructure), static and dynamic yield stresses should be considered, which can be calculated from the $G'-G''$ crossover and the intersection of σ and γ, respectively, from oscillatory measurements (Dinkgreve et al., 2016). From a structural viewpoint, in an idealized fat crystal network, σ_y can be envisioned as the force necessary to move interacting microstructural elements apart so that their interaction energy becomes zero. This force depends on the solid volume fraction ϕ, fractal dimension of the system D, crystal–melt interfacial tension δ, and primary particle size a (Marangoni and Rogers, 2003). Fig. 6.3 illustrates σ_y of roll-in and all-purpose shortenings. Despite their distinct end use, they share comparable σ_y (defined as the maximum stress). On closer inspection of the strain–stress curves, striking differences are observed during yielding, i.e., the material response of an all-purpose shortening resembles a brittle-like solid ("catastrophic" yielding and strong stress relaxation), whereas a roll-in shortening resembles a ductile-like solid ("smooth" yielding and stress plateau). Such differences are consistent with the macroscopic behavior of these materials (Fig. 6.2A).

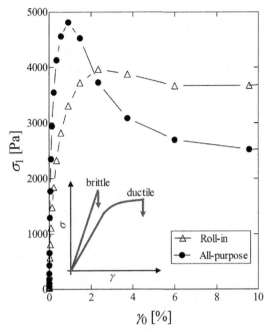

Figure 6.3 Stress σ_1 as a function of strain γ_0 for roll-in and all-purpose shortenings. Idealized brittle and ductile failure of solid materials (inset) share qualitative resemblance with the failure of fats. Data were collected at $T = 16°C$ and $\omega = 3.6$ rad/s.

The nonlinear rheological behavior of fats is per se not new; similar observations have been made in the past through compression and start–up shear tests. Nevertheless, it has not been widely investigated with oscillatory shear tests (de Bruijne and Bot, 1999; Goh and Scanlon, 2007; Gonzalez-Gutierrez and Scanlon, 2013). To the best of my knowledge, there are only two early studies that explicitly addressed nonlinear viscoelasticity of fatty materials (e.g., butter, margarine, cheese based spread) under oscillatory shear tests (Elliot and Ganz 1971; Prentice 1993). It is noteworthy that other rheometric tests can also exert nonlinear deformations, but their use is limited by several shortcomings. For example, steady shear and start–up of steady shear are not well suited for "stiff" materials with sensitive microstructures and highly prone to experimental artifacts (e.g., wall slip and edge fracture) such as fats. Step shear (creep or relaxation) imposes sudden and large displacements, which typically lead to highly heterogeneous deformation especially at high stresses or amplitudes. Compression tests are not accurate for determining elastic properties of materials with yield strains below 2% (Kloek et al., 2005). In contrast, nonlinear oscillatory shear tests possess many advantages that rationalize their use over the aforementioned techniques. LAOS tests provide easier and controlled flow (i.e., gradual yielding) and simultaneous survey of energy storage and loss mechanisms in a wide range of timescales ($1/\omega$) and amplitudes (γ or σ). The resultant material response is represented in an experimental test space known as the "Pipkin" diagram. Moreover, LAOS enables better SN ratio because oscillations lock in on each frequency and amplitude probed. Having set the stage, in the following sections we provide a summary of LAOS fundamentals, the two options for performing LAOS tests: strain-controlled and stress-controlled along with their corresponding material functions, guidelines for experimental acquisition, analysis and interpretation of the data, the various ontological frameworks provided for data analysis, and their application in fat crystal networks and fat-containing products (e.g., cheese). The goal here is not to provide an exhaustive description of the technique, recently reviewed by (Hyun et al., 2011), but rather prove its potential for understanding the "functionality" of fat materials.

Fundamentals

During LAOS, nonlinear viscoelastic behavior arises when the output responses become a measurable function of the imposed amplitude (i.e., deformation or load). Consequently, linear viscoelastic moduli or compliances are not uniquely defined (Ewoldt et al., 2008). These nonlinearities manifest as incrementing distortions of the waveform output (i.e., nonsinusoidal response) as deformation/load increases beyond the linear viscoelastic region (LVR). The time-periodic nonlinear responses encode rich material information related to its structure, function, processing, and usability (Ewoldt, 2013a). There are two possible types of LAOS characterization: strain-controlled (hereafter denoted LAOStrain) and stress-controlled (hereafter denoted LAOStress), which provide complements, but not conjugates, material functions, i.e., viscoelastic moduli (G' and G'') and compliances (J' and J''). The natural and most common method to quantify a time-periodic nonlinear response is a Fourier series. Several quantitative frameworks have been proposed to analyze nonlinear waveforms, including Fourier series, stress decomposition method, Chebyshev polynomials, or Chebyshev stress decomposition method, time-dependent moduli, among other methods (Cho et al., 2005; Ewoldt et al., 2008; Rogers and Lettinga, 2012; Wilhelm, 2002).

Fourier Series

A natural representation of a time-periodic oscillating response is a Fourier series. Fourier transform (FT) rheology constitutes the basis for most of the frameworks proposed to analyze nonlinear viscoelasticity (Wilhelm, 2002). For the case of strain-controlled experiments, a sinusoidal strain input $\gamma(t) = \gamma_0 \sin(\omega t)$ gives a stress response that can be captured by a Fourier series written in two alternative forms to highlight elastic or viscous scaling. The elastic scaling of the series is expressed as follows:

$$\sigma(t; \omega, \gamma_0) = \gamma_0 \sum_{n:odd} \left\{ G'_n(\omega, \gamma_0)\sin n\omega t + G''_n(\omega, \gamma_0)\cos n\omega t \right\} \qquad (6.2)$$

$$\sigma(t; \omega, \gamma_0) = \gamma_0 \sum_{n:odd} \left\{ G_n'(\omega, \gamma_0)\sin n\omega t + G_n''(\omega, \gamma_0)\cos n\omega t \right\} \qquad (6.3)$$

From which, a viscous scaling may be obtained by factoring out $\dot{\gamma}_0 = \gamma_0\omega$, giving the coefficients $\eta_n'' = G_n'/\omega$ and $\eta_n' = G_n''/\omega$.

Alternatively, the stress response may be expressed in terms of amplitude and phase angles:

$$\sigma = \gamma_0 \sum_{n:odd} \sigma_n(\omega, \gamma_0)\sin(n\omega t + \delta_n(\omega, \gamma_0)) . \qquad (6.4)$$

These mathematical descriptions only include odd harmonics assuming the stress response is an odd function with respect to the direction of the shear strain γ or shear rate $\dot{\gamma}_0$, i.e., the stress is independent of the shear direction, and its sign changes as the shear direction changes. Such assumption can be violated (i.e., even harmonics will appear) due to loss of rotational symmetry about the origin attributed to transient responses, edge failure, and secondary flows (Hyun et al., 2011; Li et al., 2009).

In the linear regime, the rheological response is only described by the first ($n = 1$) or "fundamental" harmonic occurring at the frequency of excitation ω because by definition, there is no stress distortion (i.e., the response is sinusoidal) (Fig. 6.4A). In the nonlinear regime, the stress output is not purely sinusoidal, i.e., the stress response contains higher-order harmonics that grow unboundedly ($n = 1,3,5...$) (Fig. 6.4B). In principle, nonlinear responses manifest as distortions of the elliptical Lissajous–Bowditch loops observed in the linear regime (see Figs. 6.5, 6.8, 6.11). Lissajous–Bowditch loops are parametric plots of γ versus σ (elastic perspective) or $\dot{\gamma}$ versus σ (viscous perspective). To quantify nonlinear viscoelastic responses appearing in such loops, the amplitudes of all harmonic contributions or their ratios ($I_{n/1}$), and more often the third harmonic

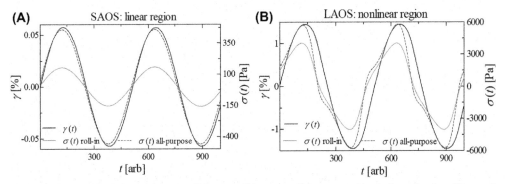

Figure 6.4 Raw waveforms obtained from roll-in and all-purpose shortenings in the (A) small amplitude oscillatory shear (SAOS) (sinusoidal strain and stress) regime and (B) large amplitude oscillatory shear (LAOS) (sinusoidal strain, distorted stress) regime, respectively.

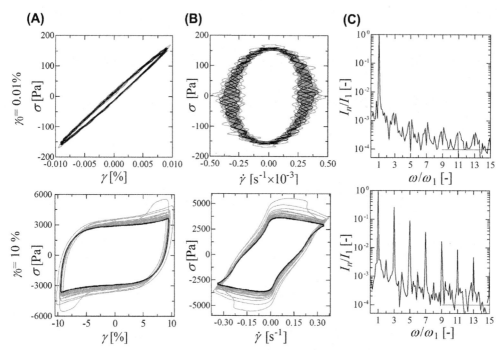

Figure 6.5 Raw Lissajous—Bowditch plots of (A) stress versus strain (elastic representation) and (B) stress versus shear rate (viscous representation). The *gray lines* correspond to transients, whereas the *solid black lines* represent the steady state data. (C) Frequency-domain Fourier spectra. The data on the top row were collected in the small amplitude oscillatory shear regime ($\gamma_0 = 0.01\%$), whereas the data on the down row were obtained in the LAOS regime ($\gamma_0 = 10\%$), for a roll-in shortening at $T = 16°C$ and $\omega = 3.6$ rad^{-1}.

normalized by the fundamental frequency ($I_{3/1}$), are used. The $I_{3/1}$ is the leading order and most intense compared with the rest of harmonic contributions (Wilhelm, 2002). For weakly nonlinear stress response (i.e., strains "slightly" beyond the LVR), the harmonic response is only dependent on oscillatory frequency (not strain amplitude) and scales with a standard integer power law of ~ 2 ($I_{3/1} \propto \gamma_0^2$). This regime may be referred as to medium amplitude oscillatory shear (MAOS), but here we do not make distinction of such regime. The intensity of the $I_{3/1}$ is highly sensitive to morphological features (e.g., shape, dimension, size distribution), structural changes induced by flow (e.g., shear-induced orientation in lamellar block copolymers), or formulation (e.g., fillers in polymer matrix), yielding in complex fluids and soft matter (e.g., polymeric, colloidal). Recently, the yielding of fat shortenings was monitored by the growth of $I_{3/1}$ to quantitatively differentiate between their commercial "functionality" (e.g., all-purpose, roll-in, cake and icing). The evolution of $I_{3/1}$ was described by a descriptive function with minimum degree of free parameters (Wilhelm, 2002):

$$I_{3/1} = A[1 - \exp(- (\gamma_0 - \gamma_L)/k)], \gamma_0 > \gamma_L, \qquad (6.5)$$

where A reflects the maximum possible contribution of $I_{3/1}$ ($A \leq 1/3$) at an idealized infinite shear strain amplitude γ_0, γ_L describes the maximum shear amplitude within the linear region, and k (inverse slope) describes the change in intensity as a function of γ_0. The parameter k has shown to be heavily influenced by structure morphology (e.g., molecular topology). From the fitting, there were only differences in the parameter k. Roll-in shortenings had higher inverse slopes ($k = 0.6-1.3 \times 10^{-2}$) than shortenings used for other applications ($k = 0.3-0.4 \times 10^{-2}$), reflecting a more gradual transition into the nonlinear region for roll-in shortenings. It was also observed that $I_{3/1}$ plateaued for roll-in shortenings, whereas it decreased abruptly for an all-purpose shortening. These findings qualitatively indicated higher load-bearing ability for the primer than for the latter. The similar γ_L indicated that yielding initiates at similar strain amplitudes for all samples irrespective of their end use.

Higher-order Fourier harmonics provide mathematically robust and sensitive indicators of nonlinearities. However, they do not confer physical meaning to the material response. To address this shortcoming, several ontological frameworks have been developed that augment physical meaning while providing locally defined material measures. Among these, the sequence of physical processes, and the Chebyshev stress decomposition method are among the most important ones (Ewoldt et al., 2008; Rogers, 2012). In the following section, we focus on the fundamentals of the latter method, its application in strain-controlled and stress-controlled experiments.

LAOStrain Chebyshev Framework

Cho et al. (2005) used a geometrical interpretation of viscoelasticity to decompose the nonlinear response. For a deformation-controlled test, the nonlinear stress response to a periodic sine-strain input is given by $\sigma(t) = \sigma'(x) + \sigma''(y)$ and is decomposed into a superposition of elastic stress $\sigma^e(x)$ and the viscous stress $\sigma^v(y)$:

$$\sigma^e(x) \equiv \frac{\sigma(\gamma, \dot{\gamma}) - \sigma(-\gamma, \dot{\gamma})}{2} = \gamma_0 \sum_{n:odd} G'_n(\omega, \gamma_0) \sin n\omega t \qquad (6.6)$$

$$\sigma^v(y) \equiv \frac{\sigma(\gamma, \dot{\gamma}) - \sigma(\gamma, -\dot{\gamma})}{2} = \gamma_0 \sum_{n:odd} G''_n(\omega, \gamma_0) \cos n\omega t, \qquad (6.7)$$

where x and y are the normalized strain and normalized strain rate, respectively, $x = \gamma/\gamma_0 = \sin\omega t$ and $y = \dot{\gamma}/\dot{\gamma}_0 = \cos\omega t$. This decomposition is based on the concept that $\sigma^e(x)$ is odd with respect to x and even with respect to y, and $\sigma^v(y)$ is even with respect to x and odd with respect to y. The right-hand side of Eqs. (6.6) and (6.7) shows the mathematical equivalence of the geometrical decomposition with the Fourier decomposition of the stress signal. Plots of elastic stress σ^e versus γ or viscous stress σ^v versus $\dot{\gamma}$ result in single-valued functions rather than closed loops formed when total stress σ is in the ordinate. Cho et al. (2005) proposed fitting the elastic and viscous curvatures to simple polynomial regression of x or y. This approach as simple as it appears is nonetheless unsuitable because the selection of the regression fit is arbitrary and hence the associated material properties are not exclusive either. For example, for a smooth stress function $F(x)$ in the nonlinear regime, a third-order polynomial $F = a_0 + a_1 x + a_2 x^2 + a_3 x^3$ naturally leads to better fitting and different regression coefficients a_0, a_1, a_2, and a_3 than a first-order polynomial $F = a_0 + a_1 x$, and fitting higher-order terms affects the lower-order terms. Building on the work of Cho et al. (2005), Ewoldt et al. (2008) proposed fitting $\sigma^e(x)$ and $\sigma^v(y)$ to a set of Chebyshev polynomials of the first kind. This basis set were selected because they are (1) orthogonal over the finite domain $[-1, 1]$ (i.e., the fitting coefficients are independent of each other), (2) bound for higher-order contributions, and (3) directly related to the Fourier harmonics. Using these basis functions, the elastic and viscous components of the stress signal can be described by the following equations:

$$\sigma^e(x) = \gamma_0 \sum_{n:odd} e_n(\omega,\, \gamma_0)\, T_n(x) \tag{6.8}$$

$$\sigma^v(y) = \dot{\gamma}_0 \sum_{n:odd} v_n(\omega,\, \gamma_0)\, T_n(y), \tag{6.9}$$

where $T_n(x)$ and $T_n(y)$ correspond to nth-order of the Chebyshev polynomials of the first kind, γ_0 and $\dot{\gamma}_0$ represent the input strain and strain, respectively, and $e_n(\omega,\, \gamma_0)$ and $v_n(\omega, \gamma_0)$ are the orthonormal elastic and viscous coefficients (independent of each other), respectively. Chebyshev coefficients directly correspond one-to-one to Fourier coefficients in the time domain as follows:

$$e^n = G'_n(-1)^{(n-1)/2} \quad n:odd \tag{6.10}$$

$$v^n = G''_n/\omega = \eta'_n \quad n:odd. \tag{6.11}$$

Thus, similar to the third-order Fourier harmonics, the third-order Chebyshev coefficients e_3 and v_3 signal the departure from nonlinearity while also providing physical interpretation. In the linear regime, $e_3/e_1 \ll 1$ and $v_3/v_1 \ll 1$ because the effective contribution of higher-order terms is negligible. Eqs. (6.8) and (6.9) then reduce to the linear viscoelastic measures $e \to G'_{LVE}$ and $v_1 \to G''_{LVE}/\omega$ (G' and G'' are analogous to the first-harmonic moduli G'_1 and G''_1, the common output of most rheometers). In the nonlinear regime, the first-order coefficients (or viscoelastic moduli) e'_1 and v''_1 describe average, global, or "intercycle" responses (i.e., the basis function is linear but changes for each cycle of increasing strain amplitude), whereas the third-order reveal local or "intracycle" responses (i.e., the basis function indicates relative nonlinearities within a single cycle). Nonlinearities manifest as curvatures (see Figs. 6.5, 6.8, and 6.11), characterized by "upturns" or "bends" dependent mainly on the contribution of e_3 and v_3. Positive values of the first-harmonic nonlinearities, $e_1 > 0$ and $v_1 > 0$ signify intercycle elastic stiffening and viscous thickening (visually observed as counterclockwise rotation or "increasing slope"), whereas negative values $e_1 < 0$ $v_1 < 0$ mean intercycle elastic softening and viscous thinning (visually observed as clockwise rotation or "decreasing slope"). In the same manner, for positive contributions of the third-harmonic nonlinearities $e_3 > 0$ and $v_3 > 0$, the response is referred as to intracycle strain stiffening and intracycle shear thickening, respectively (visually observed as "upturns," i.e., higher stress at maximum strain or shear rate). For negative contributions $e_3 < 0$ and $v_3 < 0$, the response is referred as to intracycle strain softening and intracycle shear thinning, respectively (visually observed as "downturns" at maximum strain or shear rate) (Ewoldt et al., 2008). An alternative interpretation of the local coefficients e_3 and v_3 is that they indicate the predominant cause driving global nonlinearities of e_1 and v_1. For example, for e_1 (or G'_1) decreasing, and $e_3 > 0$, large instantaneous strain rates are responsible for the average elastic softening. In the same way, for v_1 (or G''_1) decreasing and $v_3 < 0$, large instantaneous strain rates also drive average viscous thinning (Ewoldt and Bharadwaj, 2013). Apart from the interpretation of Chebyshev coefficients Ewoldt and Bharadwaj (2013) proposed a set of clearly defined and meaningful viscoelastic moduli in the nonlinear regime to capture local elastic and viscous effects at minimum and large instantaneous strains or rates of deformation:

$$G'_M \equiv \frac{d\sigma}{d\gamma}\bigg|_{\gamma=0} = \sum_{n:odd} nG'_n = e_1 - 3e_3 + ..., \tag{6.12}$$

$$G'_L \equiv \frac{\sigma}{\gamma}\bigg|_{\gamma=\pm\gamma_0} = \sum_{n:odd} G'_n(-1)^{(n-1)/2} = e_1 + e_3 + ..., \tag{6.13}$$

$$\eta'_M \equiv \frac{d\sigma}{d\dot{\gamma}}\bigg|_{\dot{\gamma}=\pm\dot{\gamma}_0} = \frac{1}{\omega}\sum_{n:odd} nG''_n(-1)^{\frac{n-1}{2}} = v_1 - 3e_3 + ..., \tag{6.14}$$

$$\eta'_L \equiv \frac{\sigma}{\dot{\gamma}}\bigg|_{\dot{\gamma}=\pm\dot{\gamma}_0} = \frac{1}{\omega}\sum_{n:odd} G''_n = v_1 + v_3 + ..., \tag{6.15}$$

where G'_M is the minimum strain or tangent modulus at $\gamma_0 = 0$ and G'_L is the large strain or secant modulus at $\gamma = \gamma_0$. In the same manner, η'_M is the minimum-rate viscosity at

$\dot{\gamma} = 0$, and η'_L is the large-rate viscosity at $\dot{\gamma} = \dot{\gamma}_0$. These material functions reduce to G'_1 and $G''_1 (\eta = G''_1/\omega)$ in the linear regime. A graphical depiction of some of these measures in elastic and viscous Lissajous–Bowditch projections is offered in Fig. 6.11. Relative differences between intracycle elasticities (or local dynamic viscosities) at large strain (or shear rate) and at minimum strain (or shear rate) can be also expressed as strain-stiffening and shear-thickening ratios, respectively:

$$S \equiv \frac{G'_L - G'_M}{G'_L} = \frac{4e_3 + \dots}{e_1 + e_3 + \dots} \tag{6.16}$$

$$T \equiv \frac{\eta'_L - \eta'_M}{\eta'_L} = \frac{4v_3 + \dots}{v_1 + v_3 + \dots}, \tag{6.17}$$

where simple linear viscoelastic responses yield $S = 0$ and $T = 0$. Instead, nonlinear responses yield $S > 0$ and $T > 0$ to indicate intracycle strain stiffening and intracycle strain-rate thickening, respectively, and $S < 0$ and $T < 0$ to indicate intracycle strain softening and shear thinning, respectively.

LAOStress Chebyshev Framework

A similar ontological framework has been proposed for LAOStress (Dimitriou et al., 2013; Ewoldt, 2013; Läuger and Stettin, 2010). For stress-controlled test, an imposed cosine oscillating stress $\sigma(t) = \sigma_0 \cos \omega t$ gives a strain response. Using geometry arguments, the total strain can be decomposed into its elastic and viscous components defined $\gamma(t) = \gamma^e(t) + \gamma^v(t)$. An identical decomposition can be considered for strain rate by replacing γ by $\dot{\gamma}$. The decomposition is based on the concept that γ^e and γ^v are instantaneous single-value functions of the stress input.

Using this idea, the Chebyshev representation is given by the following equations:

$$\gamma^e(t) = \sigma_0 \sum_{n:odd} J'_n(\omega, \sigma_0) \cos n \omega t = \sigma_0 \sum_{n:odd} J'_n(\omega, \sigma_0) T_n(x) \tag{6.18}$$

$$\gamma^v(t) = \sigma_0 \sum_{n:odd} n\omega J''_n(\omega, \sigma_0) \cos n \omega t = \sigma_0 \sum_{n:odd} n\omega J''_n(\omega, \sigma_0) T_n(x), \tag{6.19}$$

where $J'_n(\omega, \sigma_0) = c_n(\omega, \sigma_0)$ and $n\omega J''_n(\omega, \sigma_0) = f_n(\omega, \sigma_0)$ are the interrelations between the Chebyshev and the Fourier coefficients. The coefficients c_n and f_n represent Chebyshev compliance and fluidity coefficients, respectively.

Likewise, nonlinear elastic and viscous metrics have been proposed for minimum and maximum stress:

$$J'_M \equiv \frac{d\gamma}{d\sigma}\bigg|_{\sigma=0} = \sum_{n:odd} (-1)^{(n-1)/2} n J'_n = \sum_{n:odd} (-1)^{(n-1)/2} n c_n \tag{6.20}$$

$$J'_L \equiv \frac{\gamma}{\sigma}\bigg|_{\sigma=\sigma_0} = \sum_{n:odd} J'_n = \sum_{n:odd} c_n \tag{6.21}$$

$$\phi'_M \equiv \frac{d\dot{\gamma}}{d\sigma}\bigg|_{\sigma=0} = \sum_{n:odd} (-1)^{(n-1)/2} n^2 \omega J''_n = \sum_{n:odd} (-1)^{(n-1)/2} n f_n \tag{6.22}$$

$$\phi'_L \equiv \frac{\dot{\gamma}}{\sigma}\bigg|_{\sigma=\sigma_0} = \sum_{n:odd} n\omega J''_n = \sum_{n:odd} f_n, \tag{6.23}$$

where J'_M is the minimum-stress compliance at $\sigma = 0$ and J'_L is the large-stress compliance at $\sigma = \sigma_0$. In the same manner, ϕ'_M is the minimum-stress fluidity at $\sigma = 0$, and ϕ'_L is the large-stress fluidity at $\sigma = \sigma_0$. Fluidity denotes "propensity to flow" and is simply the inverse of viscosity $\phi = 1/\eta$. In addition to these measures, a relative ratio of the change of intracycle compliance can be defined as

$$R \equiv \frac{J'_L - J'_M}{J'_L} = \frac{4c_3 + \ldots}{c_1 + c_3 + \ldots}, \tag{6.24}$$

when $R = 0$ indicates linear viscoelastic responses, and $R > 1$ suggests strong nonlinearities due to apparent softening.

Experimental: Data Collection and Analysis

Major experimental considerations to keep in mind when performing LAOS experiments include selection of the most suitable experiment (i.e., LAOStrain vs. LAOStress), the form of data collection, arising shear rheometric artifacts before processing and analyzing the data.

Selection of Type of Experiment

The selection of LAOStrain versus LAOStress techniques and the measurement of corresponding material functions depend on three main factors: instrument specifications, material under investigation and its application, and structure-rheology models (Ewoldt, 2013). Instrument specifications include rheometer design, performance, and sensitivity. Depending on the rheometer design, stress control or strain control may be more suitable one over the other. Likewise, "extra" intrinsic nonlinearities may arise such as for instruments that use active deformation control to measure strain (Merger and Wilhelm, 2014). However, minor intrinsic nonlinearities might not be as important for materials with high moduli (e.g., fats) whose nonlinear responses dominate the response signal. Microstructure "sensitivity" to deformation or loads will also influence the selection of the protocol. Stress-controlled test leads to large jumps in strain in the nonlinear region where the moduli decrease. This has been shown in fat crystal

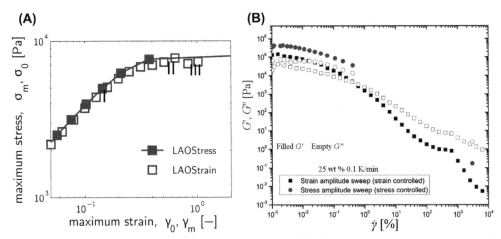

Figure 6.6 (A) LAOStrain and LAOStress experiments of full-fat cheese, measure the same ratios of maxima in the stress and maxima in strain, defined by σ_{max}/γ_0 and σ_0/γ_{max}, respectively. However, the LAOStrain measurements more readily allow probing the material response deeper into the fluid regime. The roman numerals correspond to the specific LAOS cycles are examined in more detail in Fig. 6.11. (A) Data were collected at $T = 25°C$ and $\omega = 5$ rad/s. (B) Viscoelastic moduli from LAOStrain and LAOStress modes of 25 wt%. ((A) Reproduced from Faber, et al., 2017. J. Food. Eng 210, 1–18. Copyright (2017), with permission from Elsevier Ltd. (B) Reproduced from Ramamirtham, et al., 2017. Rheol. Acta 56 (12), 971–982. Copyright (2017), with permission from Springer-Verlag GmbH Germany.)

networks and emulsion-filled gels (e.g., cheese containing fat) where LAOStrain allow better input control and deeper probing into the nonlinear regime (Fig. 6.6) (Faber et al., 2017; Ramamirtham and Madivala, 2002). The application also defines material properties of interest, e.g., butter, spreads, and shortenings, undergo nonlinear shear rates during processing or usability, namely during lamination and spreading and thus LAOStrain better describe this situation. It has also been recently shown that textural attributes (e.g., firmness, rubberiness) described in terms of rheological material functions and calculated from LAOStrain are independent and have consistent units compared to LAOStress measures (Faber et al., 2017). Finally, the selection of the input function also depends on the sensitivity of structure-rheology models to specific material functions, which help infer molecular-, nano-, or microscale structures (Ewoldt et al., 2013).

Data Collection

LAOS requires the collection of raw oscillatory waveforms, which is typically achieved by two ways. The first involves collecting strain—stress raw signals using standard capabilities included in the software of commercial rheometers (i.e., TA Orchestrator software, Raw data LAOS waveform) (Ewoldt et al., 2008; Läuger and Stettin, 2010). The second consists in installing additional hardware that enable collection and conversion of the analog voltage outputs of torque and motor displacements into digital signals using an analog-to-digital converter card, to provide oversampling and superior sensitivity (Wilhelm, 2002). We have used the first method because it does not necessitate additional instrumental setup and provides adequate sampling rates and SN ratios (Ewoldt et al., 2008). Beyond the LVR region, in the LAOStrain regime, the raw stress begins to show transient decay in amplitude attributed to slow intercycle thixotropic or aging. Because any LAOS analysis using FT or stress decomposition Chebyshev representation requires periodic stress waveforms (i.e., steady- or quasi—steady-state responses), multiple periods of oscillations at a specific set of $\{\omega, \gamma_0\}$ need to be applied, so transients die out. The achievement of a steady state (or alternance) can be verified by observing overlapping of the Lissajous—Bowditch loops or equilibrium of the first-harmonic viscoelastic moduli G' and G'' or peak value of the shear stress over time (e.g., in a time sweep) at certain set of $\{\omega, \gamma_0\}$. The material response might also contain prolonged transient behavior due to aging and shear rejuvenation (e.g., in waxy crude oils). In such a case, an instantaneous LAOS approach such as that developed by Rogers (2012) (a time-dependent moduli) will be more desirable because transient and even incomplete responses are all amenable to analysis.

The Importance of Good Data (Potential Experimental Artifacts)

In general, rheological measurements can be affected by several experimental artifacts, which "at best" change slightly the absolute values of apparent material functions and "at worst" communicate false mechanical behavior. Ewoldt et al. (2015) provided an excellent treatise of experimental challenges and tools to avoid "bad data" in shear rheology, many of which are highly relevant to fat-based materials Macias-Rodriguez and Marangoni (2017). It is important that the practitioner is aware of these issues so as to mitigate/minimize their occurrence, ensure "good data" and select an appropriate experimental window during LAOS measurements. The most important ones include slip, edge fracture, gap underfilling (e.g., for "in situ" crystallization). Slip is a prevalent artifact that is largely overlooked. Given the self-lubricated nature of fat or fat-containing materials, it may be rather rare the case when slips do not occur. Key indications of slip include the following: (1) gap dependence of the apparent stress and strain rate, (2) reduced flow stress (Ewoldt et al., 2015), (3) irregular fluctuations in the transient response of the stress—strain signal in the Lissajous—Bowditch curves, (4) occurrence of broadband noise in the FT of the stress—strain response, (5) asymmetric and open intracycle Lissajous—Bowditch loops, (6) relative intensity of $I_2/I_1 > 0.1$, (7) secondary loops in Lissajous—Bowditch curves also indicating slip in some cases, and (8) "free motion" of the sample between the contacting boundaries and even migration outside the gap when slip is strongly present. Approaches to check for and minimize slip include looking for any of the signatures abovementioned, collecting data at different gaps, performing marker tests (Chakrabarti, 2005), testing rheological behavior with different measuring geometries or modified surface geometry, and selecting appropriate experimental window (i.e., define frequencies and amplitudes that prevent/minimize slip) (Ewoldt et al., 2015; Macosko, 1994). Sandpaper is the most commonly used material used to prevent slip though filter paper—modified surfaces (i.e., filter paper glued to the measuring geometry) also enables good adhesion of fat samples to contacting boundaries (Macias-Rodriguez and Marangoni, 2016; Prentice, 1993). Application of mild normal force control (e.g., NFC = 0—0.5 N) helps to maintain sample-geometry contact during measurement and thus minimizing slip (Prentice, 1993). Under certain experimental conditions, slip may be unavoidable (Tariq et al., 1998). Edge failure is another potential artifact particularly ubiquitous in stiff fats (e.g., cocoa butter, shortenings) and can be detected by visual observation of the edge of the sample, marker tests, reduction of the apparent stress due to decrease of effective contact area and gap dependence of the data. Sample underfilling may occur in fats or other fat-based materials due to volume contraction on crystallization or sol-to-gel transition (e.g., such as in "organogels") in the rheometer. Underfilling manifests as the development of a large negative normal force (Mao et al., 2016). Viable ways to eliminate this issue include applying zero normal force control during crystallization or using internal gap adjustment procedures to compensate for sample contraction (Läuger and Stettin, 2010; Mao et al., 2016; Morales-Rueda et al., 2009). Any of these artifacts will contribute to erroneous estimation of the viscoelastic moduli, particularly in the LAOS regime where strong waveform distortion and decrease of the shear stress may be observed (e.g., edge failure reduces the effective amount of sample at the plate periphery) (Li et al., 2009).

Data Processing and Analysis

Here we describe the data analysis for the LAOStrain framework already applied to fats and fat-based materials. Data processing and analysis is performed with a custom-written freeware MATLAB routine (The "MITlaos" software) (Ewoldt et al., 2009). A screenshot of the software user interface and a flowchart describing the program sequence are shown in Fig. 6.7. The software requires time series signals of strain and stress along with user-specified parameters for analysis. As alluded to earlier, data processing of any time-periodic signal via FT and Chebyshev stress decomposition method requires inputting steady or quasi—steady stress cycles. Inputting multiple oscillatory cycles to the discrete FT increases the SN ratio. The software also smoothen the total stress signal by filtering out even and noninteger harmonics associated with random noise and broken shear symmetry, e.g., with responses that have not yet reached the time-periodic state. A cutoff frequency is selected to further remove noise attributed to higher-frequency components masking the real signal. Here, the user has to be careful as to not select "too many" harmonics that introduce random noise but neither "too little" that underrepresent the real signal and incorporate additional nonlinear features. The output of the software

(A)

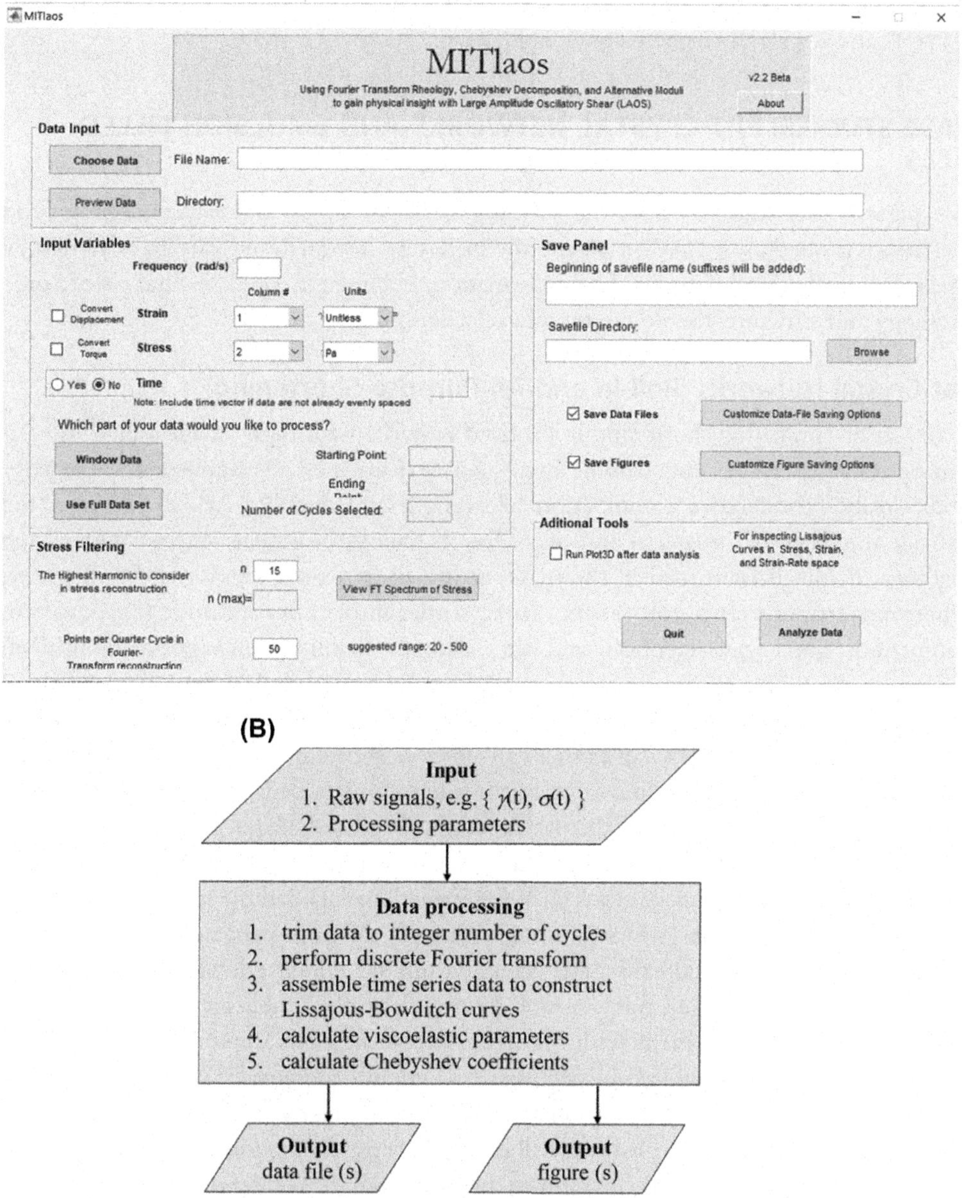

(B)

Figure 6.7 MITlaos data analysis software (request from: mitlaos@mit.edu) to assist in implementing the framework for nonlinear viscoelasticity. (A) Screenshot of the main user interface and (B) flow chart of the processing sequence.

includes Fourier and Chebyshev spectrum of the stress response, intercycle and intracycle viscoelastic material measures, and time series signals of elastic and viscous stresses, $\sigma'(\gamma(t))$ and $\sigma'(\dot{\gamma}(t))$, respectively (Ewoldt et al., 2009).

CASE STUDIES: FAT CRYSTAL NETWORKS AND EMULSION-FILLED GELS

We present case studies where the yielding behavior of fat and fat-containing food products is probed using LAOStrain and the stress response analyzed with the Chebyshev stress decomposition method. The information obtained is used to draw structure-rheology and structure-rheology—texture relationships.

Fat Crystal Networks (Roll-In and All-Purpose Shortening)

A roll-in or laminating shortening is a soft fat material with high "malleability" (i.e., it can be shaped without catastrophic breakage) or toughness. Toughness is a property traditionally considered a combination of strength and ductility, which indicates the damage tolerance of a material and quantifies its ability to dissipate energy under large loadings. A roll-in fat is used in the manufacture of croissants, Danish and puff pastry, where it contributes crucial processing characteristics and sensory attributes (e.g., uniform lamination, good gas retention and lift volume, flakiness, and good mouthfeel) (Ooms et al., 2015). During processing, a roll-in fat is extruded, squeezed, and shaped into micron-width films, i.e., the material undergoes large deformations, without breaking catastrophically. By contrast, an all-purpose shortening is used in multiple bakery applications mainly cake and icing but not in laminates because it performs poorly. The baker is well familiar with this fact and usually discriminates a "good" roll-in fat from a "bad" roll-in fat by tactile perception, i.e., using the so-called "finger" or thumb test. To mimic this test, indentations were performed on each sample (presented earlier as macrographs in Fig. 6.2A), which illustrate strikingly different behavior. On one side, a roll-in fat behaves as a ductile (invisible cracks within the length scale of observation) solid material, on the other, an all-purpose fat behaves as a brittle (visible cracks) solid material. This behavior is in agreement with the strain—stress diagrams shown earlier (Fig. 6.3). From fundamental and applied perspectives, it is desirable to pinpoint suitable rheological functions that describe the yielding behavior and "functionality" of these materials. From a health standpoint, this information will enable reverse engineering of roll-in fats with no *trans* and lower saturates. To investigate these materials, LAOStrain deformations were applied, and the stress responses analyzed and interpreted using Lissajous—Bowditch curves and the Chebyshev stress decomposition method (Figs. 6.8 and 6.9). In both material classes, linear viscoelasticity dominates the stress response at $\gamma_0 \approx 0.05\%$ (i.e., tight elliptical Lissajous—Bowditch loops). Above strains $\gamma_0 \approx 0.09\%$, Lissajous—Bowditch distort because of periodic nonlinear variations in the stress response particularly visible

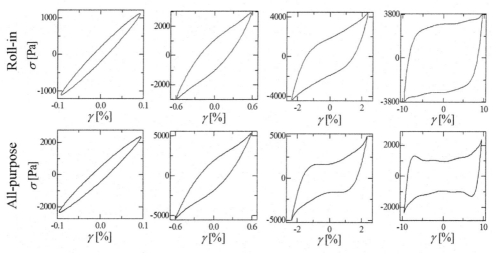

Figure 6.8 Lissajous—Bowditch curves for roll-in and all-purpose shortenings at selected strain amplitudes.

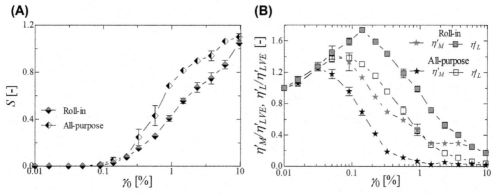

Figure 6.9 Nonlinear (A) elastic and (B) viscous measures for roll-in and all-purpose shortenings calculated from LAOS data at $\omega = 3.6$ rad/s. (B) Measures are parametrized by the linear dynamic viscosity η'_{LVE} at $\gamma_0 = 0.01\%$.

in local points of the stress response. In general, both samples display similar qualitative features, e.g., stress upturns at $\gamma=\gamma_0$ (i.e., lower elasticity or energy storage at $\gamma = 0$ or $\dot{\gamma}_0$) in the elastic perspective and "bends" at $\dot{\gamma} = \dot{\gamma}_0$ (i.e., lower viscosity or energy dissipation at $\dot{\gamma} = 0$ or γ_0) in the viscous perspective indicating intracycle strain stiffening and intracycle shear thinning, respectively. At high strains $\gamma_0 \geq 6\%$, self-intersections and secondary loops occur at $\dot{\gamma}_0$ in the viscous curves (not shown here) attributed to stronger and quicker unloading of instantaneous elastic stresses (which translates to stress overshoots as more pronounced in an all-purpose shortening), competition between network destruction and formation at high shear rates and even slippage in some cases (Ewoldt and McKinley, 2010; Jacob et al., 2014). Within an elastic LAOS cycle (Fig. 6.8), e.g., at $\gamma_0 \approx 6\%$, the peak in stress demarks two regions: a nearly linear region preceding the overshoot that extends roughly from the lower reversal point to the stress overshoot (Fig. 6.8) and a "flow" region after the overshoot. This linear stress region has been associated with residual elasticities by other researchers (Kim et al., 2014; Rogers et al., 2011; van der Vaart et al., 2013). In the "flow" region, as shear rate increases (or strain decreases), the stress decreases reaching a minimum. Subsequently, as the strain increases (or shear rate decreases), the stress increases again indicating thixotropy or network restructuring as reported previously (Macias-Rodriguez and Marangoni, 2016). This progresses until the end of the half-cycle ($\gamma_0 = 0$ or $\dot{\gamma} = \dot{\gamma}_0$), and then the sequence is repeated during flow reversal. Moreover, larger areas enclosed by the elastic Lissajous–Bowditch curves at high strains ($\gamma_0 \geq 6\%$), indicate increased plastic response, more evident in a roll-in shortening. Similar intra-cycle LAOS responses have been reported in model fat systems (tribehenin dispersed in isopropyl myristate) (Kim et al., 2014). Using local nonlinear measures, previously introduced, augment the insight gained from Lissajous-Bowditch curves. Local measures indicate that a roll-in and all-purpose shortening store energy in a similar manner (i.e., comparable G'_M and G'_L); however, they dissipate energy in a strikingly different way (i.e., different η'_M and η'_L). A roll-in shortening dissipates $\sim 2 \times$ energy of an all-purpose shortening. To link the observed nonlinear mechanical response to structure, nano-to-micro scale structures were investigated. Data from ultra-small angle X-ray scattering data fitted to the unified fit model revealed two major structural differences: (1) the number of levels making up the fat hierarchy and (2) the morphology and size of nanoplatelets (Macias-Rodriguez and Marangoni, 2017). A roll-in shortening had three structural levels [each level identified as a power law regime in a $\log(q)-\log I(q)$ plot], whereas an all-purpose shortening had only two structural levels. A roll-in shortening had nanoplatelets characterized by "smooth" surfaces and average sizes of 50 nm, whereas an all-purpose shortening had nanoplatelets with "rough" surfaces and average size of 400 nm (i.e., nanoplatelets of a roll-in shortening were about an order of magnitude smaller than those of an all-purpose shortening). Scanning electron microscopy further revealed differences in the morphology and spatial distribution of crystal aggregates making up the crystal network. A roll-in shortening had homogenous and layered-like structures, each layer made up of well-defined crystal aggregates (~ 4 μm length), whereas an all-purpose shortening displayed heterogeneous, distorted flakelike structures with no particular orientation. Based on rheology and the structural insight, it can be inferred that a roll-in shortening dissipates more effectively shear deformations due to the following characteristics: (1) an "extra" hierarchy that allows better energy allocation (an all-purpose shortening has one less hierarchy available for energy dissipation) and (2) control sliding motion of the microscopic layerlike crystal aggregates in which the liquid oil serves as a lubricant. Furthermore, the fact that several compositions meet the unique rheological "fingerprint" (i.e., higher normalized dynamic viscosities) indicates that the structure—function framework rather than the exact bulk composition determines the desired performance (Macias-Rodriguez and Marangoni, 2017).

Emulsion-Filled Food Gels (High-Fat and Low-Fat Semihard Cheeses)

From a material perspective, cheese can be regarded as an emulsion-filled gel (i.e., filled by the fat emulsion), whose quality and acceptability are largely dependent on rheology and texture. Rheology and texture are largely governed by the fat filler, its volume fraction, and its physicochemical properties. Fat acts as a 'perfect' filler, as it tailors textural attributes, such as firmness, springiness, rubberiness, and also influences cheese processability. Any reformulation effort to produce low-fat cheese requires careful consideration of rheology and texture to anticipate quality and sensory trade-offs. In this case study, relationships between rheology and structure and texture are presented for foiled, ripened Gouda cheeses formulated with zero-fat ($\phi_{fat} = 0$) and full-fat content ($\phi_{fat} = 0.3$), as investigated by Faber et al. (2017) (For additional formulations refer to the original study). By using LAOStrain, the effect of the fat filler on the firm-to-fluid transition of cheese (i.e., its yielding) is elucidated. Faber et al. (2017) showed that the intercycle damage progression and failure of full-fat cheese differs from that of zero-fat cheese (Fig. 6.10). The roman numbers denote the various stages of stress response for both cheeses: I, elastic (i.e., linear growth rate of the intercycle stress and recoverable strain); I—II, elastoviscoplastic (irrecoverable strain due to initiation of microcrack formation); II, global stress maximum $\sigma'_{max}(\gamma_0)$ or elastic failure stress; and III, decay of elastic stress (due to percolation of crack into larger fractures). The main differences are that a full-fat cheese shows larger initial growth rate (steeper slope) and larger elastoviscoplastic response (i.e., plastic response toward lower strain amplitudes I—II, and broad plateau before failure III) than a zero-fat cheese because fat plasticizes the cheese matrix (Faber et al., 2017). Faber et al. (2017) indicate that $\sigma'_{max}(\gamma_0)$ (its magnitude and rate of decline) is a strong indicator of "brittleness" (i.e., "the tendency to break under the condition of minimal previous plastic deformation"), a similar observation made in the first case study. To quantify these differences, Lissajous—Bowditch curves and locally defined measures are inspected (depicted in Fig. 6.11). Lissajous—Bowditch curves reveal qualitative differences between a full-fat cheese and a zero-fat cheese, e.g., areas enclosed by the elastic loops of full-fat cheese are larger than those of zero-fat cheese and viscous loops of full-fat cheese display a longer plateau above $\dot{\gamma} = \dot{\gamma}_0$ ($\gamma = 0$) than those of zero-fat cheese. These features are indicative of more plastic flow in full-fat cheese. Quantitatively, the local metrics G'_M and G'_K

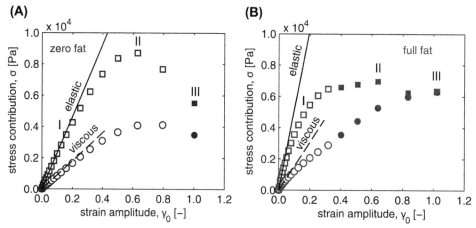

Figure 6.10 Evolution of the intracycle maxima of the elastic stress $\sigma'_{max}(\gamma_0)$ (hollow squares) and viscous stress $\sigma''_{max}(\gamma_0)$ (hollow circles) as a function of the strain amplitude γ_0 for zero-fat and full-fat cheese measured at $T = 25°C$ and a frequency $\omega = 5$ rad/s. The *continuous and dashed lines* represent the predictions of the linear viscoelastic constitutive model. Both zero-fat and full-fat cheese display an intercycle maximum of the elastic stress σ'_f (indicated by the numeral II) at a failure strain amplitude $\gamma_f \approx 0.7$. We define this intercycle maximum as the failure criterion for the food gel. The full-fat cheese curve (B) displays a broad plateau in the maximum elastic stress σ'_{max} and a small decrease beyond the failure point. By contrast, the zero-fat cheese, shown in (A), displays a more clearly pronounced peak in the σ'_{max} curve. The filled symbols in Fig. 6.5A and B indicate the strain amplitudes at which mild slip was observed. *(Reproduced from Faber, et al., 2017. J. Food. Eng. Copyright (2017), with permission from Elsevier Ltd.)*

measure the onset of plastic flow and accumulation of damage in the elastic network and η'_M and η'_K, the associated viscous responses. The level of "fluidization" (i.e., the extent of "solid-to-fluid" transition) in an elastic perspective is defined as

$$\Phi \equiv \frac{G'_K - G'_M}{G'_K} = \frac{12e_3 + 20e_5 + \ldots}{e_1 + 9e_3 + 25e_5 + \ldots} \qquad (6.25)$$

and in a viscous perspective is defined in analogy with the "thickening" ratio T (Ewoldt et al., 2008) as

$$\Theta \equiv \frac{\eta'_K - \eta'_M}{\eta'_K} = \frac{12v_3 + 20v_5 + \ldots}{v_1 + 9v_3 + 25v_5 + \ldots}. \qquad (6.26)$$

These definitions are analogous to "stiffening" and "thickening" ratios previously described, with the only difference being that both G'_K and η'_K are tangential (not secant) measures that describe elastic modulus and viscosity at maximum strain/strain rate, respectively (Ewoldt et al., 2008; Faber et al., 2017). Fig. 6.12 shows Φ and Θ for zero-fat and high-fat cheeses where three regimes (A, B, and C) encompassing a sequence

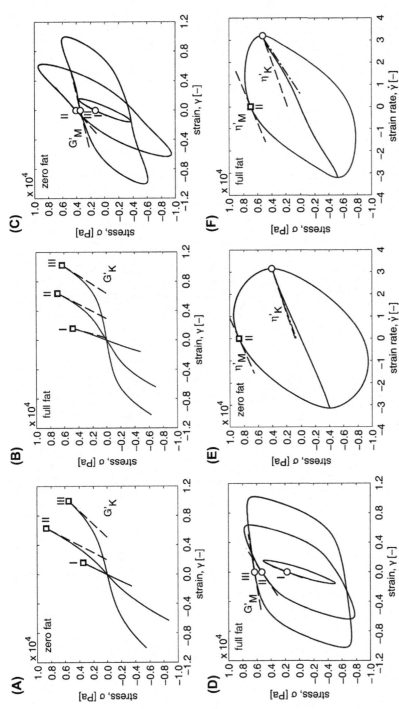

Figure 6.11 The solid–fluid transition of zero-fat and full-fat cheese at $T = 25°C$ and $\omega = 5$ rad/s depicted using the Lissajous representation. (A,B) The *continuous line* represent the intracycle evolution of the elastic stress σ'. The hollow squares are the intracycle maxima σ'_{max} of the elastic stress as a function of strain amplitude γ_0. The *dashed lines* indicate the tangent modulus G'_K to the curve of the elastic stress at $\gamma = \gamma_0$ and provide a measure of the loss of strength of the elastic network. (C,D) Lissajous curves of the total stress response at three strain amplitudes $\gamma_0 = 0.2, 0.6, 1.0$. The hollow circles show the intracycle maxima in the viscous stress $\sigma''_{max}(\gamma_0)$. The *dashed lines* through these points are the tangent modulus G'_M at maximum strain rate $\dot{\gamma} = \dot{\gamma}_0$. The slope of these *tangents* represents the resistance of the material to plastic flow. (E,F) Lissajous plots of cycle II from the viscous perspective with the viscous stress σ''_{max} plotted as black continuous lines. *The dashed lines* represent the minimum and maximum strain rate dynamic viscosity η'_M and η'_K, respectively. (F) Full-fat cheese displays a more pronounced (fifth-order) nonlinearity. Truncating the description of η'_K at third-order overpredicts its magnitude as shown by the *dot-dashed line*. *(Reproduced from Faber, et al., 2017. J. Food. Eng. Copyright (2017), with permission from Elsevier Ltd.)*

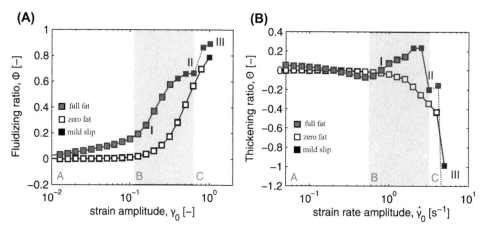

Figure 6.12 Strain sweeps showing the evolution in (A) the fluidizing ratio ϕ and (B) the thickening ratio Θ, of zero-fat cheese and full-fat cheese measured at a temperature $T = 25°C$ and $\omega = 5$ rad/s. (A) Both cheese formulations show comparable ultimate magnitudes of the fluidizing ratio; however, the rise of ϕ of full-fat cheese is more gradual and sets in at lower strains. (B) The non-Newtonian fluid properties of full-fat and zero-fat cheese are characterized by the evolution of the thickening ratio, which reveals three flow regimes A, B, and C. Zero-fat cheese displays continuous intercycle thinning, whereas full-fat cheese shows some initial thinning, followed by thickening and thinning. Beyond cycle II, the sample of full-fat cheese is no longer homogeneous, indicated with a *dotted line* in (A) and (B). The strain and strain-rate amplitudes at which mild slip is observed are indicated using filled symbols in each figure. *(Reproduced from Faber, et al., 2017. J. Food. Eng. Copyright (2017), with permission from Elsevier Ltd.)*

of physical processes are distinguished: (1) predominant linear viscoelastic response without plastic flow $\Theta \approx 0$ for zero-fat cheese *versus* narrow linear viscoelastic followed by mild fluidization $\Phi > 0$ and intracycle thinning $\Theta < 0$ due to strain localization in the emulsified fat component, (2) start of fluidization/thinning $\Phi > 0$ and $\Theta > 0$ due to microcrack nucleation and propagation for zero-fat cheese *versus* continued fluidization $\Phi > 0$ and change in response from intracycle viscous thinning to intracycle viscous thickening $\Theta > 0$ (akin to rubber toughening in thermoplastic composites) for full-fat cheese, and (3) extreme fluidization/thinning $\Phi > 0$ and $\Theta > 0$ attributed to failure, which is driven by crack propagation and sample-spanning fractures for both zero-fat and high-fat cheeses. Additionally, Faber et al. (2017) proposed that "firmness" $(\widehat{F})$ and "rubberiness" $(\widehat{R})$ ("the resistance to yield") are time-dependent linear viscoelastic and nonlinear elastoviscoplastic properties, respectively, quantified by the complex modulus G^* and the yield strain amplitude γ_0:

$$\widehat{F} \equiv \left| G^*(\omega_f) \right| \tag{6.27}$$

$$\widehat{R} \equiv \gamma_{0,\gamma} \tag{6.28}$$

$$\left| \frac{\left| G_1^*(\omega, \gamma_0) \right| - \left| G^*(\omega) \right|}{\left| G^*(\omega) \right|} \right| > \gamma, \qquad (6.29)$$

where G^* is the complex modulus and ω_f is the test frequency. This definition of firmness is similar to that suggested for fats by Narine and Marangoni (1999b) with the only obvious difference that firmness only depends on the elasticity of the network G'. The magnitude of the critical strain and hence rubberiness depends on the selection of γ (Faber et al. chose $\gamma = 0.01$, meaning that γ_0 corresponds to strain amplitudes at which G^* drops its linear value by 1%). Based on these definitions, increase of fat content in cheeses result in higher firmness (i.e., fat increases elastic modulus) and lower rubberiness (i.e., fat reduces the yield strain and induces fluidization).

CONCLUSIONS

Nonlinear viscoelastic properties of soft materials under oscillatory shear have garnered renewed interest in the last decade partly due to their relevance in material processing and performance. During use and consumption, fats and fat-containing materials undergo large amplitudes, which trigger interesting and complex rheological phenomena. A wide range of traditional methods are available to access to the nonlinear viscoelastic regime; however, their use is constrained by several limitations (e.g., poor sensitivity). Instead, LAOS tests are suitable for simultaneously characterizing the time-dependent and amplitude-dependent nonlinear viscoelastic properties and elucidate the underlying physical processes responsible for the collapse of the microstructure during yielding as described in this chapter. The LAOS technique and the set of nonlinear measures introduced here may serve as a novel analytical tool for structure-texture engineering of fats and fat-containing soft materials.

ACKNOWLEDGMENT

I would like to express my gratitude to Prof. Randy H. Ewoldt at The Department of Mechanical Science and Engineering, University of Illinois at Urbana-Champaign for providing the MITlaos freeware, his input and guidance in using the LAOStrain protocol.

REFERENCES

Bird, R.B., Armstrong, R.C., Hassager, O., 1987. Dynamics of Polymeric Liquids In: Fluid Mechanics, second ed., vol. 1. John Wiley & Sons, Inc., New York.

Chakrabarti, S., 2005. Probing ingredient functionalities in food systems using rheological methods. In: Gaonkar, A.G., McPherson, A. (Eds.), Ingredient Interactions: Effects on Food Quality. CRC Press, Taylor and Francis Group, Boca Raton, pp. 49−87.

Cho, K.S., Hyun, K., Ahn, K.H., Lee, S.J., 2005. A geometrical interpretation of large amplitude oscillatory shear response. J. Rheol. 49, 747−758.

Coussot, P., 2007. Rheophysics of pastes: a review of microscopic modelling approaches. Soft Matter 3, 528—540.

de Bruijne, D.W., Bot, A., 1999. Fabricated fat-based foods. In: Rosenthal, A.J. (Ed.), Food Texture: Measurement and Perception. Aspen Publishers, Gaithersburg, pp. 185—227.

De Graef, V., Dewettinck, K., Verbeken, D., Foubert, I., 2006. Rheological behavior of crystallizing palm oil. Eur. J. Lipid Sci. Technol. 108, 864—870.

Dealy, J.M., Wissbrun, K.F., n.d. Meltrheology and its role in plastic processing: theory and applications. New York, Van Nostrand Reinhold.

Dimitriou, C.J., Ewoldt, R.H., McKinley, G.H., 2013. Describing and prescribing the constitutive response of yield stress fluids using large amplitude oscillatory shear stress (LAOStress). J. Rheol. 57, 27—70.

Dinkgreve, M., Paredes, J., Denn, M.M., Bonn, D., 2016. On different ways of measuring "the" yield stress. J. Nonnewton. Fluid Mech. 238, 233—241.

Elliott, J.H., Ganz, A.J., 1971. Modification of food characteristics with cellulose hydrocolloids I: Rheological Characterization of an Organoleptic Property (Unctuousness). J. Texture Stud. 2, 220—229.

Ewoldt, R.H., 2013. Defining nonlinear rheological material functions for oscillatory shear. J. Rheol. 57, 177—195.

Ewoldt, R.H., Bharadwaj, N.A., 2013. Low-dimensional intrinsic material functions for nonlinear viscoelasticity. Rheol. Acta 52, 201—219.

Ewoldt, R.H., McKinley, G.H., 2010. On secondary loops in LAOS via self-intersection of Lissajous—Bowditch curves. Rheol. Acta 49, 213—219.

Ewoldt, R.H., Hosoi, A.E., McKinley, G.H., 2008. New measures for characterizing nonlinear viscoelasticity in large amplitude oscillatory shear. J. Rheol. 52, 1427—1458.

Ewoldt, R.H., Hosoi, A.E., McKinley, G.H., 2009. Nonlinear viscoelastic biomaterials: meaningful characterization and engineering inspiration. Integr. Comp. Biol. 49, 40—50.

Ewoldt, R.H., Johnston, M.T., Caretta, L.M., 2015. Experimental challenges of shear rheology: how to avoid bad data. In: Spagnolie, S.E. (Ed.), Complex Fluids in Biological Systems. Springer, New York, pp. 207—241.

Faber, T.J., Van Breemen, L.C.A., McKinley, G.H., 2017. From firm to fluid — structure-texture relations of filled gels probed under large amplitude oscillatory shear. J. Food Eng. 210, 1—18.

Ferry, J.D., 1980. Viscoelastic Properties of Polymers. Wiley, New York.

Goh, S.M., Scanlon, M.G., 2007. Indentation of lipid-based particle gels: an experimental, theoretical and numerical study. Acta Mater. 55, 3857—3866.

Gonzalez-Gutierrez, J., Scanlon, M.G., 2013. Strain dependence of the uniaxial compression response of vegetable shortening. J. Am. Oil Chem. Soc. 90, 1319—1326.

Hyun, K., Wilhelm, M., Klein, C.O., Cho, K.S., Nam, J.G., Ahn, K.H., Lee, S.J., Ewoldt, R.H., McKinley, G.H., 2011. A review of nonlinear oscillatory shear tests: analysis and application of large amplitude oscillatory shear (LAOS). Prog. Polym. Sci. 36, 1697—1753.

Jacob, A.R., Deshpande, A.P., Bouteiller, L., 2014. Large amplitude oscillatory shear of supramolecular materials. J. Nonnewton. Fluid Mech. 206, 40—56.

Kim, J., Merger, D., Wilhelm, M., Helgeson, M.E., 2014. Microstructure and nonlinear signatures of yielding in a heterogeneous colloidal gel under large amplitude oscillatory shear. J. Rheol. 8, 1359—1390.

Kloek, W., van Vliet, T., Walstra, P., 2005. Large deformation behavior of fat crystal networks. J. Texture Stud. 36, 516—543.

Läuger, J., Stettin, H., 2010. Differences between stress and strain control in the non-linear behavior of complex fluids. Rheol. Acta 49, 909—930.

Li, X., Wang, S.-Q., Wang, X., 2009. Nonlinearity in large amplitude oscillatory shear (LAOS) of different viscoelastic materials. J. Rheol. 53, 1255—1274.

Macias-Rodriguez, B., Marangoni, A.G., 2016. Rheological characterization of triglyceride shortenings. Rheol. Acta 55, 767—779.

Macias-Rodriguez, B.A., Marangoni, A.A., 2017. Linear and nonlinear rheological behavior of fat crystal networks. Crit. Rev. Food Sci. Nutr. 8398.

Macosko, C.W., 1994. Rheology: Principles, Measurements and Applications. Wiley, New York.

Mao, B., Divoux, T., Snabre, P., 2016. Normal force controlled rheology applied to agar gelation. J. Rheol. 60, 473–489.

Marangoni, A.G., Narine, S.S., 2002. Identifying key structural indicators of mechanical strength in networks of fat crystals. Food Res. Int. 35, 957–969.

Marangoni, A.G., Rogers, M.A., 2003. Structural basis for the yield stress in plastic disperse systems. Appl. Phys. Lett. 82, 3239–3241.

Merger, D., Wilhelm, M., 2014. Intrinsic nonlinearity from LAOStrain—experiments on various strain- and stress-controlled rheometers: a quantitative comparison. Rheol. Acta 53, 621–634.

Møller, P.C.F., Mewis, J., Bonn, D., 2006. Yield stress and thixotropy: on the difficulty of measuring yield stresses in practice. Soft Matter 2, 274.

Morales-Rueda, J.A., Dibildox-Alvarado, E., Charó-Alonso, M.A., Toro-Vazquez, J.F., 2009. Rheological properties of candelilla wax and dotriacontane organogels measured with a true-gap system. J. Am. Oil Chem. Soc. 86, 765–772.

Narine, S.S., Marangoni, A.G., 1999a. Mechanical and structural model of fractal networks of fat crystals at low deformations. Phys. Rev. E. Stat. Phys. Plasmas. Fluids. Relat. Interdiscip. Topics 60, 6991–7000.

Narine, S.S., Marangoni, A.G., 1999b. Microscopic and rheological studies of fat crystal networks. J. Cryst. Growth 198–199, 1315–1319.

Ooms, N., Pareyt, B., Brijs, K., Delcour, J.A., 2015. Ingredient functionality in multilayered dough-margarine systems and the resultant pastry products: a review. Crit. Rev. Food Sci. Nutr. 8398.

Prentice, J.H., 1993. Rheology and texture of dairy products. J. Texture Stud. 3, 415–458.

Ramamirtham, S., Madivala, S.A., 2002. Controlling the yield behavior of fat-oil mixtures using cooling rate. Rheol. Acta 56, 1–12.

Rogers, S.A., 2012. A sequence of physical processes determined and quantified in (LAOS): an instantaneous local 2D/3D approach. J. Rheol. 56, 1129–1151.

Rogers, S.A., Lettinga, M.P., 2012. A sequence of physical processes determined and quantified in large-amplitude oscillatory shear (LAOS): application to theoretical nonlinear models. J. Rheol. 56, 1.

Rogers, S.A., Erwin, B.M., Vlassopoulos, D., Cloitre, M., 2011. A sequence of physical processes determined and quantified in LAOS: application to a yield stress fluid. J. Rheol. 55, 435–458.

Rohm, H., Weidinger, K.H., 1993. Rheological behaviour of butter at small deformations. J. Texture Stud. 24, 157–172.

Tariq, S., Giacomin, A.J., Gunasekaran, S., 1998. Nonlinear viscoelasticity of cheese. Biorheology 35, 171–191.

Thareja, P., 2013. Rheology and microstructure of pastes with crystal network. Rheol. Acta 52, 515–527.

Thareja, P., Street, C.B., Wagner, N.J., Vethamuthu, M.S., Hermanson, K.D., Ananthapadmanabhan, K.P., 2011. Development of an in situ rheological method to characterize fatty acid crystallization in complex fluids. Colloids Surf. A Physicochem. Eng. Asp. 388, 12–20.

Thareja, P., Golematis, A., Street, C.B., Wagner, N.J., Vethamuthu, M.S., Hermanson, K.D., Ananthapadmanabhan, K.P., 2013. Influence of surfactants on the rheology and stability of crystallizing fatty acid pastes. J. Am. Oil Chem. Soc. 90, 273–283.

Tschoegl, N.W., 1989. The Phenomenological Theory of Linear Viscoelastic Behavior: An Introduction. Springer-Verlag, Berlin.

van den Tempel, M., 1961. Mechanical properties of plastic-disperse systems at very small deformations. J. Colloid Sci. 16, 284–296.

van der Vaart, K., Rahmani, Y., Zargar, R., Hu, Z., Bonn, D., Schall, P., 2013. Rheology of concentrated soft and hard-sphere suspensions. J. Rheol. 57, 1195–1209.

Wilhelm, M., 2002. Fourier-transform rheology. Macromol. Mater. Eng. 287, 83–105.

CHAPTER 7

Modeling Interactions in Edible Fats

David A. Pink

St. Francis Xavier University, Antigonish, NS, Canada

INTRODUCTION

The intent of this chapter is not only to be practical but also to cast a wider gaze around the modeling and simulation landscape in order that those interested in undertaking modeling of food systems can set their goals within a context of what was attempted and what has been achieved in related fields of condensed matter physics.

Many food systems are aqueous dispersions containing inorganic ions and aggregates of fats, polysaccharides, and proteins, ranging in size from nanometers to microns, along with larger-scale phases (Walstra, 1996). Aggregates can arise via low-energy "physical" molecular interactions due to Coulomb forces. Broadly speaking, physical interactions may be taken as comprising the ubiquitous (charge-fluctuation-induced) London—van der Waals type interactions (Lifshitz, 1956; Landau and Lifshitz, 1968; Israelachvili, 2006; Leckband and Israelachvili, 2001; Parsegian, 2005) and "electrostatic" interactions—interactions between those moieties that carry electric charges on sufficiently large (long) spatial (time) scales and that can be considered as "permanent" (Cevc and Marsh, 1987; McLaughlin, 1989; Cevc, 1990). The last-named encompass electric multipole (e.g., dipole—dipole) interactions. Although the distance dependence of electric dipole—dipole interactions can be similar to that of van der Waals interactions, the origins are different. Hydrogen bonding is a third, weaker, Coulombic effect, which is specific to certain proton-containing moieties and is short-range (<0.3 nm) and directional (Israelachvili, 2006). In nonquantum molecular dynamics, hydrogen bonding is treated as a combination of van der Waals forces together with attractive electrostatic interactions. Electrical charges are ubiquitous in food systems, and this leads to the importance of electrostatic interactions that are isotropic and long-range.

From a point of view of their physical attributes, four properties characterize food systems: (1) many of them involve aqueous solutions; (2) they are composed of salts, proteins, fats, and polysaccharides, some of which are in their ionic form; (3) they can be in the form of emulsions or other ordered structures, or simply heterogeneous mixtures that involve different amounts of solid and liquid phases; and (4) their mechanical properties can range from "hard" materials to that of soft condensed matter. Accordingly, their composition is such that their structure can involve separation into many coexisting phases—possibly microphases—separated by interfaces of different characteristics and complexities.

Structure-Function Analysis of Edible Fats, Second Edition
ISBN 978-0-12-814041-3, https://doi.org/10.1016/B978-0-12-814041-3.00007-1

This chapter deals largely with the physical interactions between crystals, or aggregates of them such as fractal structures, in a liquid oil environment. The interactions are all Coulombic of which the major components are van der Waals interactions; although in cases where hydrogen bonding might play a role, electrostatic interactions become important. Other effective interactions exist of which, perhaps, the most important are those brought about by depletion effects. This is an entropic interaction and requires at least two components comprising objects of different sizes. No description of physical forces is complete without an outline of how to calculate their effects in a given system. In an oil environment, the system is necessarily complex, a mix of assorted solids in a liquid, which may exhibit complex (micro)phase separation. The possibility that such a system can be realistically modeled by analytically solvable mathematical models is unlikely and, in recent years, computer simulation has provided a practicable avenue to obtain useful results. Accordingly, this chapter will include a section that describes the principle computer simulation techniques and will restrict itself to practical "how-to" advice. The philosophy of this chapter is practical and can be summed up in the following sentences: (1) What are the fundamental interactions in an oil environment? (2) How does one go about modeling these interactions? This chapter comprises sections on van der Waals interactions, followed by electrostatics. These lead into sections on computer simulation techniques made up of atomic-scale molecular dynamics (AMD), Monte Carlo methods, and dissipative particle dynamics. We shall end by describing a technique of which more use should be made for certain food systems: photon and neutron scattering—not to study crystalline structure via identifying Bragg peaks, but to analyze the scattering intensity as a function of scattering vector magnitude ranging over many decades. The advantages of this technique are as follows: (1) one can study a system in vivo—the way the oil system is when in use, (2) one can study, simultaneously, the structures in it over many spatial scales, and (3) one can study either time-dependent or steady-state effects. Examples of calculations will be given. The level here is that of a graduate student or postdoctoral fellow in food science with a basic knowledge of physics, and the intent is that this chapter can be used as a general reference and as a basis for the theoretical section of a graduate course in Modeling the Physics of Food Systems.

There is one last point that should be raised even though it presents a daunting mountain to climb. Below, we shall refer to crystalline nanoplatelets (CNPs) as appearing to be the fundamental components from which many systems of fats in oils are composed (Acevedo et al., 2011; Acevedo and Marangoni, 2010a,b). These objects are typically highly anisotropic structures measuring ~ 500 nm on a side and being ~ 50 nm thick. The question that is raised is that of the self-assembly of structures from CNPs. Much work has been done on the self-assembly of membranes, vesicles, and micelles from phospholipids (Patel and Velikov, 2011; McClements and Li, 2010; Fathi et al., 2012; Weiss et al., 2009; Antunes et al., 2009; Larsson, 2009; Sagalowicz and Leser, 2010), and work has been done on the related area of polar food lipids (Leser et al., 2006). However, the question of what structures arise from the interaction of CNPs with each other and with the oils in which they are embedded, and the pathways through which self-assembly on multimicron scales comes about, is essentially unresolved. Below, we shall see that although a first step might have been taken along this route, there is still a fascinating universe to explore.

VAN DER WAALS INTERACTIONS

Van der Waals forces refer to interactions between materials that are brought about by interactions between atomic electric dipoles induced in them by fluctuations in the electromagnetic field. This was shown by the work of Dzyaloshinskii et al. (1961) and subsequently applied to problems in soft condensed matter by the work of Parsegian (2005) and references therein. A good, very readable introduction is that of Lee-Desautels (2005). What has come to be known as "Lifshitz theory," describes the induced dipole—dipole interaction in terms of electric permittivities of the two materials as well as those of any materials separating them. Those permittivities arise reflect the fact that Lifshitz theory involves many-body correlations. This is in contrast to Lennard-Jones 6—12 potential theory, which considers two infinitesimal volumes of material ("bodies") and writes the van der Waals interaction energy, $u(r)$, between them as a pairwise attractive potential in $1/r^6$ and a repulsive potential in $1/r^{12}$, where r is the distance from one body to the other (Parsegian, 2005; Israelachvili, 2006),

$$u(r) = -\frac{C_6}{r^6} + \frac{C_{12}}{r^{12}}.$$ (7.1)

The total interaction between the two materials would then be the sum of all these pairwise interactions,

$$U = \sum_j \sum_k u(r_{jk}),$$ (7.2)

where the sum is over all pairs of infinitesimal volumes labeled j and k, and one must not count interactions twice. Let us now restrict our attention only to the attractive part, $-C_6/r^6$, of the 6—12 potential because the short-range repulsion is merely a phenomenological interaction (below). The picture of the attractive interaction being described as a two-body effect is clearly inadequate as can be seen in Fig. 7.1.

The creation of fluctuating dipoles anywhere inevitably take part in creating fluctuating dipoles throughout the materials so that the total energy is not simply a sum of pairwise energies (Eq. 7.2) but is a sum of energies due to all many-body correlations,

$$\widetilde{U} = \sum_j \sum_k \widetilde{u}_{jk} + \sum_j \sum_k \sum_l \widetilde{u}_{jkl} + \sum_j \sum_k \sum_l \sum_m \widetilde{u}_{jklm} + \cdots,$$ (7.3)

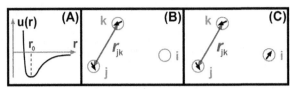

Figure 7.1 (A) The 6—12 potential (Eq. 7.1). (B) Pairwise interaction between fluctuating dipoles (*black arrows*) at bodies *j* and *k* created by the fluctuating electromagnetic field. The sum of all these yields *U* (Eq. 7.2). (C) Fluctuating dipoles anywhere create fluctuating dipoles everywhere: the fluctuating dipoles at *j* and *k* take part in creating a fluctuating dipole at *i*. This is a three-body interaction.

where we have indicated the summation up to four-body correlations (fourth "order"). This is the key difference between models that use the 6—12 potential approach and those that recognize that correlations to all orders are required. The two-body interactions of Eq. (7.1) clearly do not involve correlations higher than two-body correlations in contrast to Lifshitz theory (below), which involves correlations of all orders.

The attractive term in the 6—12 potential is derivable from second-order perturbation theory (below), which makes it clear why only two-body interactions enter. The repulsive term involving $1/r^{12}$ was chosen as a convenient phenomenological short-range repulsion between electron "clouds." Lifshitz theory has provided many insights (French et al., 2010), and one might think that the pairwise 6—12 potential theory has been essentially superseded by Lifshitz theory, except for one fact: 6—12 potential theory is used in AMD and it works: AMD is widely used to make predictions for gases and liquids, which have been successfully compared with experiments. It is the intent of this section to describe the Lifshitz and the 6—12 potential theories and remark on the coupled dipole method (CDM). At the end, the question, "why does the 6—12 potential work?" will be raised.

One should note that, in the van der Waals energy, there are three terms describing the attractive interaction between two objects a distance *r* apart, all of which exhibit a dependence of $1/r^6$: the London dispersion term due to the interaction between induced fluctuating dipoles, the Debye interaction between such a fluctuation and a permanent electric dipole, and the Keesom term that describes the interaction between a pair of electric dipoles. The Lifshitz approach includes all these terms in the expression for this electromagnetic field fluctuation-induced interaction between two objects, but the 6—12 potential method includes only the London dispersion term, and the other two have to be explicitly included.

Some recent work that could be relevant for food environments are Hanna et al. (2006) and Parsegian (2005) who calculated the distance dependence of the van der Waals interaction between (soft) polymeric or similar surfaces. Li et al. (2006) considered the van der Waals adhesion between rough surfaces. The equation for the van der Waals interaction between two spheres is especially useful because, under certain conditions, larger structures can be synthesized as a close-packed aggregate of spheres. For many years, the Derjaguin approximation (Butt et al., 2003) has been used for this purpose, but one should be aware of the Papadopoulos and Cheh (1984) approximation, which is considered to be an improvement.

It can be noted that, although the attractive part of the van der Waals interaction between atomic-scale moieties depends on $1/r^6$, it can be shown that when one considers two mesoscale soft surfaces at an average distance, D, apart, the resulting interaction is much different from that between two hard surfaces, which is proportional to $1/D^2$. It has recently been shown that if the molecular motion leading to the resulting averaging of the two soft surfaces is much faster than the rate at which this average distance, D, is changing then the attractive van der Waals interaction between these two mesoscale soft interfaces is proportional to $ln(1/D)$ (Hanna et al., 2006).

Lifshitz theory is an elegant approach for calculating the free energy of a mesoscale system. It assumes that one knows the electric permittivities as functions of frequency of all the components of a (possibly complex) system. Electric permittivity is defined as a parameter specifying the average electric dipole moment vector (polarization), $\overrightarrow{P}$, set up in some material by the application of an electric field, $\overrightarrow{E}$,

$$\overrightarrow{P} = [\varepsilon - \varepsilon_0]\overrightarrow{E} = \varepsilon_0[\varepsilon_r - 1]\overrightarrow{E}, \qquad (7.4)$$

where the permittivity is $\varepsilon = \varepsilon_0\varepsilon_r$, with ε_r being the relative permittivity and ε_0 being the permittivity of "free space," i.e., of the vacuum. Accordingly, the volume of space over which permittivity is a meaningful concept must be large enough that one samples sufficient atoms to define an average polarization, $\overrightarrow{P}$. "Large enough" will then depend on the size and complexity of the molecules in that volume. Huang and Levitt (1977) modeled the dielectric properties of the hydrocarbon chain region of a phospholipid bilayer membrane and found that the permittivity for an electric field applied perpendicular to the local membrane plane differed from that of bulk liquid hydrocarbons by a few percent. They concluded that the permittivity (perpendicular to the membrane plane) of such a bilayer membrane could be taken as that of bulk liquid hydrocarbons. This point is raised here because it might be thought that such a result could be applied to oils composed of triacylglycerols (TAGs), but the following words of warning should be heeded: the thickness of the hydrocarbon chain region of such a bilayer membrane with palmitoyl chains (which were taken to be in their *all-trans* conformations) and oriented perpendicular to the plane of the membrane, is ~4.5 nm. This is almost precisely the distance, d, separating two CNPs (Acevedo and Marangoni, 2010a), where the attractive van der Waals interaction begins to depart from the familiar $1/d^2$ dependence for oil density oscillations to appear (Pink unpublished). Accordingly, use of bulk values of permittivity when an oil is being studied in confined nanospaces should be viewed with caution.

There is another note of caution that is, perhaps, even more important in food systems. Lifshitz theory can be happily used in systems with fixed components such as solids (French et al., 2010), or at the mesoscale in systems with a one-component fluid. However, for a many-component fluid, in which phase separation can take place, the final system might involve fluid mixtures of composition that has to be determined by the minimum in the free energy. A fluid mix of unknown a priori composition will necessarily possess an unknown permittivity. In such a case, the use of Lifshitz theory could be problematical because the free energy determines the various phases and depends on a knowledge of the permittivities. But the permittivities depend on the composition of the liquid phases so that although, in principle, it can be used to model phase separation, nonetheless, one would have to identify the composition of the various phases and deduce or measure the permittivity of each of the possible phases into which the system might separate. Furthermore, if some of the phases are "nanophases" involving nanoscale layers of fluid around solids, then one might have difficulties in specifying a permittivity of a nanoscale layer, as described in the last paragraph. With these reservations, let us give a very simplified version of Lifshitz theory.

Lifshitz theory calculates the interaction energies of meso- or macrophases and is based on temperature-dependent fluctuations in the electromagnetic field, which, in turn, give rise to fluctuations in atomic or molecular electric dipole moments. Let us consider a linear isotropic material characterized by an (angular) frequency-dependent permittivity, $\varepsilon(\omega)$. From the relationship between an electric dipole, $\overrightarrow{P}$, induced by an electromagnetic field, $\overrightarrow{E}$, $\overrightarrow{P} = [\varepsilon - \varepsilon_0]\overrightarrow{E}$, we obtain the frequency-dependent polarization induced by an electromagnetic field of angular frequency, ω, $\overrightarrow{P}(\omega) = [\varepsilon(\omega) - \varepsilon_0]\overrightarrow{E}(\omega)$. This means that the frequency-dependent induced polarization, $\overrightarrow{P}(\omega)$, is proportional to the permittivity, $\varepsilon(\omega)$. It is for this reason that Lifshitz theory relates the free energy to the frequency-dependent permittivities of the components of a system. Let us consider two solids, 1 and 3, possessing relative permittivities $\varepsilon_{r1}(\omega) = \varepsilon_{r1}$ and $\varepsilon_{r3}(\omega) = \varepsilon_{r3}$ immersed in a liquid phase, 2, of relative permittivity $\varepsilon_{r2}(\omega) = \varepsilon_{r2}$. The boundary conditions on the electric and displacement field vectors involve these relative permittivities.

Consider the solids to be two macroscopic homogeneous solid slabs. If we approximate them as two solids with their faces smooth and parallel to each other, and filling entire half-spaces, $(-\infty < x,y < \infty)$ with one of the solids occupying $0 < z < \infty$ and the other occupying $-\infty < z < -d$, and the space between them, $(-\infty < x,y < \infty)$ and $-d \leq z \leq 0$ occupied by a material labeled 2, as shown in Fig. 7.2, then the free energy per unit area of the solid faces is given by the Lifshitz expression,

$$G(d) = -\frac{A_{H123}}{12\pi d^2}. \tag{7.5}$$

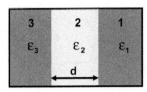

Figure 7.2 The system of two solid slabs, 1 and 3, modeled as filling two half-spaces and separated by a distance, d, with the intervening space, 2, occupied by an oil. The permittivities are frequency-dependent with $\varepsilon_j = \varepsilon_0 \varepsilon_{rj}(\omega)$, $j = 1, 2, 3$.

Here A_{H123} is the Hamaker coefficient given by (Hamaker, 1937; Dzyaloshinskii et al., 1961; Parsegian, 2005; French et al., 2010),

$$A_{H123} = -\frac{3}{2}k_B T \sum_{m=0}^{\infty} \int_0^{\infty} u\,du\,\ell n\left[1 - \Delta_{32}(i\xi_m)\Delta_{12}(i\xi_m)e^{-u}\right]$$

$$\approx \frac{3}{2}k_B T \sum_{m=0}^{\infty} \Delta_{32}(i\xi_m)\Delta_{12}(i\xi_m). \tag{7.6}$$

k_B is Boltzmann constant, T is the absolute temperature, and where we have approximated $\ell n[1 - x] \approx -x$, and used that $\int_0^{\infty} u e^{-u} du = 1$, with

$$\Delta_{ab}(i\xi_m) = \frac{\varepsilon_{ra}(i\xi_m) - \varepsilon_{rb}(i\xi_m)}{\varepsilon_{ra}(i\xi_m) + \varepsilon_{rb}(i\xi_m)}. \tag{7.7}$$

The sum is over imaginary (Matsubara) frequencies, $i\xi_m = 2\pi i k_B T m/\hbar$, where $\hbar = h/2\pi$ with $h =$ Planck constant, and the term for $m = 0$ has a prefactor of $^1/_2$. In the case that the system responds predominantly only in the ultraviolet region of the electromagnetic spectrum, media 1 and 3 are identical, and the permittivities for media 1 and 2 are sufficiently similar, Eq. (7.6) can be approximated (Israelachvili, 2006) by

$$A_{H123} = \frac{3}{4}k_B T \left(\frac{\varepsilon_{r1} - \varepsilon_{r2}}{\varepsilon_{r1} + \varepsilon_{r2}}\right)^2 + \frac{3h\nu_{UV}}{16\sqrt{2}} \frac{(n_1^2 - n_2^2)^2}{(n_1^2 + n_2^2)^{3/2}}, \tag{7.8}$$

where n_k ($k = 1, 2$) are the refractive indices, $\nu_{UV} = \omega_{UV}/2\pi$ is the electronic frequency in the UV region, and h is Planck constant. However, in general, a system will not respond only in the ultraviolet: systems also respond in both the microwave and infrared regions, and these terms must be calculated and included in (7.8). Parsegian and Ninham (1970) represented the permittivity for a system as a set of harmonic oscillators,

$$\varepsilon(i\xi_m) = 1 + \sum_j \frac{C_j}{1 + (\xi_m/\omega_j)^2}. \tag{7.9}$$

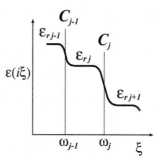

Figure 7.3 $\varepsilon(i\xi)$ as a function of ξ. The coefficient C_j is essentially the decrease in $\varepsilon(i\xi)$ as ξ increases from below ω_j to above ω_j.

The intent would be to substitute Eqs. (7.9) into (7.6) and carry out the summation. A plot of Eq. (7.9) is shown in Fig. 7.3.

Substituting Eq. (7.9) in Eq. (7.6), we can go beyond the approximations that lead to Eq. (7.8) and calculate the effect of including infrared or microwave terms in the expression for the Hamaker coefficient. To do this, we need to know about the coefficients of (7.9), the oscillator strengths $\{C_k\}$. The permittivities $\{\varepsilon_{rj}\}$ are related to the oscillator strengths $\{C_k\}$ by $C_j = \varepsilon_{rj} - \varepsilon_{rj+1}$ (Hough and White, 1980). Thus, if the system responds only in the infrared and the visible-ultraviolet, then, as $\varepsilon_r \propto n^2$, there are only two terms in (7.5): $C_{IR} = \varepsilon_r(0) - n_{UV}^2$ and $C_{UV} = n_{UV}^2 - n_\infty^2 = n_{UV}^2 - 1$.

Although the original formulation by Dzyaloshinskii et al. (1961) was for spatially homogeneous systems, it has been extended to more complex geometries and spatial dependences (Parsegian, 2005). The complexity increases, and it is not trivial to treat a system in which the permittivities might be continuously varying functions of spatial position. A relatively simple case is the Hamaker coefficient for the interaction between two spheres in the case that the permittivity of the spheres is a function of the radial distance from the centers of the spheres (Parsegian, 2005, p.79).

Except for the work by Johansson and Bergenståhl (1992), who used Eq. (7.8), Lifshitz theory has not been applied to fats and oils, even the simplest cases of a single type of fat in a known polymorphic form, in a single-component oil. The reason might be that one needs a knowledge of the frequency-dependent permittivities for both the solid fat and the oil, and it is not trivial to make appropriate samples of the former to measure these quantities.

The Lennard-Jones (L-J) 6—12 potential, $u_{AB}(r)$, describing the London—van der Waals dispersion energy of interaction between two spherically symmetric atomic moieties, A and B is (Israelachvili, 2006)

$$u_{AB}(r) = -\frac{C_6^{AB}}{r^6} + \frac{C_{12}^{AB}}{r^{12}}, \tag{7.10}$$

where we have denoted the energy between two atomic moieties, the smallest objects that go to make up larger systems, by a lower-case u. Here r is the center-to-center distance between A and B. Although the attractive $1/r^6$ term can be justified from first principles, the form of the repulsive, $1/r^{12}$ term, representing the electron "overlap" repulsion between different atoms, has been chosen for computational convenience. Eq. (7.10) has also been used to represent the interaction energy between two infinitesimal "chunks of matter" and has been used in mean field theories, such as the Derjaguin–Landau–Verwey–Overbeek (DLVO) theory of colloid interactions (Israelachvili, 2006). The attractive part of (7.10) for atom or molecules in their ground state can be derived from quantum mechanical second-order perturbation theory involving the dipole–dipole interaction as the perturbation. This yields (e.g., Wennerström, 2003)

$$u_{AB}(r) = -\frac{1}{24(\pi\varepsilon_0)^2}\frac{1}{r^6}\sum_l\sum_n\frac{\left|\langle A,l|\overrightarrow{P}|A,0\rangle\right|^2\left|\langle B,n|\overrightarrow{P}|B,0\rangle\right|^2}{(E_{A,l}-E_{A,0})+(E_{B,n}-E_{B,0})} = -\frac{C_6^{AB}}{r^6}. \quad (7.11)$$

Here, $|A,l\rangle$ is a statevector for the A molecule in state l, belonging to energy $E_{A,l}$, with $l=0$ representing the ground state with energy $E_{A,0}$. A similar definition holds for the statevectors of the B molecule in state n. $\overrightarrow{P}$ is the electric dipole moment operator, and r is the distance between the molecules. That one obtains this from second-order perturbation theory means that it involves only pairs of molecules so that the interaction between two objects composed of such molecules will be the sum of all the pairwise interactions. Nonadditive perturbation terms arise only in third-order perturbation theory and are known as the Axilrod–Teller effect (Axilrod and Teller, 1943). Wennerström (2003) has shown the equivalence of the Hamaker and Lifshitz approaches for nonpolar molecules and has established the relationship between the two approaches.

If one knows the functional form of the number densities, $\Phi_i(\overrightarrow{r})$, of the solids ($i=1$, 3) and the oil ($i=2$), then one can use the Lennard-Jones 6–12 potential to compute the total energy of the system and, from it, deduce the Hamaker coefficient. The energy of the system of Fig. 7.1 interacting via the 6–12 potential is (Israelachvili, 2006)

$$U_{123} = \sum_{j\neq i}\int_i d^3\overrightarrow{r}\int_j d^3\overrightarrow{r}'\left[\frac{-C_6^{ij}}{|\overrightarrow{r}-\overrightarrow{r}'|^6}+\frac{C_{12}^{ij}}{|\overrightarrow{r}-\overrightarrow{r}'|^{12}}\right]\Phi_i(\overrightarrow{r})\Phi_j(\overrightarrow{r}'), \quad (7.12)$$

where the sum is over $i,j=1,2,3$, $j\neq i$, and the double counting of energies is disallowed. The coefficients, C_n^{ij}, describe the interaction energy ($n=6,12$) between substances i and j. One has to make some assumptions about the number densities, and the most common assumption is that they are constant, independent of the solid separation, d. However, if one requires that the chemical potential (free energy per molecule) of the slabs in Fig. 7.1 is equal to the chemical potential of a pure oil system (the "bulk") in contact with that slab system, then one need not assume that the number densities are constant. From this, one can obtain an average number density of oil between the two nanocrystals as a function of d.

If we consider the system of Fig. 7.2 to be a pair of TAG crystals, 1 and 3, separated by a TAG oil with the same chain lengths, then it is reasonable to assume that the coefficients C_6 and C_{12} are equal for solid—solid, solid—liquid, and liquid—liquid interactions, and, assuming that the number densities are constant, then by a trivial integration, we obtain the energy per unit cross section of the two solids,

$$U_{123}(d) = \left[-\frac{\pi C_6}{12} \frac{1}{d^2} + \frac{\pi C_{12}}{360} \frac{1}{d^8} \right] (\Phi_1 - \Phi_2)^2, \qquad (7.13)$$

which yields an attractive term of

$$U_{123}(d) = -\frac{A_{H123}}{12\pi d^2} + \frac{\pi C_{12}}{360 d^8} (\Phi_1 - \Phi_2)^2, \qquad (7.14)$$

where the Hamaker coefficient is $A_{H123} = \pi^2 C_6 (\Phi_1 - \Phi_2)^2$.

The CDM (Kim et al., 2007; Verdult, 2010) was developed to calculate van der Waals forces—in the spirit of Lifshitz theory—by including all correlations between dipoles but without invoking a mesoscale variable such as the permittivity. The method begins by considering N vibrating atoms, {1,2,3...}, which can be polarized by an electric field and which interact with each other via the Coulomb interaction. Atom j experiences an electric field set up at its position by all the other atomic dipoles: the dipole of atom j is created (induced) by the net electric field acting on it due to all the other dipoles. This picture is similar to that described by Fig. 7.1C. One obtains a set of N coupled equations involving the unknown induced dipole moments $\{\vec{p}_j\}$ and the known atomic polarizabilities $\{\alpha_{j0}\}$ for $j = 1,...,N$ and any externally applied electric field, $\vec{E}_0$. One then solves for the induced dipole moments via matrix inversion.

To derive the energy of interaction between two solids, one writes down the quantum mechanical Hamiltonian operator, $\mathcal{H}$, for the atomic harmonic oscillators of each solid in the usual way. We assume that there are N_k atoms ($k = 1,2$) in the two solids labeled 1 and 2, and we assume, for simplicity, that all atoms are identical so that we obtain

$$\mathcal{H} = \sum_i^{N_1} \frac{p_i^2}{2m} + \sum_j^{N_2} \frac{p_j^2}{2m} + \frac{m\omega_0^2}{2} \left[\sum_i^{N_1} u_i^2 + \sum_j^{N_2} u_j^2 \right] - \frac{q^2}{2\varepsilon_0} \sum_i^N \sum_j^N \vec{u}_i \cdot T_{ij} \cdot \vec{u}_j, \qquad (7.15)$$

where the first four terms comprise the usual Hamiltonian for $N_1 + N_2$ simple harmonic oscillators (e.g., Sakurai and Tuan, 1994; Messiah, 1999) and the last term describes the electromagnetic coupling between the atomic dipoles. Here the vector $\vec{u}_k$ describes the polarization vector of the kth atom, T_{ij} is a second-rank tensor that defines the electric field set up at atom j due to the atomic dipole i and $N = N_1 + N_2$. One then diagonalizes $\mathcal{H}$, or uses perturbation theory, to obtain the eigenvalues (energies) of (7.15). The fact that T_{ij} is proportional to r_{ij}^{-3}, where $r_{ij} = \left| \overrightarrow{r_i} - \overrightarrow{r_j} \right|$ is the distance between atom i and atom j and introduces the distance between, and relative orientation of, solids 1 and 2.

The advantage of the CDM approach is that it does not introduce a mesoscale variable such as the permittivity as Lifshitz theory does but depends on atomic polarizabilities. One disadvantage is that it needs to obtain the eigenvalues, approximate or exact, for the Hamiltonian operator of the system. For the case of two solids in a vacuum, as done above, the problem can be solved. If, however, we have to introduce a fluid in which the two solids are immersed, the problem becomes more complicated, and no such applications have yet been realized. This is, however, an interesting approach that could find applications to fluids in confined nanospaces.

ELECTROSTATIC INTERACTIONS (FRANKL, 1986; LORRAIN ET AL., 1988)

Electrostatic forces in biological systems, from which foods are derived, are ubiquitous. Electrostatic interaction is defined as "short range," if the force depends on distance, r, from the charged moiety such as r^{-n} where $n > d$, the dimensionality of the system. Thus, an electric charge ($n = 2$) and a point dipole ($n = 3$) give rise to "long-range" forces in a three-dimensional space. Biological molecules possess a plethora of electrical charges, many of which appear as electric dipoles (e.g., the peptide bond). However, in an aqueous solution, the presence of free ions can bring about Debye screening of charges, which, under certain circumstances, restricts that range of an electrostatic force. In addition, the presence of many electrical dipoles in close proximity can create a force, which is effectively short range and which can be highly anisotropic. A suitable combination of dipoles can thus create complex, short-range electrostatic fields that can be tuned for the purposes of selecting and binding particular molecules.

Even if the effects of a point charge is screened by the presence of free monovalent ions, the range of the field, using a linearized approximation to the Poisson–Boltzmann (P–B) equation, depends on $e^{-\kappa r}/r^2$, where κ is determined by the free ions and the temperature, and the length, κ^{-1}, is the Debye screening length. The force from this field can have an effect beyond the range of a hydrogen bond or van der Waals interactions. In an edible oil lacking any freely mobile charges, electric fields are unscreened and cannot be approximated by a short-range interaction. This statement is true for any edible oil in three dimensions. Below, a description of how to handle long-range electrostatics will be described.

There is, thus, a significant difference between electrostatic effects in water-based systems, such as milk, or in pure edible oil systems. (1) In both water-based systems and in pure edible oils, if boundary conditions at surfaces are not taken into account sufficiently correctly, then the behavior of macromolecules at such an interface could be modeled incorrectly. (2) In water-based systems, the effect of monovalent ions can, under certain conditions, be represented by the Debye screening length, κ^{-1}. The same is not generally true for multivalent ions, which cannot be treated as part of an average Debye screening function, although they do carry out screening. However, not only do they contribute to screening but also, equally importantly, divalent ions, such as for example Ca^{2+}, can take part in ion bridging. Ion bridging, in particular, divalent ion bridging, as the name suggests, might be thought of as static configurations, on the timescales of the system being studied, in which a divalent ion bridges a pair of monovalent moieties. This view is not tenable. Multivalent ion bridging can be a dynamical process in which the bridging ions move, on a timescale possibly faster than molecular motion of the bridged moieties, between local energy minima and in which their entropy plays an important role. If multivalent ion bridging was to be modeled as a static process, it is possible that incorrect phenomena would be predicted. (3) In the presence of charged multicomponent polymers in which soft interfaces are changing on a timescale similar to that of the movement of the various charged species present, the correct application of the boundary conditions is essential. (4) Another point concerns the use of buffers, which might be thought of as simply a pH-controlling technique. Buffer molecules, however, become ionized and give rise to multivalent molecular ions, which, in sufficiently high concentrations, can change the screening in a dielectric. Such possible effects must at least be considered when electric fields are calculated. These are some of the effects that electrostatics in water-based systems must address.

Electrostatic interactions in edible oils frequently involve permanent dipoles of polarized chemical bonds or charged moieties, which arise via the oxidation of edible oils. Each mono-, di-, or triglycerol molecule possesses one, two, or three such atomic electrical dipoles at the carbonyl group. Fig. 7.4A shows partial charge distributions on selected groups of triolein, whereas Fig. 7.4B shows partial charges, which can arise when one of the oleic chains is oxidized. Total charge neutrality is preserved, but this does not take away from the fact that, locally, there can be a significant net charge. The numbers shown are in units of the magnitude of the electronic charge. This system is similar to that modeled by Khandelia and Mouritsen (2009).

Poisson's equation. The electric field, $\overrightarrow{E}(\overrightarrow{r})$, at any point in space can, in the absence of time-dependent magnetic fields, be derived from the electric potential, $\psi(\overrightarrow{r})$, via $\overrightarrow{E}(\overrightarrow{r}) = -\overrightarrow{\nabla}\psi(\overrightarrow{r})$. The interaction of an electrical charge with other charges is proportional to the charge itself and to the electric potential created by other charges. Accordingly, the problem of charge—charge interactions reduces to the correct

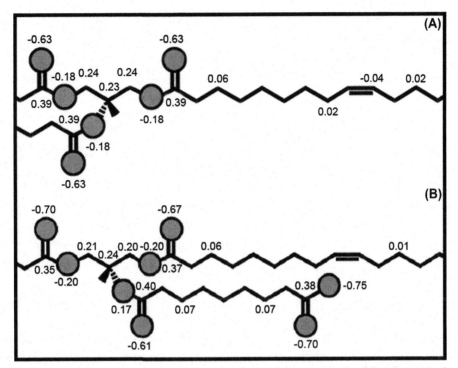

Figure 7.4 Partial charges on a triolein molecule in units of the magnitude of the electronic charge. (A) Triolein. Gray spheres represent oxygen atoms. (B) Triolein with one chain oxidized. Charges computed using http://davapc1.bioch.dundee.ac.uk/prodrg The GlycoBioChem PRODRG2 Server. A somewhat different charge distribution can be obtained using http://moose.bio.ucalgary.ca/index. php?page=Structures_and_Topologies

calculation of the electric potential distribution for the system of interest. If the charges are all at fixed positions, then the calculation need only be done once. If the charges are "free" (to move), then one has a self-consistent problem: the movement of the charges is determined by the electrical potential and vice versa. In the latter case, the potential has to be calculated at each instant of time to determine how the accelerations — and hence positions — of the charges change with time. The integral and differential forms of Gauss's law,

$$\int_S \vec{E}(\vec{r}) \cdot d\vec{s}(\vec{r}) = \frac{Q}{\varepsilon} \tag{7.16a}$$

$$\nabla^2 \psi(\vec{r}) = -\frac{\rho(\vec{r})}{\varepsilon} \tag{7.16b}$$

can be used to determine the electric potential (distribution). Here Q is the total charge contained inside the volume bounded by the surface S over which the integral of the electric field $\vec{E}(\vec{r})$ is carried out, $\rho(\vec{r})$ is the electric charge density at point $\vec{r}$, and $\varepsilon = \varepsilon_0 \varepsilon_r$ is the permittivity, where ε_0 is the permittivity of free space and ε_r is the relative permittivity describing the polarizability of the dielectric. Eq. (7.1) is used to calculate the electric field when the system possesses sufficiently high symmetry; for example, when $\vec{E}(\vec{r})$ has a constant magnitude over the surface S.

Uniqueness theorem. The electrical potential is unique if it satisfies Poisson's equation and the boundary conditions. This very powerful statement means that we may use *any* technique whatsoever (analytical calculation, computer simulation, inspired guess, message in a fortune cookie) to obtain the electric potential, subject only to those two requirements.

Boundary conditions, polarization, induced charges, and energy. To obtain a correct solution for the potential from Poisson's equation, it is necessary to satisfy the boundary conditions. In the case of a charge Q located at the origin in a uniform dielectric, these are $\lim_{r \to \infty} \psi(\vec{r}) = 0$ and $\lim_{r \to \infty} \frac{\partial \psi(r)}{\partial r} = 0$. In this case, we obtain as a solution to Eq. (7.16b),

$$\psi(r) = \frac{1}{4\pi\varepsilon} \frac{Q}{r}. \tag{7.17}$$

The boundary conditions above are not valid for the case of a nonuniform space, such as a system possessing an interface. Examples of this include emulsions, bacterial surfaces, and micellar delivery systems. Indeed, it is unlikely that any food system does *not* possess interfaces within it. Interfaces can become electrically polarized, which amounts to inducing electrical dipoles at the interface, and it is therefore necessary to determine the boundary conditions at interfaces. To apply the uniqueness theorem at an interface adjacent to regions 1 and 2, we need to satisfy the boundary conditions, which are the continuity of the potential at the interface, $\psi_1(\vec{r}_S) = \psi_2(\vec{r}_S)$, and the continuity of the tangential component of the electric field, $\vec{E}$, at the interface,

$$\left[\vec{E}_2(\vec{r}_S) - \vec{E}_1(\vec{r}_S) \right] \times \hat{n}(\vec{r}_S) = 0 \quad \text{or} \quad E_{2T}(\vec{r}_S) = E_{1T}(\vec{r}_S), \tag{7.18}$$

and the continuity of the normal component of the electric displacement $\vec{D}$,

$$[\vec{D}_2(\vec{r}_S) - \vec{D}_1(\vec{r}_S)] \cdot \hat{n}(\vec{r}_S) = \sigma_F(\vec{r}_S) \quad \text{or} \quad \vec{D}_{2N}(\vec{r}_S) - \vec{D}_{1N}(\vec{r}_S)$$
$$= \sigma_F(\vec{r}_S), \tag{7.19}$$

where

$$\vec{D}_m(\vec{r}) = \varepsilon_m \vec{E}_m(\vec{r}) = \varepsilon_0 \vec{E}_m(\vec{r}) + \vec{P}_m(\vec{r}) \tag{7.20}$$

is the electric flux density. Here, the vector $\overrightarrow{r}_S$ locates a position on the interface in question, $E_{mT}(\overrightarrow{r}_S)$ is the component of the electric field in region m that is tangential to the interface at position $\overrightarrow{r}_S$, $\overrightarrow{D}_m(\overrightarrow{r})$ and $\overrightarrow{P}_m(\overrightarrow{r})$ are the electric displacement and polarization vectors in region m at $\overrightarrow{r}$, $\hat{n}(\overrightarrow{r}_S)$ is a unit vector perpendicular to the interface at $\overrightarrow{r}_S$ and pointing outward from the interface, $\overrightarrow{D}_{mN}(\overrightarrow{r}_S)$ is the component of the electric displacement in region m perpendicular to the interface at $\overrightarrow{r}_S$, $\sigma_F(\overrightarrow{r}_S)$ is the free charge density at $\overrightarrow{r}_S$, and ε_m is the permittivity of region m.

The boundary conditions in Eqs. (7.18)–(7.20) can be satisfied by finding the polarization induced at every interface and inside the bulk of every material making up the system. For a linear isotropic dielectric, the electric polarization $\overrightarrow{P}(\overrightarrow{r}_S)$ at point $\overrightarrow{r}_S$ at an interface that is induced by an electric field $\overrightarrow{E}(\overrightarrow{r})$ is given by

$$\overrightarrow{P}(\overrightarrow{r}_S) = (\varepsilon - \varepsilon_0)\overrightarrow{E}(\overrightarrow{r}_S). \tag{7.21}$$

The induced "bound" surface charge density $\sigma_b(\overrightarrow{r}_S)$ can be obtained from

$$\sigma_b(\overrightarrow{r}_S) = \overrightarrow{P}(\overrightarrow{r}_S)\cdot\hat{n}(\overrightarrow{r}_S). \tag{7.22}$$

The surface charge density is referred to as the "bound" surface charge density because it arises from the polarization of atoms or molecules essentially fixed in space. It is thus distinguished from "free" charges that can move in space.

For a given charge distribution, the electrostatic energy of the system inside a volume v is

$$U = \frac{1}{2}\int_{v} \rho_f(\overrightarrow{r}')\psi(\overrightarrow{r}')dv(\overrightarrow{r}'), \tag{7.23}$$

where $\rho_f(\overrightarrow{r}')$ is the "free" charge density at $\overrightarrow{r}'$, $\psi(\overrightarrow{r}')$ is the electric potential at $\overrightarrow{r}'$, $dv(\overrightarrow{r}')$ is the infinitesimal volume at $\overrightarrow{r}'$ and the integration is over v. Eq. (7.23) is, strictly speaking, formally correct only for continuous charge distributions $\rho_f(\overrightarrow{r}')$. When discrete charges are included, care must be taken to exclude the infinite self-energies that arise from the unphysical interaction of a discrete charge with its own electrostatic potential. The potential at any point $\overrightarrow{r}'$ is

$$\psi(\overrightarrow{r}') = \frac{1}{4\pi\varepsilon_0}\int_{v}\frac{\rho_f(\overrightarrow{r}'') + \rho_b(\overrightarrow{r}'')}{|\overrightarrow{r}' - \overrightarrow{r}''|}dv(\overrightarrow{r}'') + \frac{1}{4\pi\varepsilon_0}\int_{S}\frac{\sigma_f(\overrightarrow{r}'') + \sigma_b(\overrightarrow{r}'')}{|\overrightarrow{r}' - \overrightarrow{r}''|}dA(\overrightarrow{r}''), \tag{7.24}$$

where $\rho_f\left(\overrightarrow{r}''\right)$ and $\rho_b\left(\overrightarrow{r}''\right)$ are the "free" and "bound" volume charge densities in the infinitesimal volume $dv\left(\overrightarrow{r}''\right)$, and $\sigma_f\left(\overrightarrow{r}''\right)$ and $\sigma_b\left(\overrightarrow{r}''\right)$ are the "free" and "bound" surface charge densities in the infinitesimal surface area $dA\left(\overrightarrow{r}''\right)$. The integration is over the volume v and the surface(s) S.

Image charges. This method involves an approach that avoids having to calculate induced bound surface charges although it can be used, in practice, only for sufficiently simple interfaces. The *method of image charges* involves finding the (unique) electrostatic potential that satisfies Poisson's equation, together with the appropriate boundary conditions, by the artifice of introducing fictitious "images" of the free charges. Consider a single free electric charge Q_1 located at position $\overrightarrow{r}_1$ at a perpendicular distance d_1 from a plane interface that separates the space into two half-spaces, regions 1 and 2, which have permittivities ε_1 and ε_2, respectively. The charge Q_1 is in region 1, and the perpendicular to the interface is taken to be the z-axis. Using the boundary conditions, it is easy to see that the potential at any point $\overrightarrow{r}$ *in region 1* can be calculated by placing a fictitious image charge Q'_1 at position $\overrightarrow{r}'$ *in region 2*, at a perpendicular distance d_1 from the interface (Fig. 7.5A), so that

$$\overrightarrow{r}_1 - \overrightarrow{r}'_1 = 2d_1\,\widehat{z}, \tag{7.25}$$

and taking Q'_1 to be

$$Q'_1 = \frac{\varepsilon_1 - \varepsilon_2}{\varepsilon_1 + \varepsilon_2}Q_1. \tag{7.26}$$

The electric potential at point $\overrightarrow{r}$ in region 1 is then

$$\psi_1(\overrightarrow{r}) = \frac{1}{4\pi\varepsilon_1}\left(\frac{Q_1}{\left|\overrightarrow{r} - \overrightarrow{r}_1\right|} + \frac{Q'_1}{\left|\overrightarrow{r} - \overrightarrow{r}'_1\right|}\right). \tag{7.27}$$

There are two contributions to the electrostatic potential $\psi_1(\overrightarrow{r})$ in Eq. (7.27). The first term is the source potential [see Eq. 7.17] for the free charge Q_1 located at position $\overrightarrow{r}_1$. The second term is, in reality, the electric potential produced by the bound surface charge density $\sigma_b(\overrightarrow{r})$ at the interface $z = 0$ that is induced by the free charge Q_1. Remarkably, *in region 1*, this second term is equal to the potential [see Eq. 7.17 again] that would be produced by a fictitious image charge Q'_1 located at position $\overrightarrow{r}'_1$ in region 2 (Fig. 7.5).

The free charge density for a discrete free charge Q_1 located at position $\overrightarrow{r}_1$ is given by

$$\rho_f(\overrightarrow{r}) = Q_1\delta\left(\overrightarrow{r} - \overrightarrow{r}_1\right), \tag{7.28}$$

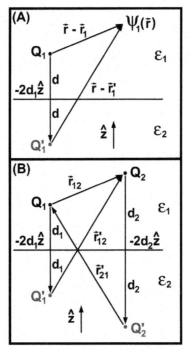

Figure 7.5 (A) A charge Q_1 located at $\vec{r}_1$ in region 1, at a perpendicular distance d_1 from an interface (*horizontal line*) that separates region 1 from region 2, for which the permittivities are ε_1 and ε_2, respectively. The electric potential $\psi_1(\vec{r})$, at $\vec{r}$ in region 1 can be calculated by introducing an image charge Q_1' located at an equal distance from the interface inside region 2. The unit vector $\hat{z}$ is perpendicular to the interface. (B) Geometry of two charges, Q_1 and Q_2 in region 1 near an interface (*horizontal line*) with $\vec{r}_{12} = \vec{r}_2 - \vec{r}_1$. The potential anywhere in region 1 can be calculated by introducing two image charges, Q_1' and Q_2', in region 2 as shown. ((B) *Adapted with permission from Pink, D.A., Hanna, C.B., Quinn, B.E., Levadny, V., Ryan, G.L., Filion, L. and Paulson, A.T., 2006 Modelling electrostatic interactions in complex soft systems. Food Res. Int. 39, 1031−1045.)*

where $\delta(\vec{r})$ is Dirac delta function. Note that only the free charge Q_1 contributes to the free charge density ρ_f, not the fictitious image charge Q_1'. It follows from Eqs. (7.23)−(7.28) that the electrostatic energy resulting from introducing the free charge Q_1 is

$$U(d_1) = \frac{1}{2} \frac{1}{4\pi\varepsilon_1} \frac{Q_1 Q_1'}{\left|\vec{r}_1 - \vec{r}_1'\right|} = \frac{Q_1^2}{16\pi\varepsilon_1 d_1}\left(\frac{\varepsilon_1 - \varepsilon_2}{\varepsilon_1 + \varepsilon_2}\right),\qquad(7.29)$$

where the infinite self-energy due to the unphysical interaction of the free charge Q_1 with its own potential has been omitted. Eq. (7.29) is equal to *one-half* of Q_1 times the potential at Q_1 due to its image charge Q_1'.

Note that if $\varepsilon_1 < \varepsilon_2$, then $U < 0$, and the charge Q in region 1 is attracted to region 2. This is the essential physics behind, for example, the attraction of dry bread crumbs to an electrically charged object in dry air, such as a charged rubber comb. But if $\varepsilon_1 > \varepsilon_2$, then $U > 0$, which means that the charge Q in region 1 will be repelled by region 2. This latter situation is relevant, for example, to lipid membranes in aqueous solutions. The permittivity of the hydrophobic region of a cell membrane is $\varepsilon_{mem} \approx 5\varepsilon_0$, whereas that of an aqueous solution is $\varepsilon_{water} \approx 81\varepsilon_0$. This means that nonorganic ions located in solution are repelled from a membrane interface so that the concentration of the ions close to a membrane carrying zero net charge is less than that in a bulk solution.

Note that the image charge Q_1' required for the calculation of the potential in region 1 must be located in region 2. This is necessary in order that the boundary conditions and Poisson's equation be satisfied. This suggests that if we want to calculate the potential in region 2, then we should locate an image charge in region 1. This is true. If we replace the original charge Q_1 by a charge Q_1'' with

$$Q_1'' = \frac{2\varepsilon_2}{\varepsilon_1 + \varepsilon_2} Q_1, \tag{7.30}$$

then the electrical potential at any point $\vec{r}$ in region 2 is given by

$$\psi_2(\vec{r}) = \frac{Q_1''}{4\pi\varepsilon_2} \frac{1}{|\vec{r} - \vec{r}_1|}. \tag{7.31}$$

We emphasize that the method of image charges is only one way, albeit a very convenient one, of satisfying the boundary conditions at interfaces. Despite the fact that this method is applicable only to a limited class of boundary conditions in which the interfaces possess relatively high symmetry, the method of images can be very useful in modeling and understanding the physics underlying electrostatics in food systems.

The P–B equation. Although electrical charges are localized on atoms or molecules, we can approximate the charge distribution by a continuous function of position, if the spatial scale of the system is such that the distances between the charges are very much less than the characteristic spatial scale in which we are interested. Even if we are interested in spatial scales comparable with the characteristic distance between charges, it is possible to approximate their locations by an average distribution if the timescale of their movement and redistribution is much shorter than that of the system for which we are calculating the electric potential. An example of this is the electrical potential due to a relatively stationary charge located on a surface or on a polymer in the presence of ions (free charges) in the surrounding aqueous solution. One assumes that the free charges in solution are moving sufficiently rapidly that the stationary charge sees an average "smeared-out" distribution of free charges characteristic of some equilibrium. In many cases, the aqueous solution is in thermodynamic equilibrium, characterized by an ambient temperature T, so that the spatial distribution of the free charge density is determined by the Boltzmann factor $\exp(-\beta U)$, where U is the energy and $\beta = 1/k_B T$. Taking into account the requirement of total electroneutrality, this converts the (linear) Poisson Eq. (7.16b) into the P—B equation (Israelachvili, 2006), which is nonlinear because ψ appears in the exponent:

$$\nabla^2\psi = \frac{e}{\varepsilon_r\varepsilon_0}\sum_i z_i n_i^0 \exp\left(\frac{z_i e\psi}{k_B T}\right). \tag{7.32}$$

Here e is the elementary charge, and n_i^0 and z_i are the concentration and valence of ith type of ion in solution, respectively.

The linearized P—B equation and the Debye screening length. In cases where entropic effects (sufficiently high temperature T) dominate the energetics, i.e., when the potential ψ is sufficiently small so that $\psi \ll k_B T/e$ (~ 25 mV at room temperature), then the exponential term in Eq. (7.32), $\exp(z_i e\psi/k_B T)$, can be expanded to first order in $z_i e\psi/k_B T$. This results in approximating the nonlinear P—B equation by a linearized differential equation (the linearized P—B equation) for which an analytical solution can be obtained. In this case, the P—B equation becomes, using the condition of total charge neutrality $\sum_i z_i n_i^0 = 0$,

$$\nabla^2\psi = \psi\frac{e^2}{\varepsilon k_B T}\sum_i z_i^2 n_i^0 = \kappa^2\psi, \tag{7.33}$$

where

$$\kappa^2 = \frac{e^2}{\varepsilon k_B T}\sum_i z_i^2 n_i^0 = \lambda^{-2} \tag{7.34}$$

and n_i^0 and z_i are the concentration and the valence of the ith type of ion in the bulk solution. The units of κ are clearly inverse length (e.g., nm^{-1}). The distance λ can be treated as the effective radius of a perturbed volume in a neutral solution that contains mobile ions and is called the "Debye (screening) length" (Israelachvili, 2006). In the simplest case of a 1:1 electrolyte, where $n_1^0 = n_2^0 = n_0$, Eq. (7.34) becomes

$$\kappa^2 = \lambda^{-2} = \frac{2e^2 n_0}{\varepsilon k_B T}. \tag{7.35}$$

The Debye screening length of a 100 mM solution at room temperature (300 K) is ~ 1 nm.

Debye and Gouy—Chapman solutions. There are two classical cases of the application of the linearized P—B equation: (1) the case of spherical geometry (so that the solutions to Eq. 7.33 are functions of $r = |\vec{r}|$), where the radius of curvature of the surface is comparable with or smaller than the Debye length $\lambda = \kappa^{-1}$ and (2) the case of plane geometry (where the solutions to Eq. (7.33) are functions only of the perpendicular distance, z, from the plane), where the radius of curvature of the surface is much larger than the

Debye length. The solution of Eq. (7.33) in the first case is the basis of Debye's theory of solutions and is called the "Debye equation" (Israelachvili, 2006; Hunter, 2001):

$$\psi = \frac{Q}{4\pi\varepsilon} \frac{\exp(-r/\lambda)}{r}. \tag{7.36}$$

The solution of Eq. (7.33) in the second case is the basis of the Derjaguin—Landau—Verwey—Overbeek (DLVO) theory of the stability of solutions of colloidal particles (Derjaguin and Landau, 1941; Verwey and Overbeek, 1948; Hunter, 2001) and is called the "Gouy—Chapman equation" (Israelachvili, 2006; Hunter, 2001),

$$\psi = \psi_0 \exp(-z/\lambda) = \frac{\sigma\lambda}{\varepsilon}\exp(-z/\lambda), \tag{7.37}$$

where ψ_0 is the surface potential of the particle and σ is its surface charge density (Israelachvili, 2006; Hunter, 2001).

Handling long-range interactions. Ewald summation (Darden et al., 1993; Hockney and Eastwood, 1988; Sadus, 2002; Lee and Cai, 2009; Ballenegger et al., 2012). Consider the total electrostatic force acting on a given charge. One might conceivably truncate the electrostatic forces acting on this charge if all the charges are sufficiently screened, but, in general, this cannot be done. This is especially true in the cases of edible oils because there are, in general, no freely moveable screening charges. This consideration should hold special interest for those concerned with structures arising in edible oils, both solid and liquid. The point is that, in nonaqueous phases, electrostatic forces can retain their importance for all distances from a charge. In such cases, it might appear that one must simply carry out the summation of electrostatic forces to all distances from a charge until further contributions can be ignored. However, such a limit might never be reached. It is for this reason that the Ewald technique was implemented.

The principle underlying this approach is as follows. The space is divided into two regions: (1) a short-range region in which all (pairwise) electrostatic interactions are calculated explicitly and (2) the remainder of the space in which the long-range forces are Fourier-transformed and their sum computed in reciprocal space yielding a force as a function of k-vector. This sum in k-space converges relatively rapidly. This force is then transformed, via an inverse Fourier transformation, into direct space where it is added to the short-range force. This technique was first extended to the Particle—Mesh (PM) Ewald and, subsequently, to the Particle—Particle—Particle—Mesh (P^3M) technique. The last-named carries out a particle—particle sum for particles that are sufficiently close to each other, in addition to the PM calculation for those separated by greater distances.

In most molecular dynamics simulation packages, these techniques are built into the software and need only be called.

OTHER MODELS

There are two other models of direct relevance to fats aggregation. One was developed to relate mechanical properties of fat aggregates to their structure, whereas other deduces the value of the Hamaker coefficient from the disjoining energy of a fats crystal.

The fractal model. Vreeker et al. (1992) studied tristearin aggregates in olive oil using light scattering and deduced that they formed a fractal structure with dimension $D = 1.7$, which increased to $D = 2.0$ as the sample aged. They also studied the storage modulus, G' as a function of solid fraction (solid fat content), ϕ, and observed a power law dependence in accord with a fractal dimension, $D = 2.0$. These experiments were carried out for low solid fat content (<10%), the "strong-link" regime, in which the interaction between the aggregating objects is stronger than that between the subunits of which the aggregates are composed. The value of $D = 1.7$ is characteristic of diffusion-limited cluster–cluster aggregation (DLCA). They did not, however, identify the objects that were aggregating, nor did they report on how large the aggregating objects were. The first to attempt to make models of the process was Van den Tempel (1961), but it was not until the work of Marangoni and Rousseau (1999) followed by Narine and Marangoni (1999a,b) and subsequently by Marangoni (2000) and Marangoni and Rogers (2003) that the fractal structure of aggregation was modeled.

From rheological experiments, as well as PLM observations, fractal models were developed (Rogers et al., 2008), and Fig. 7.6 shows the interpretation of fat crystal networks with high solid fat content. It is composed of "flocs," which are themselves formed from fractal aggregates of smaller units.

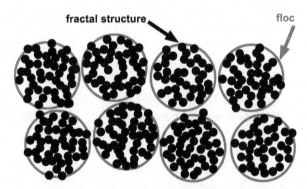

fractal structure   floc

Figure 7.6 Schematic diagram of the weak-link fractal model. Small units (black) self-assemble into fractal structures, of fractal dimension and form flocs (gray). The weak-link describes the weak interaction between flocs compared with the stronger interaction between the small units of the fractal structures. *(From reference Rye, G.G., Litwinenko, J.W., Marangoni, A., 2005. Fat crystal networks. In: Shahidi, F. (Ed.), Bailey's Industrial Oil and Fat Products. Edible Oil and Fat Products: Products and Applications, sixth ed., ed., vol. 4. John Wiley & Sons, Inc. (Chapter 4), Fig. 7.8.)*

Whereas the low value of ϕ used by Vreeker et al. is the "strong-link" regime, the system studied by Marangoni's group is in the "weak-link" regime, corresponding to the large value of ϕ used: the interactions between the flocs (large gray spheres) are stronger (strong-link) or weaker (weak-link) than those within the fractal structures, composed of small spheres (black), that make up the flocs. The size of the flocs can be calculated from images (Rye et al., 2005, Fig. 7.8). If the interpretation of Fig. 7.6 is correct, then those calculations should be in accord with estimates of the floc size deduced from structure—function analysis described below. The size of the units making up flocs can also be deduced from the "Bragg peaks."

The model deduces that the storage modulus, G', is related to the Hamaker coefficient, A_H, by

$$G' = \frac{A_H}{6\pi a E d_0^2}\phi^{1/(3-D)} \tag{7.38}$$

$$A_H = 6\pi a E \lambda d_0^2. \tag{7.39}$$

Here, λ is the slope of $ln[G']$ with respect to $ln[\phi]$ (Narine and Marangoni, 1999b; Marangoni, 2000), where ϕ is the solid fat content, D is the fractal dimension of the structure of which the floc is composed, a is the diameter of a fat floc, E is the extensional strain at the limit of linearity, and d_0 is the equilibrium distance between the floc surfaces.

For the cases of fully hydrogenated canola oil (predominantly tristearin) and high oleic sunflower oil (predominantly triolein), $E = 1.9 \ 10^{-4}$ (Hiemenz and Rajagoplalan, 1997), $\lambda = 1.17 \ 10^7$ Pa (Ahmadi et al., 2008) and $a = 150$ nm (Acevedo and Marangoni, 2010b). However, the value of d_0 is unknown.

Work has been done that could have relevance to modeling solids in oils. Of special note is that of Babick et al. (2011), who presented a scheme for calculating the van der Waals interaction between a pair of fractal aggregates. They make the point that, normally, it is assumed that the fractal aggregates are spherical, and they present results of simulations for which this is not so. The point here is that, if the rotation time of an aggregate is very much smaller than its characteristic lateral movement time, then the spherical approximation could be valid. Otherwise, one cannot assume that it is even approximately valid: fractal structures can assume many different shapes. Other works of potential relevance are those of Dhont and Briels (2003)) concerned with the hydrodynamics and ordering of long rods and Li et al. (2011) concerned with the hydrodynamics of nanorods and nanoplatelets. Another work of possibly less relevance is that of Kontogiorgos et al. (2006) who modeled the fractal dimension of a polysaccharide gelling system modeled as connected spheres. The references contained in these papers should not be ignored.

The semiclassical model relates the Hamaker coefficient A_H to the disjoining free energy of a fat crystal (Hiemenz and Rajagoplalan, 1997), separated into two blocks at their equilibrium separation, d_e:

$$A_H = 24\pi\gamma d_e^2, \tag{7.40}$$

where γ is the interfacial free energy per unit area. As in the case of the equilibrium distance between flocs, d_0, of the fractal model, the value of d_e is unknown.

From Eqs. (7.39) and (7.40), we have two unknowns: the equilibrium distance between flocs, d_0, and the equilibrium distance between two blocks of fat crystals, d_e.

COMPUTER SIMULATION

There are many very good reviews of computer simulation techniques. Those listed reflect a cross section of the many excellent works (Feig, 2010).

Ergodicity. An important concept in statistical mechanics and one that underpins successful implementations of computer simulation techniques is ergodicity. Consider a system with a set of generalized coordinates—variables that define the state of a system. The set of all the variables define the coordinates of phase space. At any time, the values of the variables identify where in phase space the system is. As the values of these variables change, the system moves on a trajectory through phase space. The ergodic hypothesis states that for certain systems the time average of their properties is equal to the average over all of phase space. A system is ergodic if the average values of its variables have been obtained by sampling all of space. That is, the computer simulation has been run for sufficiently long so that all of phase space has been adequately sampled.

Coarse-grained approaches—1 (Müller et al., 2006; Bennnun et al., 2009; Monticelli et al., 2008; Marrink et al., 2007; Laradji and Kumar, 2011). How we model a system depends on what spatial scale interests us. As far as the Coulomb interaction in atomic systems is concerned, we could model a molecule by treating the nucleus as a point object and taking into account the electronic states, most of which exhibit nonspherical distributions, as well as electronic polarizabilities, and spin and orbital angular momentum states. However, many of these degrees of freedom are irrelevant to, for example, nanoplatelet interactions. One intent of modeling must be to eliminate those system coordinates that undergo motions on timescales much less than the timescales in which we are interested. We then replace the detailed dynamics of the fat coordinates by average values. Such a procedure is called "coarse graining." Although the references, above, describe coarse graining as applied to phospholipid bilayer membranes, they describe coarse-modeling of hydrocarbon chains and carbonyl groups, which is of relevance to oils. In addition, coarse-grained models have been used to model the outer leaflet of the outer membranes of gram-negative bacteria (Oliveira et al., 2009, 2010; Schneck et al., 2009, 2010). Only one paper of which we are aware has used coarse-grained models and dissipative particle dynamics (below) to model the liquid phase of TAGs (Pink et al., 2010).

Monte Carlo techniques. The Metropolis Monte Carlo (MMC) method involves only the interaction, and internal, energies of the model being simulated. It is not concerned with kinetic energies and so does not involve forces, accelerations, or velocities. What this implies is that because we have no forces in this approach, we cannot, in general, use MMC techniques to model the dynamics of a system: the path through phase space generated by the method is not necessarily the one that would be generated in response to the forces of the system (Binder and Heermann, 2010; Binder, 1997) if one were to use a method of molecular dynamics (below). The MMC method uses the following procedure:

1. The system is set up in any initial state that is convenient. This is defined by the values of a set of variables, $a_1, a_2, \ldots, a_K$. One MMC step involves trying to change each of these values to a different value. At any point in the simulation, let the state of the system be labeled A, possessing energy E_A, and let us try to change the system to state B, possessing energy E_B by attempting to change one or more of the variables, $a_1, \ldots, a_K$. Define $\Delta E = E_B - E_A$.

2. If $\Delta E \leq 0$, then the change is accepted, and the state of the system becomes B.

3. If $\Delta E > 0$, then choose a random number R: $0 \leq R < 1$.

4. If $R \leq \exp(-\beta \Delta E)$, where $\beta = 1/k_B T$ with T the absolute temperature and Boltzmann constant.

5. When we have tried to change all variables, 1, 2,...,K, then one MMC step has been carried out.

6. There are two aspects to a simulation: (1) initializing the system that entails permitting the system to come to equilibrium after the start of the simulation and (2) calculating average values while the system samples equilibrium states.

   a. After the simulation is begun, let us keep track of the energies at each MMC step. Let these energies be $E(0)$, $E(1)$, $E(2)$, ... We will notice that these change in a regular way, apart from fluctuations and that they approach a value $<E>$, which does not change as the simulation proceeds further. This is the average equilibrium value of the energy.

   b. The simulation is continued, and the average values of other variables are calculated by carrying out a further S MMC steps and computing the average of the set $\{a_n, n = 1, \ldots K\}$. Let the sequence of values obtained from these S MMC steps be $a_n(1)$, $a_n(2)$, ... $a_n(S)$. Then, the average over the S MMC steps is,

$$< a_n > = \frac{1}{S} \sum_{r=1}^{S} a_n(r). \tag{7.41}$$

7. It has been shown that if a simulation is carried out for a sufficient number of MMC steps, then the system being simulated will come to thermal equilibrium at the temperature selected for the simulation and will exhibit average values and fluctuations characteristic of thermal equilibrium.

Born−Oppenheimer atomic-scale molecular dynamics (AMD) is concerned with simulating the motion of atomic nuclei, or atoms as a whole, in contrast to other methods, such as the Car−Parinello technique (Car and Parinello, 1985), which includes electronic degrees of freedom. In AMD, electronic "motion" is assumed to take place on a timescale much faster than the motion of the atom as a whole so that an average electronic distribution can be assumed. This is the Born−Oppenheimer approximation. AMD is concerned with weak physical interactions and is not concerned with chemical reactions involving energies that can break or form covalent bonds. The technique defines sets of forces ("force fields") that act between pairs of atoms (C, O) or small atomic moieties (CH, CH_2, CH_3, OH, H_2O). The atoms or atomic moieties are characterized by fixed (average) point partial electric charges located inside a spherical volume, which defines the size of the atomic moiety. Distinction is made between "bonded" and "nonbonded" atoms: a pair of bonded atoms possesses a permanent covalent bond between them. All other pairs of atoms are "nonbonded." The forces to be used do not act between pairs of "bonded" atoms. The net force, $\vec{F}(t)$, acting at time t on an atom with mass m defines its acceleration, $\vec{a}(t)$, via Newton's equation, $\vec{a}(t) = \vec{F}(t)/m$, which is then integrated twice using, for example, the velocity Verlet numerical algorithm, to eventually obtain the velocity, $\vec{v}(t + \Delta t)$, and position, $\vec{r}(t + \Delta t)$, after a preselected elapsed "time step," Δt. From the new position, $\vec{r}(t + \Delta t)$, one calculates the net force, $\vec{F}(t + \Delta t)$, acting on the atom at time $t + \Delta t$ and repeats the procedure. The magnitude of Δt used in AMD is typically $\sim 1 - 10 \times 10^{-15}$ s (1−10 fs). Simulations are generally carried out under one of the following conditions ("ensembles"): microcanonical (constant number of particles, N; constant volume, V; constant energy, E), canonical (constant N; constant V; constant temperature, T) or the isothermal−isobaric (constant N; constant T; constant pressure, p). There is also the grand canonical ensemble (constant temperature T; constant volume V; constant chemical potential μ).

We should also note that one can go beyond Born−Oppenheimer AMD and include excited electronic states and so treat chemical bonding. A "first-principles" quantum mechanical method such as density functional theory (Sholl and Steckel, 2009), an example of an ab initio quantum mechanical technique, can be used but, as one would expect, this is more compute-intensive than Born−Oppenheimer AMD. The Car−Parinello molecular dynamics package, also an ab initio method, is based on density functional theory, but here is not the place to go into this.

Other ensembles can be defined and used. In what follows we are concerned primarily, but not exclusively, with describing the procedure followed by the simulation package GROMACS (GROningen MAchine for Chemical Simulations) (Hess et al., 2008). The general approach for running an AMD simulation is as follows:

1. Define the questions about a particular phenomenon or event for which you wish answers.

2. Decide whether AMD can provide these answers or elucidate some parts of the question so that you can formulate another method of answering your question.

3. If AMD is applicable, determine what tools are needed to perform your AMD simulation.

   a. What software is available for you to use and how easy is it to get help if you run into problems?

   b. Do you have a coordinate file to describe your molecule of interest? Have you identified and specified the location of all the atoms of the molecules that you want to model? Do you know which pairs of atoms are chemically bonded to each other via covalent bonds? If not, do you have the ability or resources to answer these questions? Can you create the files in your computer that adequately describe the molecules of interest?

   c. What force fields are appropriate for the phenomenon under study?

4. Generate topologies for the molecules under study by identifying and specifying the relative locations of all the atoms of the molecules that you want to model and specifying which pairs of atoms are bonded to each other.

5. Define the simulation box by specifying its dimensions and whether you are using periodic boundary conditions or not.

6. Insert the appropriate number of molecules in the appropriate orientations into the simulation box.

7. Fill the simulation with "solvent." This can be TAGs in the case of oils, or, more commonly, water. It is likely that you will have to create the TAGs, but libraries of water molecules exist. Simulations can also be carried out in vacuum.

8. Determine the appropriate simulation parameters for the equilibrium simulation that are consistent with the chosen force field and ensemble. For example:

   a. Choose the cutoff length for the van der Waals and electrostatic interactions. These are the distances beyond which the interaction is set equal to zero because it is very small, or it is handled by a procedure that is not a force field. Thus, electrostatic interactions beyond their cutoff can be handled by the Ewald summation procedure (see above and York et al., 1993, Darden et al., 1993, Petersen, 1995, Essmann et al., 1995).

   b. Temperature

   c. Pressure

   d. Magnitude of the simulation timestep. This must be small enough so that the system satisfies the requirement of ergodicity.

9. Run the simulation such that the system relaxes sufficiently as shown by the energy coming to equilibrium and not changing further.

10. Run production simulations for a length of time such that it is sufficient samples phase space, i.e., that it satisfies the requirement of ergodicity. This journey through phase space—the space defined by the dependent variables of the simulation—is called the "trajectory." It is at this step that we carry out averaging.

11. Analyze the trajectory to obtain information about the observed phenomenon or event.

AMD is probably the best technique to study the structure and dynamics of food systems on the nanoscale. Many modeling studies have been carried out on TAG systems (Yan et al., 1994; Sum et al., 2003; Hall et al., 2008; Hsu and Violi, 2009). An open problem is to model the condensation of TAG molecules from a melt into a crystal and the appearance of the various polymorphic forms (Sato, 2001; Sato and Ueno, 2011; Himawan et al., 2006). Work by Vatamanu and Kusalik (2007) and Razul and Kusalik (2011) presents a technique that should be tried.

Example 1. Aggregation of TAG CNPs. The smallest components of fats appear to be CNPs (Acevedo et al., 2011), and we begin with considering them as "fundamental components." These highly anisotropic objects have dimensions of hundreds of nanometers along two, approximately, perpendicular axes and less than 100 nanometers in the third dimension and are composed of planes of TAG molecules. To model CNPs, we represented the flat nanocrystals by a close-packed lattice of three-dimensional solid unit structures composed of crystalline TAGs. It is convenient to represent each CNP as a close-packed structure of spheres because the attractive part of the dispersion interaction between spheres has been established[7]. Each sphere, represented a continuum of TAG molecules with a density characteristic of the crystalline phase of interest, possesses radius R, and Fig. 7.7A shows a representation of a CNP. Such a model has been used elsewhere (Glotzer et al., 2005).

The interaction between two identical homogeneous spheres, each of radius R, a center-to-center distance, r, apart is the Hamaker hybrid form (Parsegian, 2005 Table S.3, p.155)

$$V_d(r) = -\frac{A_H}{6}\left[2R^2\left(\frac{1}{s^2+4Rs}+\frac{1}{(s+2R)^2}\right)+\ell n\left(\frac{s^2+4Rs}{(s+2R)^2}\right)\right] \quad r \geq 2R \quad (7.42)$$

where $s = r - 2R$ is the surface-to-surface separation of the two spheres and A_H is the Hamaker coefficient.

Aggregation of CNPs were modeled in an $L \times L \times L$ simulation box with periodic boundary conditions. One MC step involved translating and rotating CNPs and translating and rotating clusters of CNPs with respect to their centers of mass. We permitted movement of all clusters with the translational step size as well as the angle of rotation

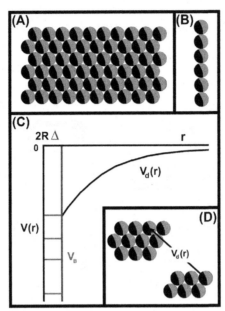

Figure 7.7 (A) Model of a crystalline nanoplatelet (CNP). It is defined by the set of integers $\{m,n,\ell\}$ specifying the number of spheres in each row (m) and column (n) of a lateral plane and the number of planes (ℓ). Here $m = 10$ and $n = 6$. (B) Model of a structure with cylindrical symmetry, for which $m = n = 1$ and $\ell = 6$. (C) The van der Waals potential, $V_d(R,r)$, between two spheres belonging to different model CNPs. The spheres are a center-to-center distance, r, apart and $\Delta = 2R + \delta$, where R is the radius of the spheres and δ is a small distance. When $r \leq 2R + \delta$ the spheres become bound with an energy $V_B < 0$ with two examples shown. We define the "physically realistic" value of V_B to be $V_B = V_d(R,\Delta)$. (D) Illustration of the interaction between CNPs. The total interaction between the two CNPs shown is the sum of all pairwise interactions, $V_d(R,r)$.

around a randomly chosen axis through the center of mass, proportional to $M^{-1/2}$, where M is the mass of the cluster. All CNPs were of size $m \times m \times 1$, We chose the radius of the spheres making up the CNPs to be $R = 0.5$ in arbitrary units. We carried out two kinds of simulation which would lead to a total of N CNPs taking part in aggregation, although not necessarily forming a single cluster: (1) This case had all N CNPs present initially. (2) The other case argued that CNPs are formed as the temperature dropped below the freezing point of those TAGs which went to form the nanoplatelets or spherical structures. This creation of CNPs is described by a characteristic time τ_{create}. As soon as they come into existence; however, such structures can begin to aggregate and relax with a characteristic time of τ_{relax}. This process was simulated by choosing an initial number of CNPs, N_I, and incrementing this by a number, ΔN, every ΔK MC steps. After a total of $(N - N_I)/\Delta N$ MC steps, all N CNPs would have been created. N is determined by the concentration of TAGs going to make up CNPs in the system.

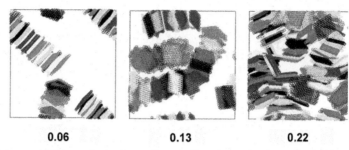

0.06 **0.13** **0.22**

Figure 7.8 Representative configurations of crystalline nanoplatelets of size (10 × 10 × 1) with sphere radius of $R = 0.5$ for three mass concentrations, 0.06, 0.13, and 0.22.

Let us choose the total number of spheres to be $N = 8000, 16,000$ or $24,000$ yielding 80, 160 or 240 CNPs which corresponds to CNP concentrations of 0.06. 0.13 and 0.22.

For the example of case (1) and shown in Fig. 7.8, it can be seen that, at low concentrations, the model CNPs form multilayer "sandwiches." We also considered the case in which the CNP sizes followed a Gaussian distribution, $\{m \times m \times 1\}$, with average, $m = 10$, and variance, $\sigma^2 = 5$. We found the same results as shown in Fig. 7.8. We note that, for higher concentrations of CNPs, it is possible that the multilayer sandwiches could appear as needles and that there is evidence for needles in Fig. 7.1 of Acevedo et al. (2011).

Example 2. Oils in confined nanospaces (Razul et al., 2014). Simulations have been carried out to investigate the validity of the assumption that the liquid oil was homogeneous when the distance, d, between two CNPs, as shown in Fig. 7.2, was a few nanometers. The radial interactions between nonbonded TAG atomic moieties were modeled using GROMACS (Berendsen et al., 1995; Kutzner et al., 2007). TAGs possess only CH, CH_2, CH_3, O and C=O moieties and the force fields given by Berger et al. (1997) were used.

Although the NpT ensemble (constant number of molecules, pressure and temperature) is the "natural" ensemble to use, the problem of simulating two sufficiently large slabs immersed in a liquid oil (bulk) is excessively compute-intensive. Instead, the bulk and the fats particles were represented separately, representing the fats particles as in Fig. 7.1 of Razul et al. (2014), and required that the chemical potential of the fats particles system be equal to that of the bulk. The fats particles were taken to be tristearin nanocrystals and the oil as triolein. Because the density of a bulk triolein fluid, ~ 930 kg/m^3, is known, the NVT (constant number of molecules, volume and temperature) ensemble for both the bulk and the system of Fig. 7.2 with periodic boundary conditions in the planes perpendicular to the interfaces was used. The solids were represented as continua interacting with each other by replacing the individual molecules by average values of the 6−12 theory parameters. The coefficients, C_6^{ij} and C_{12}^{ij} ($i,j = 1, 2$), were replaced by their averages over the atomic moieties making up the solids. Fig. 7.9 shows some of

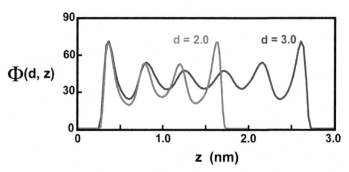

Figure 7.9 Triacylglycerol number density, $\Phi(d,z)$, as a function of position, z (in nm) between the two solid slabs (Fig. 7.2), for given slab separation, d.

the results for the TAG number density, $\Phi(d,z)$, as a function of position, z, between the two solid slabs of Fig. 7.2, for given slab separation, d. The result from this is that the number density, and therefore the density, is oscillatory and not constant as assumed by the Lifshitz and the 6−12 potential theories, as well as any theories which make use of these, such as DLVO theory.

However, does this matter? Does this result lead to a better understanding of the structural properties of a system of tristearin nanocrystals in triolein oil? Let us integrate the number density over z to obtain the area under the two curves of Fig. 7.9. If we divide the integral by the slab separation distance, d, the resulting number is the average number density between the two slabs, $\langle \Phi(d) \rangle$. We then discover that, for example, $\langle \Phi(2.0) \rangle < \langle \Phi(3.0) \rangle$. In fact, $\langle \Phi(d) \rangle$ exhibits a general decrease as $d \to 0$. What this means is that triolein oil between two tristearin slabs will begin to leak away for finite values of d until the two slabs come into molecular contact: for the system of tristearin CNPs in triolein, the oil-binding capacity is low.

Why does the 6−12 potential work for AMD? In the introduction to the section on van der Waals interactions, it was recognized that, even though the pairwise attractive potential is inadequate to describe the many-body interactions; nonetheless, it works in AMD computer simulations. The reason is likely to be that the C_6 coefficients used in the two-body attractive interactions of AMD arise from the "decoupling" of the many-body interactions of Eq. (7.3), a technique that has long been used in, for example, solid- and liquid-state theories. This amounts to breaking up many-body interactions into sums of two-body interactions weighted by average values, indicated by brackets <...>,

$$
\begin{aligned}
\tilde{U} &= \sum_j \sum_k \tilde{u}_{jk} + \sum_j \sum_k \sum_l \tilde{u}_{jkl} + \ldots \\
&\approx \sum_j \sum_k \tilde{u}_{jk} + \sum_j \sum_k \sum_l \left[<\tilde{u}_l> \tilde{u}_{jk} + <\tilde{u}_k> \tilde{u}_{jl} + <\tilde{u}_j> \tilde{u}_{kl} \right] + \ldots
\end{aligned}
\tag{7.43}
$$

In this way, three-body interactions have been approximated as a sum of three two-body interactions, which can be combined with the first term.

Coarse-grained approaches—2. The principle of Coarse Graining is that one attempts to replace the components with the fastest timescales by average values and average fluctuations. Naturally, correlations will be lost but in certain cases, e.g., when one is not near a (generally continuous) phase transition, such a replacement might be justified. The most compute-intensive component of many food systems is water and any technique which can coarse grain water should be sought. One technique is **Brownian dynamics** (Ermak and McCammon, 1978) based on the Langevin equations. One replaces the water (or any small molecule) by a continuum which contributes a stochastic force to the molecular dynamics simulation. Still others simplify the motion by replacing the continuum space by a lattice on which all objects are constrained to move from site-to-site. Such models are the **lattice gas** or the **lattice Boltzmann** approaches (e.g., Rajagopalan, 2001).

Dissipative particle dynamics (DPD) (Marsh, 1998, Flekkøy et al., 2000, Español and Revenga, 2003, Schiller, 2005 is readable general account, Pink et al., 2010 applied it to TAGs) is a mesoscale model of fluid movement on whatever scale we wish to use and is thus a procedure that involves Coarse Graining. It also satisfies the requirements of momentum conservation and Galilean invariance. If one considers a continuum limit of this model, then the equations governing that flow are in accord with the Navier–Stokes equations which are the equations which describe the flow of a continuum fluid. DPD is not a technique such as Monte Carlo but is a *general model* of a, possibly complex, fluid. It can be used for a system with a flow driven by an external force which acts directly on the components of the model, such as a rheometer, or which arises through local heating.

DPD considers a system to be composed of "chunks" of matter: a chunk of water, a cluster of saccharide groups, a branched polymer, a spherulite, a microparticle. We shall call these the components, and every component is represented by a point which locates its position. The points can be of different types if they are associated with different objects such as water, polymers, microspheres. Each component interacts with other components via three forces: (1) a Conservative force in which no energy is transferred; (2) a Random force which accounts for random energy fluctuations and thermal effects imposed on the system from outside; (3) a Dissipative force which accounts for friction, which can be molecular, meso- or microscale friction in which energy is transferred from one component to another.

Each point has associated with it a set of radii which specify the range of a particular interaction. Thus a point of type j can possess radii, $R_{j1}, R_{j2}, R_{j3}, \ldots$ for hardcore interactions, hydrogen bonding, radius of polymer gyration, van der Waals or screened electrostatic interactions. Because forces are used, then, given the masses associated with he points, one can define their accelerations and use standard molecular dynamics procedures to change their velocities and positions.

Consider two points labeled, j and k, at positions given by vectors $\vec{r}_j$ and $\vec{r}_k$ with $\vec{r}_{jk} = \vec{r}_j - \vec{r}_k$, and possessing velocities $\vec{v}_j$ and $\vec{v}_k$ with $\vec{v}_{jk} = \vec{v}_j - \vec{v}_k$. Let R_{jn} and R_{km} with $n,m = 0, 1, 2...$ represent the characteristic distances over which the forces act.

The smallest sphere of radius R_0 indicates the size of the object with its center at point j. The characteristic distances over which the forces are effective are R_{j0}, R_{j1} and R_{j2}. We assume that all objects are identical so that the characteristic distances are independent of the point at which they are centered. The size is defined by a repulsive force which can be short range and strong or weak depending on whether we model a hard object such as a glass bead, or a soft object such as a "piece" of liquid oil. If $r_{jk} > 2R_0$, the three forces for the pair of points shown in Fig. 7.10B are.

The motions of the two objects are governed by their interactions with their surroundings via three pairwise forces. These forces, acting between objects j and k, depend on the radii, R_n, are (1) a conserved force, $\vec{F}_{jk}^C(r_{jk}(t),t)$, which represents elastic scattering and involves zero energy transfer, (2) a dissipative force, $\vec{F}_{jk}^D(\vec{r}_{jk}(t), \vec{v}_{jk}(t), t)$, which results in energy transfer and represents friction ('molecular' or otherwise) and (3) a random force, $\vec{F}_{jk}^R(r_{jk}(t), t)$, which represents heat transfer with the surrounding heat bath at temperature, T. The conserved force is derivable from a potential which can be made up of different potentials representing van der Waals, electrostatic or other interactions. These forces take the form,

$$\vec{F}_{jk}^C(r_{jk}(t), t) = [f_{jk}(r_{jk}) + \omega_C\alpha(2R_0 - r_{jk}(t))]\hat{e}_{jk} \tag{7.44}$$

$$\vec{F}_{jk}^D(\vec{r}_{jk}(t), \vec{v}_{jk}(t), t) = -\gamma\omega_D[2R_0 - r_{jk}(t)]^2[\vec{r}_{jk}(t) \cdot \vec{v}_{jk}(t)]\hat{e}_{jk} \tag{7.45}$$

$$\vec{F}_{jk}^R(r_{jk}(t), t) = \sigma\omega_R[2R_0 - r_{jk}(t)]\beta\Delta t^{-1/2}\hat{e}_{jk} \tag{7.46}$$

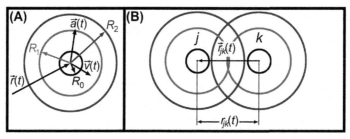

Figure 7.10 Geometry for the interaction between two points, j and k, in dissipative particle dynamics. The characteristic distances over which the forces act are R_{j0}, R_{j1}, and R_{j2}. $\hat{e}_{jk}$ is a unit vector along $\vec{r}_{jk}$, which points from k to j. R_{j0} represents the size of the object located at $\vec{r}_j(t)$.

where $\widehat{e}_{jk}$ is a unit vector pointing from k to j, and where we have assumed that the "size" of the objects, their radius, is R_0. The elapsed time, Δt, is the time between successive molecular dynamics steps. Its appearance in Eq. (7.46) comes about because if one integrates the random force from 0 to t, then one should obtain the mean diffusion, $\sigma^2 t$, but, in fact, one obtains $\sigma^2 t \Delta t$. Because this quantity cannot depend on the step size, then we introduce a factor, $\Delta t^{-1/2}$ in to Eq. (7.46). The quantity β is a randomly fluctuating variable with Gaussian statistics (appropriate for a random force due to fluctuations driven by a surrounding heat bath at temperature T) and average equal to zero. The parameters α, γ and σ are magnitudes of the three forces and ω_C, ω_D and ω_R are weighting functions which can specify the ranges of these forces, as used in the original applications of DPD. The force $f_{jk}(r_{jk})$ represents other conserved forces. It includes both the van der Waals attractive force and a phenomenological short-range repulsive force to describe solids which cannot penetrate each other. Solids are modeled as spherical particles, or as aggregates of them, because we know the analytic form of the attractive sphere—sphere (van der Waals) force (Parsegian, 2005). We shall not be modeling electrostatic or hydrogen bonding forces here because the only moieties which give rise to them are the 3 C=O groups in the glycerol core and we have no a priori information about their distribution in fats sphere. Accordingly, the total force is given by

$$f(r) = \begin{cases} \dfrac{-32AR_0^6}{3r^3\left(r^2 - 4R_0^2\right)^2} & r > 2R_0 + \delta \\[4mm] \dfrac{B}{(r - 2R_0)^7} & r \leq 2R_0 + \delta \end{cases} \tag{7.47}$$

where A is proportional to the Hamaker coefficient, B is a phenomenological repulsive force strength and δ is a small distance separating two spheres. If we require that the sum of the two potentials, which give rise to these two forces, exhibits a minimum when $r = 2R_0 + \delta$ then we obtain

$$B = \frac{AR_0\delta^5}{12}. \tag{7.48}$$

The term in α is a "soft" repulsive short-range conserved force and its purpose is to bring about particle—particle repulsion. However, we have already taken this into account for solid—solid interactions by the short-range repulsive force of Eq. (7.47). The dissipative and random forces are related via the Fluctuation-Dissipation theorem (Kubo, 1966; Español and Warren, 1995; Marconi et al., 2008) which yields

$$\sigma^2 = 2k_BT\gamma \tag{7.49}$$

$$\omega_R^2 = \omega_D$$

where k_B is Boltzmann constant. These forces serve as a thermostat, keeping the temperature constant.

Accelerations are computed from the total force, $\overrightarrow{F}$, acting on each particle, $\overrightarrow{a} = \overrightarrow{F}/m$, where m is the mass of the particle. Molecular Dynamics (Frenkel and Smit, 2002; Hess et al., 2008) is then used to obtain average quantities. The equations of motion are then integrated, using a well-behaved integrating routine such as the velocity Verlet algorithm, as modified by Groot and Warren (1997).

DPD has been used to model colloids (Bolintineanu et al., 2014). One advantage of DPD compared with, for example, computational fluid dynamics (CFD) is that, in the latter, boundary conditions have to be specified in order that the solutions to the Navier–Stokes equation represents the system of interest. The boundary conditions specify the fluid flow in that neighborhood. Difficulties can arise if the fluid flow itself creates the boundaries which, in turn specify the fluid flow. An iterative process would have to take place at every molecular dynamics time step. In the case of DPD, the forces are all local and it is they that give rise to the behavior of the fluid at boundaries. Fig. 7.11 shows a case in which a fluid (light gray spheres) is carrying (dark gray) massive spheres form left to right. The latter are falling under gravity and the interfaces between dark sphere clusters and the flowing fluid is changing with time (MacDonald and Pink, unpublished). This system is easy to treat using DPD but might give substantial difficulties if CFD was used.

Potential scattering and static structure functions. The differential cross section is the quantity measured in elastic scattering experiments, such as X-ray or neutron scattering. It is written as $d\sigma/d\Omega$ and is a measure of the number of incident particles scattered into an infinitesimal solid angle, $d\Omega$, by a set of target particles, where it is assumed that only single scattering events take place. Elementary elastic potential scattering theory shows that

$$\frac{d\sigma\left(\overrightarrow{k'}, \overrightarrow{k}\right)}{d\Omega} = \left|f\left(\overrightarrow{k'}, \overrightarrow{k}\right)\right|^2, \tag{7.50}$$

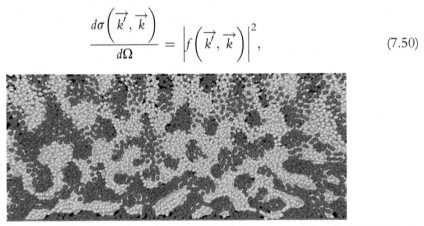

Figure 7.11 A slice through an instantaneous snapshot of a two-component flow. A fluid (*light gray spheres*) is flowing from left to right and carrying with it massive particles (*dark spheres*), which are falling to the bottom under the force of gravity. The boundaries between the flowing fluid and the massive particles are changing, as the flow distorts the shapes of the clusters of massive particles.

where the incident and scattered wavevectors are $\vec{k}$ and $\vec{k'}$, respectively, with $E = \hbar^2 k^2/2m = \hbar^2 k'^2/2m$ and m is the reduced mass of the two-body scattering system. An equivalent definition can be made if the scatterers are photons possessing mass equal to zero. The energy is then related to that of the photon in the usual way. Thus, the X-ray and neutron scattering intensities are equal to $\left| f\left(\vec{k'}, \vec{k}\right) \right|^2$. The quantity, $f\left(\vec{k'}, \vec{k}\right)$, is the scattering amplitude.

If $V(\vec{r})$ is the interaction potential between an incident particle being scattered and the target particle doing the scattering then, in the first Born approximation (e.g., Sakurai and Tuan, 1994; Messiah, 1999) valid when the scattering is due only to single-collision events, and multiple scattering events can be ignored,

$$f\left(\vec{k'}, \vec{k}\right) \sim \iiint d^3\vec{r'}\, V\left(\vec{r'}\right) \exp\left(i\left(\vec{k} - \vec{k'}\right) \cdot \vec{r'}\right)$$
$$\sim \iiint d^3\vec{r'}\, V\left(\vec{r'}\right) \exp\left(i\vec{q} \cdot \vec{r'}\right), \tag{7.51}$$

where the wavevector $\vec{q} \equiv \vec{k} - \vec{k'}$ is proportional to the momentum transfer. In obtaining this equation, it has been assumed that the scattering potential has a finite range, i.e., $V(\vec{r}) = 0$ for $r > a$. This equation shows that (1) $f\left(\vec{k'}, \vec{k}\right)$ depends linearly on the scattering potential and (2) the scattering amplitude is the Fourier transform of the scattering potential. Because $V(\vec{r})$ will yield, for example, the structure of a crystal, then a knowledge of the scattering amplitude is of fundamental importance.

For X-ray scattering, due to the Coulomb interaction between a photon and the electrons in the target scatterer, $V(\vec{r})$ is proportional to the density of scatterers, i.e., the atomic number (Cullity and Stock, 2001). For neutron scattering, which is due to the interaction between an incident neutron and the particles in the nucleus of a target scatterer, $V(\vec{r})$ is replaced by an effective scattering potential, the scattering length (Lopez-Rubio and Gilbert, 2009). There is no simple relationship between the scattering length and the number of protons and neutrons in the nucleus. One can thus write

$$f\left(\vec{k'}, \vec{k}\right) \sim \iiint d^3\vec{r'}\, n_e\left(\vec{r'}\right) \exp\left(i\vec{q} \cdot \vec{r'}\right) \quad (\text{X} - \text{ray scattering}) \tag{7.52}$$

$$f\left(\vec{k'}, \vec{k}\right) \sim \iiint d^3\vec{r'}\, s\left(\vec{r'}\right) \exp\left(i\vec{q} \cdot \vec{r'}\right) \quad (\text{neutron scattering}), \tag{7.53}$$

where $n_e(\vec{r})$ and $s(\vec{r})$ are the atomic number and (neutron) scattering length of a scatterer at position $\vec{r}$.

It can be seen that there are two aspects to these equations: (1) the type of scatterers defined by $n_e(\vec{r})$ and $s(\vec{r})$ and which reflects the magnitude of the scattering intensity and (2) the spatial structure of the total scattering object and given by the "interference function" $\exp\left(i\vec{q}\cdot\vec{r}\right)$. Accordingly, if one wants to investigate only the total structure of the objects giving rise to the scattering, one can replace the scattering potential—or, equivalently, the atomic number or scattering length—by unity. This yields what can be called the "structure amplitude," $f_1(\vec{q})$,

$$f_1(\vec{q}) = \iiint d^3\vec{r} \exp\left(i\vec{q}\cdot\vec{r}\right) \tag{7.54}$$

for a continuous distribution of scatterers, or

$$f_1(\vec{q}) = \sum_j \exp i(\vec{q}\cdot\vec{r}_j) \tag{7.55}$$

when the scattering centers can be taken as a finite set of points.

In the second case, the static structure function is defined as

$$S(\vec{q}) = \frac{1}{N}|f_1(\vec{q})|^2 = \frac{1}{N}\left|\sum_j \exp i(\vec{q}\cdot\vec{r}_j)\right|^2$$

$$= \frac{1}{N}\sum_j \exp i(\vec{q}\cdot\vec{r}_j)\sum_k \exp i(-\vec{q}\cdot\vec{r}_k), \tag{7.56}$$

where each sum is over the N scattering centers in the system. If the scattering centers are, for example, the atomic moieties of a polymer as they are in the case of triglycerides, then $\vec{r}_j$ and $\vec{r}_k$ are their positions. Because this calculation is being done using computation methods, it is of great benefit to simplify this N^2 calculation. We start by combining the sums

$$S(\vec{q}) = \frac{1}{N}\sum_j\sum_k \exp i(\vec{q}\cdot(\vec{r}_j - \vec{r}_k)). \tag{7.57}$$

Using Euler's formula, $e^{ix} = \cos(x) + i\sin(x)$, we obtain

$$S(\vec{q}) = \frac{1}{N}\sum_j\sum_k [\cos(\vec{q}\cdot(\vec{r}_j - \vec{r}_k)) + i\sin(\vec{q}\cdot(\vec{r}_j - \vec{r}_k))]. \tag{7.58}$$

Because the sums over j and k are over the same set of values, then for each r_{jk} there is $r_{kj} = -r_{jk}$ so that the structure function reduces to

$$S(\vec{q}) = \frac{1}{N}\sum_j\sum_k [\cos(\vec{q}\cdot(\vec{r}_j - \vec{r}_k))], \qquad (7.59)$$

which we can simplify further to

$$S(\vec{q}) = \left[1 + \frac{2}{N}\sum_j\sum_{k>j}\cos(\vec{q}\cdot(\vec{r}_j - \vec{r}_k))\right]. \qquad (7.60)$$

Finally, we note that we might have a sample in which there are solid domains with different orientations. Rather than averaging over all domains, we can average over all $\vec{q}$-vectors possessing the same magnitude, $|\vec{q}| = q$. We then get

$$S(q) = \langle S(\vec{q})\rangle_{|\vec{q}|=q}. \qquad (7.61)$$

This $S(q)$ is the form of the structure function that we use here.

Finally, we note that if the structure that scatters the photons or neutrons is a fractal with fractal dimension D, then, for some range of q, we should observe that $S(q) \propto q^{-D}$ (Jullien, 1992; Beaucage, 2004), or, in a more useful form,

$$log[S(q)] \propto -Dlog[q]. \qquad (7.62)$$

Hence, if we plot $log[S(q)]$ versus $log[q]$, we would obtain a straight line with slope $-D$.

Caution should be exercised, however, in deducing, from Eq. (7.62), the fractal dimension of clusters formed via DLCA. Computer simulations performed by Lachhab et al. (1998) have shown that for finite concentrations of aggregating particles, the value obtained using Eq. (7.62) is systematically larger than the true value. A more reliable value is obtained from the particle—particle correlation function.

$S(q)$ is proportional to q^{-D}. Let us consider the structure—function for scattering from a structure possessing fractal dimension D. We must integrate over all objects in the fractal. Let us assume that the density of the structure is a constant, ρ. We now represent the fractal structure as averaged over all orientations so that the integral is an isotropic three-dimensional integral over a spherically symmetric fractal. Accordingly, the result does not involve a q-vector, $\vec{q}$, but only its magnitude, q. The integral over the fractal to yield $S(q)$ is

$$S(q) = \rho\int_0^R r^{D-1}\,dr\int_0^\pi \sin(\vartheta)\,d\vartheta\int_0^{2\pi} d\phi e^{iqr\cos(\vartheta)}, \qquad (7.63)$$

where we have chosen the local z-axis to be along $\overrightarrow{q}$. The integral over ϕ gives 2π. The integral over ϑ is

$$\int_0^\pi \sin(\vartheta)d\vartheta e^{iqr \cos(\vartheta)} = \frac{e^{iqr} - e^{-iqr}}{iqr} \tag{7.64}$$

so that

$$S(q) = \frac{2\pi\rho}{iq} \int_0^R r^{D-2}dr\left(e^{iqr} - e^{-iqr}\right) = \frac{2\pi\rho}{iq} \int_0^R \frac{(qr)^{D-2}d(qr)}{(q^{D-2})(q)}\left(e^{iqr} - e^{-iqr}\right)$$

$$= \frac{2\pi\rho}{iq^D} \int_0^{qR} x^{D-2}dx\left(e^{ix} - e^{-ix}\right) = \frac{4\pi\rho}{q^D} \int_0^{qR} x^{D-2}dx \sin(x) = 4\pi\rho f(qR, D)q^{-D}, \tag{7.65}$$

where $f(qR,D)$ is the integral, which depends on qR and D but not on q.

Example 3. CNP aggregation, TAGwoods, and structure–functions (Marangoni 2002). We modeled CNP multilayer sandwiches (Fig. 7.8)—which we called "TAGwoods"—by representing each TAGwood as a rigid linear sequence of spheres (Fig. 7.7B). TAGwoods were characterized by the radius of the spheres comprising them and the number of spheres per TAGwood, ℓ. Because they are composed of CNPs, they interacted via the sphere–sphere interaction of Eq. (7.42). Monte Carlo simulation was used to obtain the steady-state configurations after sufficiently long times and the structure–function Eqs. (7.61) and (7.62) were used to identify aggregation. Plots of $log[S(q)]$ versus $log[q]$ are shown in Fig. 7.12 (Pink et al., 2013), as

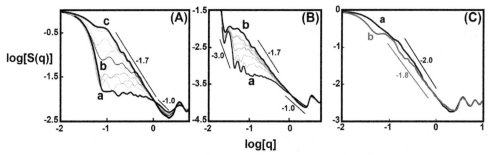

Figure 7.12 Initial TAGwood aggregation process shown by $log[S(q)]$ versus $log[q]$. (A) $\phi = 0.0005$ (SFC = 0.05%) for $\ell = 2$ in $(100)^3$ space averaged more than 27 replica simulations. Initial distribution (a), intermediate distribution (b, 600 MC steps), all 80 TAGwoods in one cluster (c, 6.6×10^4 MC steps). (B) $\phi = 0.001$ (SFC = 0.1%) of $\ell = 5$ in a $(300)^3$ space from one simulation. Initial distribution (a), all 1728 TAGwoods in one cluster (b, 1.116×10^6 MC steps). Structure functions for the intermediate cases are shown as thin solid lines. (C) Longer-time TAGwood aggregation shown by $log[S(q)]$ versus $log[q]$ when the attractive interaction energy is made weaker. Simulations were carried out in a $(100)^3$ space for TAGwood concentrations of $\phi = 0.001$ (SFC = 0.1%). **a:** $\ell = 2$. **b:** $\ell = 5$.

functions of Monte Carlo step (Fig. 7.12A and B) to see the approach to a final state. In those two figures, we can see the initial states (1) exhibiting slopes of -1 showing that we have a random distribution of one-dimensional TAGwoods. As the simulation proceeds, the TAGwoods aggregate via DLCA with the final relaxation (Fig. 7.12C, curve a) occurring via reaction-limited cluster–cluster aggregation (Binder and Heermann, 2010; Vicsek, 1999).

Example 4. How can we tell what is the spatial range of fractal behavior (Beaucage, 2004)? As a first rule of thumb, structures will scatter electromagnetic radiation with wavelength, $\lambda = 2\pi/q$, when their characteristic length, ℓ, is related to q via $\ell = 2\pi/q$. Accordingly, we can obtain an estimate of the range of sizes that a fractal encompasses by measuring the range of q-magnitudes over which the slope of $log[S(q)]$ versus $log[q]$ is a constant. Fig. 7.13 shows the plot of $log[S(q)]$ versus $log[q]$ for the case of a single spherical volume of radius, $R = 1.0$, which encloses hard cylinders arranged radially. Because they cannot overlap, we are restricted into how many can be squeezed into that space and how long they are. We chose the cylinder radius to be 0.05. We restricted the maximum cylinder length to be 0.5, and the results are shown in the figure. A stylized diagram of the sphere and the approximately randomly chosen, radially arranged cylinders is shown in the upper right side. Because the cylinders are one-dimensional, $log[S(q)]$ exhibits a straight-line segment for which the slope is -1. The range over which this entails will give some idea of the cylinder lengths. The straight-line segment possesses an approximate range, $1 < log[q] < 2$, which yields approximate lengths, $0.06 < \ell = 2\pi/q < 0.63$. This result says that the range of lengths of the cylinders lies between these two limits, and this is approximately correct.

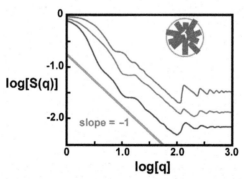

Figure 7.13 Plot of $log[S(q)]$ versus $log[q]$ for the case of a single spherical volume of radius $R = 1.0$ (shown in the upper right corner), which encloses hard cylinders arranged radially approximately randomly as shown. The three structure functions are for different numbers of cylinders inside the sphere. The slopes of the straight-line segments are a good fit to the line of slope -1.

ACKNOWLEDGMENTS

It is a pleasure to thank Erzsebet Papp-Szabo, Fernanda Peyronel, Bonnie Quinn, Dref De Moura, Charles Hanna, and Shah Razul for their comments and assistance. Thanks also to AFMnet, ACENET, and NSERC for support.

REFERENCES

Acevedo, N.C., Marangoni, A.G., 2010a. Characterization of the nanoscale in triacylglycerol crystal networks. Crystal Growth Des. 10, 3327—3333.

Acevedo, N.C., Marangoni, A.G., 2010b. Towards nanoscale engineering of triacylglycerol crystal networks. Crystal Growth Des. 10, 3334—3339.

Acevedo, N.C., Peyronel, F., Marangoni, A.G., 2011. Nanoscale structure intercrystalline interactions in fat crystal networks. Curr. Opin. Colloid Interface Sci. 16, 374—383.

Ahmadi, L., Wright, A.J., Marangoni, A.G., 2008. Chemical and enzymatic interesterification of tristearin/triolein-rich blends: chemical composition, solid fat content and thermal properties. Eur. J. Lipid Sci. Technol. 110, 1014—1024.

Antunes, F.E., Marques, E.F., Miguel, M.G., Lindman, B., 2009. Polymer-vesicle association. Advances Colloid Interface Sci. 147—148, 18—35.

Axilrod, B.M., Teller, E., 1943. Interaction of the van der Waals type between three atoms. J. Chem. Phys. 11, 299—300.

Babick, F., Schiesl, K., Stintz, M., 2011. Van der Waals interaction between two fractal aggregates. Adv. Powder Technol 22, 220—225.

Ballenegger, V., Cerdà, J.J., Holm, C., 2012. How to convert SPME to P3M: influence functions and error estimates. J. Chem. Theory Comput. 8, 936—947. See also Cerdà, J. J. Long range interactions: Direct sum and Ewald summation. See. https://www.icp.uni-stuttgart.de/~icp/mediawiki/images/e/eb/T8w.pdf.

Beaucage, G., 2004. Determination of branch fraction and minimum dimension of mass-fractal aggregates. Phys. Rev. E 70, 031401—031410.

Bennnun, S.V., Hoopes, M.I., Xing, C., Faller, R., 2009. Coarse-grained modelling of lipids. Chem. Phys. Lipids 159, 59—66.

Berendsen, H.J.C., van der Spoel, D., Vandrunen, R., 1995. GROMACS: a message-passing parallel molecular dynamics implementation. Comput. Phys. Commun. 91, 43—56.

Berger, J., Edholm, O., Jähnig, F., 1997. Molecular dynamics simulations of a fluid bilayer of dipalmitoyl-phosphatidylcholine at full hydration, constant pressure, and constant temperature. Biophys. J. 72, 2002—2013.

Binder, K., 1997. Applications of Monte Carlo methods to statistical physics. Rep. Prog. Phys. 60, 487—559.

Binder, K., Heermann, D., 2010. Monte Carlo Simulation in Statistical Physics: An Introduction. Springer, Berlin.

Bolintineanu, D.S., Grest, G.S., Lechman, J.B., Pierce, F., Plimpton, S.J., Schunk, P.R., 2014. Particle dynamics modeling methods for colloid suspensions. Comp. Part. Mech. 1, 321.

Butt, H.-J., Graf, K., Kappl, M., 2003. Physics and Chemistry of Interfaces. WILEY-VCH Verlag GmbH & Co. KGaA, Weinheim.

Car, R., Parrinello, M., 1985. Unified approach for molecular dynamics and density-functional theory. Phys. Rev. Lett. 55, 2471—2474. See also. http://www.cpmd.org/.

Cevc, G., 1990. Membrane electrostatics. Biochim. Biophys. Acta 1031, 311—382.

Cevc, G., Marsh, D., 1987. In: Phospholipid Bilayers: Physical Principles and Models. John Wiley & Sons, New York.

Cullity, B.D., Stock, S.R., 2001. Elements of X-Ray Diffraction, third ed. Prentice Hall.

Darden, T., York, D., Pedersen, L., 1993. Particle Mesh Ewald. An Nlog(N) method for Ewald sums in large systems. J. Chem. Phys. 98, 10089—10092.

Derjaguin, B., Landau, L., 1941. Theory of the stability of strongly charged lyophobic sols and of the adhesion of strongly charged particles in solutions of electrolytes. Acta Physico Chemica URSS 14, 633.

Dhont, J.K.G., Briels, W.J., 2003. Viscoelasticity of suspensions of long rigid rods. Colloids Surf. A: Physicochem. Eng. Aspects 213, 131–156.

Dzyaloshinskii, I.E., Lifshitz, E.M., Pitaevskii, L.P., 1961. General theory of van der Waals' forces. Sov. Phys. Uspekhi 73, 153–176.

Ermak, D.L., McCammon, J.A., 1978. Brownian dynamics with hydrodynamic interactions. J. Chem. Phys. 69, 1352–1361.

Español, P., Revenga, M., 2003. Smoothed dissipative particle dynamics. Phys. Rev. E 67, 026705–026712 (and references therein).

Español, P., Warren, P.B., 1995. Statistical mechanics of dissipative particle dynamics. Europhys. Lett. 30, 191.

Essmann, U., Perera, L., Berkowitz, M.L., Draden, T., Lee, H., Pedersen, L.G., 1995. A smooth particle Mesh Ewald method. J. Chem. Phys. 103, 8577–8593.

Fathi, M., Mozafari, M.R., Mohebbi, M., 2012. Nanoencapsulation of food ingredients using lipid-based delivery systems. Trends Food Sci. Technol. 23, 13–27.

Feig, M., 2010. Molecular simulation methods: standard practices and modern challenges. In: Computational Modelling in Lignocellulosic Biofuel Production, ACS Symposium Series, 1052, pp. 155–178 (Chapter 8).

Flekkøy, E.G., Coveney, P.V., De Fabritiis, G., 2000. Foundations of dissipative particle dynamics. Phys. Rev. E 62, 2140–2157.

Frankl, D., 1986. Electromagnetic Theory. Prentice-Hall, Englewood Cliffs, NJ.

French, R.H., et al., 2010. Long range interactions in nanoscale science. Rev. Mod. Phys. 82, 1887–1944.

Frenkel, D., Smit, B., 2002. Understanding Molecular Simulation: From Algorithms to Applications, second ed. Elsevier Inc., pp. 465–478 (Chapter 17).

Glotzer, S.C., Horsch, M.A., Iacovella, C.R., Zhang, Z., Chan, E.R., Zhang, X., 2005. Self-assembly of anisotropic tethered nanoparticle shape amphiphiles. Curr. Opin. Colloid Interface Sci. 10, 287–295.

Groot, R.D., Warren, P.B., 1997. Dissipative particle dynamics: bridging the gap between atomistic and mesoscopic simulation. J. Chem. Phys. 107, 4423.

Hall, A., Repakova, J., Vattulainen, I., 2008. Modelling of the triglyceride-rich core in lipoprotein particles. J. Phys. Chem. B 112, 13772–13782.

Hamaker, H.C., 1937. The London-van der Waals attraction between spherical particles. Physica 4, 1058–1072.

Hanna, C.B., Pink, D.A., Quinn, B.E., 2006. Van der Waals interactions with soft interfaces. J. Phys. Condens. Matter 18, 8129–8137.

Hess, B., Kutzner, C., van der Spoel, D., Lindahl, E., 2008. GROMACS 4: algorithms for highly-efficient, load-balanced, and scalable molecular simulation. J. Chem. Theory Comput. 4, 435–447. See also. www.gromacs.org. GROMACS users manual version 4.0, gromacs4_manual.pdf.

Hiemenz, P.C., Rajagoplalan, R., 1997. Principles of Colloid and Surface Chemistry, third ed. Marcel Dekker, New York.

Himawan, C., Starov, V.M., Stapley, A.G.F., 2006. Thermodynamic and kinetic aspects of fat crystallization. Adv. Colloid Interface Sci. 122, 3–33.

Hockney, R.W., Eastwood, J.W., 1988. Particle-particle-particle-mesh (P3M) Algorithms. Computer Simulation Using Particles. CRC Press, pp. 267–304.

Hough, D.B., White, L.R., 1980. The calculation of Hamaker constants from Lifshitz theory with applications to wetting phenomena. Adv. Colloid Interface Sci. 14, 3–41.

Hsu, W.D., Violi, A., 2009. Order disorder phase transformation of triacylglycerols: effect of the structure of the aliphatic chains. J. Phys. Chem. B 113, 887–893.

http://davapc1.bioch.dundee.ac.uk/prodrg/The GlycoBioChem PRODRG2 Server.

http://moose.bio.ucalgary.ca/index.php?page=Structures_and_Topologies.

Huang, W.-T., Levitt, D.G., 1977. Theoretical calculation of the dielectric constant of a bilayer membrane. Biophys. J. 17, 111–128.

Hunter, R.J., 2001. Foundations of Colloid Science. Oxford University Press, Oxford.

Israelachvili, J.N., 2006. Intermolecular and Surface Forces. Academic Press, New York.

Johansson, D., Bergenståhl, B., 1992. The influence of food emulsifiers on fat and sugar dispersions in oils: 2. Rheology, colloidal forces. J. Am. Oil Chem. Soc. 69, 718−727.

Jullien, R., 1992. From guinier to fractals. J. Phys. I France 2, 759−770.

Khandelia, H., Mouritsen, O.G., 2009. Lipid gymnastics: evidence of complete acyl chain reversal in oxidized phospholipids from molecular simulations. Biophys. J. 96, 2734−2743 (and references therein).

Kim, H.-Y., Sofo, J.O., Velegol, D., Cole, M.W., Lucas, A.A., 2007. Van der Waals dispersion fotces between dielectric nanoclusters. Langmuir 23, 1735−1740.

Kontogiorgos, V., Vaikousi, H., Lazaridou, A., Biliaderis, C.G., 2006. A fractal analysis approach to visco-elasticity of physically cross-linked barley b-glucan Gel networks. Colloids Surf. B: Biointerfaces 49, 145−152.

Kubo, R., 1966. The fluctuation-dissipation theorem. Rep. Prog. Phys. 29, 255.

Kutzner, C., van der Spoel, D., Fechner, M., Lindahl, E., Schmitt, U.W., de Groot, B.L., Grubmuller, H.J., 2007. Speeding up parallel GROMACS on high-latency networks. Comput. Chem. 28, 2075−2084.

Lach-hab, M., Gonzalez, A.E., Blaisten-Barojas, E., 1998. Structure function and fractal dimension of diffusion-limited colloidal aggregates. Phys. Rev. E 57, 4520−4527.

Landau, L., Lifshitz, E., 1968. Electrodynamics. Nauka, Moscow.

Laradji, M., Kumar, P.B.S., 2011. Coarse-grained computer simulations of multicomponent lipid membranes. Adv. Planar Lipid Bilayers Liposomes 14, 201−233 (Although this reference is concerned with lipid bilayers, it does contain descriptions of various simulation techniques).

Larsson, K., 2009. Lyotropic liquid crystals and their dispersions relevant in foods. Current Opin. Colloid Interface Sci. 14, 16−20.

Leckband, D., Israelachvili, J.N., 2001. Intermolecular forces in biology. Q. Rev. Biophys. 34, 105−267 (and references therein).

Lee, H., Cai, W., 2009. Ewald Summation for Coulomb Interactions in a Periodic Supercell. http://micro.stanford.edu/mediawiki/images/4/46/Ewald notes.pdf.

Lee-Desautels, R., 2005. Theory of van der Waals forces as applied to particulate materials. Educ. Reso. For Part. Techn. Available at http://www.erpt.org/051Q/leed-01.pdf.

Leser, M.E., Sagalowicz, J., Michel, M., Watzke, H.J., 2006. Self-assembly of polar food lipids. Advances Colloid Interface Sci. 123−126, 125−136.

Li, J., Forest, M.G., Wang, Q., Zhou, R., 2011. A kinetic theory and benchmark predictions for polymer-dispersed semi-flexible macromolecular rods or platelets. Physica D 240, 114−130.

Li, Q., Rudolph, V., Peukert, W., 2006. London-van der Waals adhesiveness of rough particles. Powder Technol. 161, 248−255.

Lifshitz, E.M., 1956. The theory of molecular attractive forces between solids. Sov. Phys. JETP 2, 73−83.

Lopez-Rubio, A., Gilbert, E.P., 2009. Neutron sattering: a natural tool for food science and technology research. Trends Food Sci. Technol. 20, 576−586.

Lorrain, P., Corson, D.R., Lorrain, F., 1988. Electromagnetic Fields and Waves. W.H.Freeman & Co, New York.

MacDonald, A.J., Pink, D.A. unpublished.

Marangoni, A.G., 2002. The nature of fractality in fat crystal networks. Trends Food Sci. Technol. 13, 37−47.

Marangoni, A.G., 2000. Elasticity of high-volume fraction fractal aggregate networks: a thermodynamic approach. Phys. Rev. B 62, 13951−13955.

Marangoni, A.G., Rogers, M., 2003. Structural basis for the yield stress in plastic disperse systems. Appl. Phys. Lett. 82, 3239−3241.

Marangoni, A.G., Rousseau, D., 1999. Plastic fat rheology is governed by the fractal nature of the fat crystal network and by crystal habit. In: Widlak, N. (Ed.), Physical Properties of Fats, Oils and Emulsifiers. AOCS Press, Champaign, IL, pp. 96−111.

Marconi, U.M.B., Puglisi, A., Rondoni, L., Vulpiani, A., 2008. Fluctuation-dissipation: response theory in statistical physics. Phys. Reports 461, 111−195.

Marrink, S.J., Risselada, H.J., Yefimov, S., Tieleman, D.P., de Vries, A.H., 2007. The MARTINI force field: coarse-grained model for biomolecular simulations. J. Phys. Chem. B 111, 7812−7824.

Marsh, C., 1998. Theoretical Aspects of Dissipative Particle Dynamics (D. Phil. thesis). University of Oxford and references therein.

McClements, D.J., Li, Y., 2010. Structured emulsion-based delivery systems: controlling the digestion and release of lipophilic food components. Adv. Colloid Interface Sci. 159, 213–228.

McLaughlin, S., 1989. The electrostatic properties of membranes. Ann. Rev. Biophys. Chem. 18, 113–136.

Messiah, A., 1999. Quantum Mechanics. Courier Dover Publications.

Monticelli, L., Kandasamy, S.K., Periole, X., Larson, R.G., Tieleman, D.P., Marrink, S.-J., 2008. The MARTINI coarse-grained force field: extension to proteins. J. Chem. Theory Comput. 4, 819–834.

Müller, M., Katsov, K., Schick, M., 2006. Biological and synthetic membranes: what can be learned from a coarse-grained description. Phys. Reports 434, 113–176.

Narine, S.S., Marangoni, A.G., 1999a. Fractal nature of fat crystal networks. Phys. Rev. E 59, 1908–1920.

Narine, S.S., Marangoni, A.G., 1999b. Mechanical and structural model of fractal networks of fat crystals at low deformations. Phys. Rev. E 60, 6991–7000.

Oliveira, R.G., Schneck, E., Quinn, B.E., Konovalov, O.V., Brandenburg, K., Gutsmann, T., Gill, T.A., Hanna, C.B., Pink, D.A., Tanaka, M., 2010. Crucial roles of calcium and charged saccharide moieties in survival of gram-negative bacteria revealed by combination of grazing-incidence x-ray structural characterizations and Monte Carlo simulations. Phys. Rev. E 81, 41901–41913.

Oliveira, R.G., Schneck, E., Quinn, B.E., Konovalov, O.V., Brandenburg, K., Seydel, U., Gill, T., Hanna, C.B., Pink, D.A., Tanaka, M., 2009. Physical mechanisms of bacterial survival revealed by combined grazing-incidence x-ray scattering and Monte Carlo simulation. C. R. Chimie 12, 209–217.

Papadopoulos, K.D., Cheh, H.Y., 1984. Theory on colloidal double-layer interactions. AIChE J. 30, 7–14.

Parsegian, V.A., 2005. Van der Waals Forces: A Handbook for Biologists, Chemists, Engineers and Physicists. Cambridge University Press, New York.

Parsegian, V.A., Ninham, B.W., 1970. Temperature-Dependent van der Waals forces. Biophys. J. 10, 664–674 (and references 1-3 therein).

Patel, A.R., Velikov, K.P., 2011. Colloidal delivery systems in foods: a general comparison with oral drug delivery. LWT Food Sci. Technol. 44, 1958–1964.

Petersen, H.G., 1995. Accuracy and efficiency of the particle Mesh Ewald method. J. Chem. Phys. 103, 3668–3676.

Pink, D.A., Hanna, C.B., Quinn, B.E., Levadny, V., Ryan, G.L., Filion, L., Paulson, A.T., 2006. Modelling electrostatic interactions in complex soft systems. Food Res. Int. 39, 1031–1045.

Pink, D.A., Hanna, C.B., Sandt, C., MacDonald, A.J., MacEachern, R., Corkery, R., Rousseau, D., 2010. Modelling the solid-liquid phase transition in saturated triglycerides. J. Chem. Phys. 132, 54502–54513.

Pink, D.A., Quinn, B., Peyronel, F., Marangoni, A.G., 2013. Edible oil structures at low and intermediate concentrations: I. Modeling, computer simulation and predictions for x-ray scattering. J. Applied Phys. 114, 234901.

Rajagopalan, R., 2001. Simulations of self-assembling systems. Current Opin. Colloid Int. Sci. 6, 357–365.

Razul, M.S.G., Kusalik, P.G., 2011. Crystal growth investigations of ice/water interfaces from molecular dynamics simulations: profile functions and average properties. J. Chem. Phys. 134, 014710–014713.

Razul, M.S.G., MacDougall, C.J., Hanna, C.B., Marangoni, A.G., Peyronel, F., Papp-Szabo, E., Pink, D.A., 2014. Nanoscale characteristics of molecular fluids in confined spaces: triacylglycerol oils. Roy. Soc. Chem. Food & Function 5, 2501–2508.

Rogers, M.A., Tang, D., Ahmadi, L., Marangoni, A.G., 2008. Fat crystal networks. In: Aguilera, J.M., Lillford, P.J. (Eds.), Food Materials Science. Principles and Practice. Springer Science + Business Media LLC 396–401.

Rye, G.G., Litwinenko, J.W., Marangoni, A., 2005. Fat crystal networks. In: Shahidi, F. (Ed.), Bailey's Industrial Oil and Fat Products, Edible Oil and Fat Products: Products and Applications, sixth ed., vol. 4. John Wiley & Sons, Inc. (Chapter 4).

Sadus, R.J., 2002. Molecular Simulation of Fluids. Theory, Algorithms and Object-orientation. Elsevier Science, Amsterdam.

Sagalowicz, L., Leser, M.E., 2010. Delivery systems for liquid food products. Current Opin. Colloid Interface Sci. 15, 61–72.

Sakurai, J.J., Tuan, S.F., 1994. Modern Quantum Mechanics. Addison-Wesley.

Sato, K., 2001. Crystallization behaviour of fats and lipids — a review. Chem. Eng. Sci. 56, 2255–2265.

Sato, K., Ueno, S., 2011. Crystallization, transformation and microstructures of polymorphic fats in colloidal dispersion states. Current Opin. Colloid Interface Sci. 16, 384–390.

Schiller, U.D., 2005. Dissipative Particle Dynamics: A Study of the Methodological Background (Diploma Thesis). Condensed Matter Theory Group, Faculty of Physics, University of Bielefeld and references therein.

Schneck, E., Pappne-Szabo, E., Quinn, B.E., Konovalov, O.V., Beveridge, T.J., Pink, D.A., Tanaka, M., 2009. Calcium ions induce collapse of charged O-sidechains of lipopolysaccharides from *Pseudomonas aeruginosa*. J. Roy. Soc. Interface 6, S671–S678.

Schneck, E., Schubert, T., Konovalov, O.V., Quinn, B.E., Gutsmann, T., Brandenburg, K., Pink, D.A., Tanaka, M., 2010. Quantitative determination of ion distributions in bacterial lipopolysaccharide membranes by grazing-incidence x-ray fluorescence. Proc. Natl. Acad. Sci. U.S.A. 107, 9147–9151.

Sholl, D.S., Steckel, J.A., 2009. Density Functional Theory: A Practical Introduction. J. Wiley.

Sum, A.K., Biddy, M.J., dePablo, J.J., Tupy, M.J., 2003. Predictive molecular model for the thermodynamic and transport properties of triacylglycerols. J. Phys. Chem. B 107, 14443–14451.

Van den Tempel, M., 1961. Mechanical properties of plastic disperse systems at very small deformations. J. Colloid Sci. 16, 284–296.

Vatamanu, J., Kusalik, P.G., 2007. Molecular dynamics methodology to investigate steady-state heterogeneous crystal growth. J. Chem. Phys. 126, 124703–124712.

Verdult, M.W.J., 2010. A Microscopic Approach to van-der-Waals Interactions between Nanoclusters: The Coupled Dipole Method (Master Thesis in Theoretical Physics). Utrecht University.

Verwey, E.J.W., Overbeek, J.Th.G., 1948. Theory of the stability of lyophobic colloids. Elsevier, Amsterdam.

Vicsek, T., 1999. Fractal Growth Phenomena. World Scientific, Singapore.

Vreeker, R., Hoekstra, L.L., den Boer, D.C., Agterof, W.G.M., 1992. The fractal nature of fat crystal networks. Colloids Surf. A 65, 185–189.

Walstra, P., 1996. Dispersed systems: basic considerations. In: Fennema, O.R. (Ed.), Food Chemistry, third ed. Marcel Dekker, New York, pp. 95–135 (Chapter 3).

Weiss, J., Gaysinsky, S., Davidson, M., McClements, J., 2009. Nanostructured encapsulation systems: food antimicrobials. In: Global Issues in Food Science and Technology, pp. 425–479 (Chapter 24).

Wennerström, H., 2003. The van der Waals interaction between colloidal particles and its molecular interpretation. Colloids Surf. A: Physico-Chem. Eng. Aspects 228, 189–195.

Yan, Z.Y., Huhn, S.D., Klemann, L.P., Otterburn, M.S., 1994. Molecular modeling studies of triacylglycerols. J. Agric. Food Chem. 42, 447–452.

York, D.M., Darden, T.A., Pedersen, L.G., 1993. The effect of long-range electrostatic interactions in simulations of macromolecular crystals: a comparison of the Ewald and truncated list methods. J. Chem. Phys. 99, 8345–8348.

CHAPTER 8

Oil Migration Through Fats—Quantification and Its Relationship to Structure

Farnaz Maleky
University of Guelph, Guelph, ON, Canada

OVERVIEW

The migration of oil between different phases in multicomponent food systems is of great concern to food manufacturers, as it can lead to significant changes in the structure and functionality of food materials. This phenomenon is common in chocolate-enrobed products such as chocolate-coated biscuits, filled chocolate bars or shells, wafers with fat-based cream fillings, and pralines.

Oil migration in confectionery products usually refers to a two-way diffusive process that is driven by the tiracylglycerides (TAGs) concentration gradient between a highly mobile oil-rich phase (such as a soft filling) and a crystalline fat-rich phase (Ghosh et al., 2004; Dibildox-Alvarado et al., 2004; Rousseau and Smith, 2008; Guiheneuf et al., 1997; Aguilera et al., 2004). The process will continue until a thermodynamic equilibrium state between the solid and liquid phases is reached (Timms, 2002). The migrating oil phase will dissolve certain solid TAGs, which could then migrate and recrystallize on the surface of the confectionery product resulting in a number of negative consequences such as hardening of the filling, softening of the coating, and eventually the appearance of fat bloom (the formation of grayish coating on the chocolate surface). Fat bloom negatively impacts sensory properties such as texture, color, and flavor and is one of the primary causes of consumer dissatisfaction and rejection of confectionary products (Ziegleder and Petz, 2001; Ghosh et al., 2002).

TECHNIQUES INVOLVED TO ANALYZE OIL MIGRATION

To monitor and analyze the migration of oil, various chemical and physical techniques, which range from simple visual inspection to sophisticated analytical skills, have been used. Because multiple-layer confectionery products are composed of different fats, migration has been quantified in different model systems in which several mixtures are held in direct contact with each other. Usually, a model fat system made by layering crystallized fat over a creamy filling is used, and the total uptake of migrating oil in either the solid layer or the filling is measured.

Structure-Function Analysis of Edible Fats, Second Edition
ISBN 978-0-12-814041-3, https://doi.org/10.1016/B978-0-12-814041-3.00008-3

SURFACE COLOR OBSERVATION

Product color changes relevant to migration can be assessed using commercial colorimeters, which allow surface characterization by means of a whiteness index (Bricknell and Hartel, 1998; Myung et al., 1994; Maleky and Marangoni, 2011a) or simply by visual inspection (Ali et al., 2001). Using a Hunterlab colorimeter, changes at the sample surface can be evaluated according to the Commission Intl. de l'Eclairage (CIE) system and computed as L*, lightness, a*, redness, and b*, yellowness. Converting these values to whiteness index (using Eq. 8.1) amplifies the whiteness of fat bloom from other color components and provides information about bloom formation in time.

$$WI = 100 - \left[(100 - L)^2 + a^2 + b^2\right]^{1/2}. \tag{8.1}$$

In addition, recently a computer vision system characterizing chocolate surface color was developed by Briones et al. (2006). They followed the color evolution such as white specks and gray background area on the chocolate surface to identify spatial changes and discriminate between different stages of blooming resulting from oil migration. They extracted eight features, including L*, a*, and b* values, whiteness index, chroma, hue, % bloom, and energy of Fourier from images to determine color and surface roughness.

FATTY ACIDS AND TRIACYLGLYCEROL ANALYSIS

Because multiple-layer confectionery products are composed of layers of different fats with varying chemical compositions, oil migration can be quantified by chemical analysis of the fat layers. Iodine value analysis, gas—liquid chromatography (GLC) and high-performance liquid chromatography are shown as powerful techniques to measure changes in the fatty acid (FA) and TAG compositions within the different layers when the composition of the migrating oil is sufficiently different from the receptor (sink) fat phase (Timms, 2003; Wootton et al., 1970; Ziegleder et al., 1996a; Bigalli, 1981). When the migrating and receptor (sink) fats have similar carbon number distributions (e.g., peanut oil migrates into chocolate), migration measurement can be made by FAME (fatty acid methyl esters) GC using linoleic acid in the peanut oil as a marker.

TEXTURE ANALYSIS

Oil migration can directly impact the quality attributes of fat-based confections by affecting their texture as well as their appearance and flavor. For instance, softening of the product "hard" phase or hardening of its "soft" phase is a consequence of oil movement that can be characterized by different instrumental and sensory techniques. Often, texture analyzers are used to measure the force needed to penetrate, compress, or break the phase being most affected by migration (Ali et al., 2001; Talbot, 1989; Svanberg et al., 2011).

Using maximum penetration force, Ali et al. (2001) documented chocolate softening (due to oil movement) by showing a reduction in a filled dark chocolate hardness from 3.5 to 0.5 *kg* force after 8 weeks storage at 30°C. Svanberg et al. (2011) used a mechanical testing machine and quantified the effects of oil and moisture transportation by measuring Young's modulus, tensile strength, and tensile strain of chocolate during storage. They reported about 40% reduction in tensile strength of chocolate due to oil migration in 20 days of storage.

POWDER X-RAY DIFFRACTION, DIFFERENTIAL SCANNING CALORIMETRY, AND PULSED NUCLEAR MAGNETIC RESONANCE

Liquid oil migration to a crystallized solid fat leads to recrystallization of fat into a more stable polymorphic form and a decrease in the solid fat content (SFC) of the solid phase. Hence, studies have used pulsed nuclear magnetic resonance (p-NMR) and differential scanning calorimetry simultaneously to quantify the migration process by following the variation in softening (SFC reduction) and polymorphic transitions (change in melting point) of the fatty materials (or fats) (Talbot, 1990; Tabouret, 1987; Walter and Cornillon, 2002; Ziegleder and Schwingshandl, 1998; Ziegleder, 2006). The subsequent polymorphic phase transformation can also be followed using X-ray diffraction pattern. Fat bloom occurring with oil migration is associated with phase transition from β_V to β_{VI}. One of the absolute techniques to observe this transformation unambiguously is X-ray diffraction technique (Ziegler, 2009; Sonwai and Rousseau, 2010; Sonwai and Rousseau, 2006).

RADIOLABELING

Oil migration processes can be quantified properly if the exchange between liquid and solid triacylglycerols is fully understood. To monitor the molecular exchange between the dissolved fats and fat crystals during oil migration process, Haghshenas et al. (2001) developed a new technique using radiolabeled TAGs. Although TAG radiolabeling allows the direct measurement of the molecular exchange rate and could be adapted to follow the movement of TAGs from one crystal to another, it suffers from some limitations. It is quite an expensive and laborious method. The required chemicals are costly, and only limited number of fat systems with small sample sizes can be analyzed. Moreover, the measurement takes a long time, and in many cases, the measurement frequency is not high enough (Smith et al., 2009; Lofborg et al., 2003).

SCANNER IMAGING TECHNIQUE (FLATBED SCANNER AND FLUORESCENT LIGHT MICROSCOPY)

More recently, a new method of oil migration analysis in fats was developed by Marty et al. (2005, 2009a), and the technique is used successfully by others (Maleky and Marangoni, 2011a; Hughes, 2008). These studies used a common flatbed scanner linked to image analysis software, to monitor the movement of the lipid soluble dye, Nile red stain, into crystallized fat matrices (Fig. 8.1A). The underlying assumption, which is confirmed by comparing the layers' FA profile via GLC, is that the dye migrates at the same rate as the oil. Hence the imaging data obtained via the scanner can be analyzed to quantify oil migration kinetics through confectionary fats (Marty et al., 2005). As summarized in Fig. 8.1, the position of the dye front can be determined by plotting the average pixel intensity throughout the storage time versus distance covered by the oil and graphs normalization to 100% of each data set. Where the pixel intensity is at 10% of maximum intensity, it can be used as the most reliable parameter for oil migration investigation (Marty et al., 2009a). The data describing the distance traveled by the oil

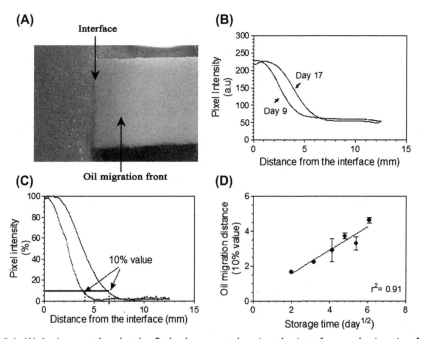

Figure 8.1 (A) An image taken by the flatbed scanner showing the interface and migration front of dyed oil. (B) An example of oil migration pixel intensity throughout the storage period: pixel intensity versus distance covered by oil. (C) The graph of normalized pixel intensity representing the pixel intensity decrease to 10% of its maximum value. Distance from interface at this 10% value was used to analyze overall migration rate. (D) An example of the linear regression of the 10% value as a function of the square root of time to determine the rate of oil migration.

(mm) as a function of the square root of time $(day^{-1/2})$ can be fitted to a linear equation to determine the rate of oil migration (Fig. 8.1D).

FLUORESCENCE RECOVERY AFTER PHOTOBLEACHING

Fluorescence recovery after photobleaching (FRAP) is a versatile technique allowing both the qualitative and quantitative analyses of labeled molecules dynamics in vivo and in vitro methods. Some studies have used FRAP to describe the diffusion of small molecules such as drugs, in macromolecular networks such as polymer solutions and gels (Meyvis et al., 1999; Kappel and Eils, 2004; Karbowiak et al., 2008; Pinte et al., 2008). The application of FRAP for studying the small-molecule diffusion in model fat matrices, such as triglyceride crystal networks was done by Marty et al. (2009c). The experiments were carried out by a confocal laser scanning microscope equipped with an *Ar/KrAr* laser, and the effective diffusion coefficient (D_{eff}) of fluorescent lipophilic molecules (Nile red) through five different fat mediums were obtained by means of Eq. (8.2):

$$D_{eff} = \frac{0.88w^2}{4t_{1/2}}, \tag{8.2}$$

where ω is bleached area radius and $t_{1/2}$ is recovery half-time. The correlation of the experimental components with the structural properties of the crystalline network (e.g., crystal mass fraction and permeability coefficient) demonstrated the potential of FRAP in quantifying the effects of TAGs' crystalline structure on the diffusivity of small lipophilic molecules through that medium.

However, it is important to mention that the study relied on some assumptions such as pure lateral (2D) diffusion and a disk-shaped bleached area with square wells. For a more accurate estimation of oil diffusivity, the authors recommended using a numerical approach instead.

MAGNETIC RESONANCE IMAGING

Besides the aforementioned analytical methods, a number of studies have introduced magnetic resonance imaging (MRI) as a powerful and accurate tool to monitor food products' temporal and spatial changes after foreign oils migrate into their networks (McCarthy and McCarthy, 2008; Guiheneuf et al., 1997; Miquel et al., 2001; Walter and Cornillon, 2002; Choi et al., 2005, 2007; Lee et al., 2010; Deka et al., 2006).

MRI is a two-dimensional technique that allows visualization of the migration process in time and provides information about the three-dimensional oil distribution within the matrix. In addition to the high accuracy of the techniques, the main advantage of MRI (compared with other techniques) is its capability to monitor and investigate the migration process in one single sample over time, which avoids the occurrence of any experimental error due to sample-to-sample variation.

The technique works by measuring the magnetic resonance (MR) signal intensity (SI) at the interface between two different phases (e.g., the interface between a "filling" and a solid matrix in confections) (McCarty and McCarthy, 2008; Altan et al., 2011; Walter and Cornillon, 2002; Maleky et al., 2012). An example of a two-dimensional MR image obtained from a two-layer sample (made of crystallized cocoa butter on the top of cream, shown in Fig. 8.2A) is presented in Fig. 8.2B. Because MRI detects brightness associated with the proton SI at each layer, the brighter part of the image with higher SI represents the cream (high in liquid lipid) and the darker part of the sample with lower SI represents the crystallized fat (low in liquid lipid). Further analysis of the MR image provides the one-dimensional SI values of the layers and illustrates the oil movement at various time points (Fig. 8.2C). Fig. 8.2C shows SI profiles at an initial time, $t = 0$ and time, $t = t$. The extent of oil migration (M_t) at any time, $t = t$, is the difference in the area under the SI curve in the lipid region at any time, $t = t$, and its area at time, $t = 0$. A representative plot for oil uptake in cocoa butter over time obtained from MRI measurement is shown in Fig. 8.2D. The amount of oil uptake (M_t) increased with time and reached a plateau at the maximum uptake.

MECHANISM OF OIL MIGRATION IN FATS

Oil migration in confectionery-type matrices has been studied for more than 50 years, yet its mechanism is subject to controversy. Although it is widely accepted that the ratio of the two lipid phases, their chemical composition, and the interactions between them may influence the migration rates, the exact mechanism behind oil movement through a fat crystalline matrix is not yet fully understood. Some studies have shown that oil migration in food products may be driven by a combination of capillary movement and diffusion process (Marty et al., 2005; Guiheneuf et al., 1997). Several other studies (Aguilera et al., 2004; Carbonell et al., 2004; Marty et al., 2009a,b; Maleky and Marangoni, 2011a; Altimiras et al., 2007) acknowledge mainly the importance of capillary forces associated with the pores in a porous material such as a chocolate on oil migration. They have even predicted the pores' size and calculated their effect on the rate of oil movement. Although these studies provide useful information, the data are not sufficient to model the proper incorporation of capillary movement. Furthermore, additional data related to the structure, and composition of the solid and liquid phases over time is needed because the solid–liquid ratio and the composition of each phase are subjected to change over time and during the migration process. Hence, the contributions of capillary flow to oil migration could be estimated more precisely when the changes in the matrix physical properties, such as the interfacial tension and contact angle, due to the oil movement are measured and considered. This lack of information could be the reason that in the

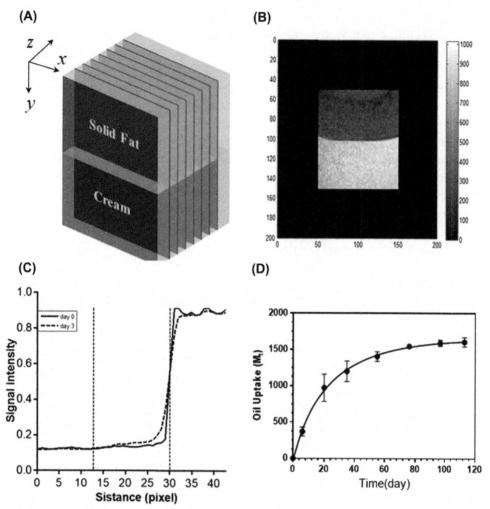

Figure 8.2 (A) Schematic 3D image of a two-layer sample (crystallized solid fat on the top of the oil source) used for oil migration experiments. (B) *Axes* represent pixel numbers and the *bar* represents the magnetic resonance signal intensity (SI) in arbitrary units. (C) SI profile of the sample in the vertical direction from top (cocoa butter) to bottom (cream) at an initial time, day 0 and day 3. (D) A representative plot obtained from MRI measurement for oil uptake by cocoa butter over time.

majority of publications a diffusive mechanism is proposed to be the prevalent mechanism of the migration. Ziegler (2009) believes that the capillary action is too fast "in the order of seconds and hours," to be considered in the timescale of oil migration, which are days and months. Galdámez et al. (2009) reports that liquid oil migration through chocolate can be explained by a diffusion-based model and that no other mass transfer mechanism is required. *Liquid oil diffusivity in solid fats:* in theory, the driving force for a diffusive process is the difference in a specific TAG concentration along the diffusive path (Ghosh et al., 2004). Moreover, the diffusion migration of liquid molecules depends on temperature, presence and type of polymers, particle concentration, and the nature of the migrating liquids (Kulkarni et al., 2003a, b). Fick's first law states that within a continuous medium and in the presence of a concentration gradient, the net migration of solute molecules due to Brownian motion occurs from a region of high concentration to one of lower concentrations. Fick's second law states that the rate at which diffusion proceeds in a diluted binary system (a single solute in a large quantity of solvent) is proportional to the variation of the slope of the concentration gradient (Fick, 1855). Fick's second law features a constant of proportionality, which depends on the nature of the migrating substances. This constant is defined as the diffusion coefficient or diffusivity, D_{eff} (Ziegleder et al., 1996a,b; Cussler, 2009; Aguilera et al., 2004), which accounts for all forms of the mass transfer occurring in the system (Aguilera et al., 2004).

Hence, to fully understand the mass transfer of liquid oil into a crystallized solid, the molecules' mobility must be examined relative to the crystallized matrix relaxation behavior. The relaxation mechanisms in the matrix are often associated with its physical changes such as swelling or phase transitions (Taub and Singh, 1998; Ghosh et al., 2002). Depending on these relative changes, the mass transfer is categorized into three main forms: Case I or Fickian diffusion, non-Fickian or anomalous diffusion, and Case II diffusion (Ghosh et al., 2002). In Fickian diffusion, the diffusion is slower compared with matrix mobility and may act as the dominating cause of migration. The deviation from Fickian behavior, known as non-Fickian diffusion, considers comparable rates of relaxation and diffusion within the system. The extreme case of this deviation is called Case II or Super Case II transport. In Case II, swelling kinetics dominates (Peppas and Brannon-Peppas, 1994), and diffusion rate is larger than matrix relaxation rate.

For a system with one-dimensional diffusion (x-direction) and a constant diffusion coefficient, Fick's second law is written as

$$\frac{\partial C}{\partial t} = D \frac{\partial^2 C}{\partial x^2}, \tag{8.3}$$

where C is concentration and t and x are the spatial and time coordinates, respectively. D is the constant of proportionality, called the diffusion coefficient (Cussler, 1990). The one-dimensional original solution of the Fick's second law in the form of trigonometric series gives Eq. (8.4), where q is the number of terms and M_t and M_∞ are the mass migrated at time t and at saturation, respectively (Peppas and Brannon-Peppas, 1994).

$$\frac{M_t}{M_\infty} = 1 - \sum_{q=0}^{\infty} \frac{8}{(2q+1)^2 * \pi^2} \exp\left[\frac{-D(2q+1)^2 * \pi^2}{l^2} t\right]. \tag{8.4}$$

Although Eq. (8.4) can be used for any ratio of M_t/M_∞ (ranges from zero at the initial time to one at a long time), using a short-time approximation to the original solution of Fick's II law summarizes Eqs. (8.4) to Eq. (8.5):

$$\frac{M_t}{M_\infty} = kt^n. \tag{8.5}$$

In Eq. (8.5), k is the uptake rate, and n is the value of the exponent, determined by fitting the experimental data into the formula. It is important to mention that this approximation, known as a power law is only valid when $M_t/M_\infty < 2/3$ (Peppas and Brannon-Peppas, 1994).

By comparing the exponent n with threshold values depending on the geometry of the matrix into which diffusion occurs (e.g., planar, cylindrical, or spherical), the migration mechanism for all categories (Fickian, non-Fickian, or Case II) can be determined through Eq. (8.5) (Ritger and Peppas, 1987). Ritger and Peppas reported different cutoff n values for Fickian diffusion corresponding to each geometry by modeling drug release from thin polymer slabs, cylinders, spheres, and discs under the perfect sink condition. For example, they reported the values of 0.5 for slabs, 0.45 for cylinders, and 0.43 for spheres. Moreover, Korsmeyer and others determined the threshold values of n from cylindrical tablets and reported $n \leq 0.45$ for Fickian, 0.45 to 0.89 for non-Fickian, 0.89 for Case II transport, and higher than 0.89 for Super Case II transport (Korsmeyer et al., 1983). Considering this explanation, the migration of liquid oil in crystallized fat matrices is described by an approximate equation (Eq. 8.6) proposed by Crank (1975) and modified by Crank (1975), Ziegleder (1985), Ziegleder et al. (1996b). This approximation, which is used by many studies Miquel et al. (2001), Aguilera et al. (2004), Marty et al. (2005, 2009a), Khan and Rousseau (2006), Maleky and Marangoni (2011a) is based on Fickian diffusion in slabs' geometry and considered a threshold n value equal to 0.5.

$$\frac{M_t}{M_\infty} = \frac{KA\sqrt{D_{eff}t}}{V}. \tag{8.6}$$

In Eq. (8.6), A is the contact area between the two phases (cm^2), V is the volume through which the diffusion takes place (cm^3), and t is the migration period (sec). D_{eff} is a time-dependent analogue of the diffusion coefficient (cm^2/s), and K is a constant specific to the two phases. In an ideal system with structural changes (e.g., swelling or eutectic effects) as a result of migration, K is equal to or greater than 1. $K < 1$ occurs when there is insufficient contact between the two phases (Ziegleder et al., 1996a,b; Khan and Rousseau, 2006; Ziegleder, 1998). Eq. (8.6) relates the extent of oil migration to the square root of storage time, and it is worth reiterating that the linear relationship between M_t/M_∞ and the square root of time is generally acceptable for $M_t/M_\infty \leq 2/3$ (Peppas and Brannon-Peppas, 1994).

Moreover, to describe the mass uptake under non-Fickian conditions, another approximation of the original solution (Eq. 8.4) is proposed by Alfrey and others (Alfrey et al., 1966). This expression (shown as Eq. 8.7) adds up contributions from both diffusion and relaxation when k_1 [$= 4 (D/\pi \, l^2)^{0.5}$] describes the diffusion coefficient of Fickian and k_2 is the relaxation constant for Case II transport (Enscore et al., 1980; Peppas and Brannon-Peppas, 1994). Similar to other short-time approximations (Eqs. 8.5 and 8.6), this assumption is also valid only when $M_t/M_\infty < 2/3$.

$$\frac{M_t}{M_\infty} = k_1 t^n + k_2 t^{2n}. \tag{8.7}$$

The generalized form of Eq. (8.7) can be modified for thin slabs (with $n = 0.5$) and used for determination of the relative contribution of diffusional (F) and relaxational (R) mechanism in different systems (Peppas and Sahlin, 1989; Paluri et al., 2015). This modification is summarized in Eqs. (8.8) and (8.9) and demonstrates the contribution ratio of the diffusivity and relaxation (R/F).

$$\frac{M_t}{M_\infty} = k_1 t^{0.5} + k_2 t^{2n} \tag{8.8}$$

$$\frac{R}{F} = \frac{k_2}{k_1} t^p. \tag{8.9}$$

Although the analysis of R/F is lacking for oil migration in lipid systems, the proportionality constant measured for water migration in crystallized lipids shows that the contribution of relaxation mechanism in moisture migration is not negligible and should be taken into consideration (Paluri et al., 2015). Depending on the structural properties of lipids, Paluri et al. (2015) documented a ratio of k_2/k_1 that ranges from 0 to 1 and indicates a shift from pure Fickian diffusion—controlled to a relaxation-controlled mechanism for water migration in fat.

Effective diffusivity of crystallized fats: in theory, the effective diffusivity is a function of molecular diffusivity, D_0; porosity, ε; and tortuosity, τ (shown in Eq. 8.10) and can be reduced by decreasing the matrix porosity and increasing its tortuosity.

$$D_{eff} = \varepsilon \frac{D_0}{\tau}. \tag{8.10}$$

Literature from various disciplines (Schaefer et al., 1995; Epstein, 1998; Grathwohl, 1998) introduces tortuosity as a reliable alternative approach, which characterizes the connectivity of pores and helps in understanding the diffusion of fluids through porous media over a long period of time. A great advantage of tortuosity is its simplicity, as it provides a simple number showing by how much diffusion will be retarded in a porous medium. In lipids, tortuosity is a function of the number of solid fat crystals, their shape, the distribution of the particles, and the platelet aspect ratio. Although affecting the tortuosity factor (diffusion length) may be the most effective strategy to alter diffusion coefficient and consequently the oil migration (Maleky and Marangoni, 2011a), tortuosity cannot be measured experimentally. Hence, in the oil migration literature, the system's tortuosity has been either ignored or speculated in past correlations between structure and oil permeability. Using Darcy's law (Eq. 8.11), Bremer et al. (1989) carried out a detailed study and illustrated the effects of all structural factors in the fat system diffusivity. They highlighted the importance of tortuosity by correlating the network permeability coefficient to the tortuosity factor of the Kozeny−Carman equation, τ (Carman, 1956; Thusyanthan and Madabhushi, 2003).

$$Q = \frac{B \cdot A_c}{\eta} \cdot \frac{\Delta P_r}{L}. \tag{8.11}$$

In these equations, Q is the volumetric flow rate, B is the permeability coefficient, A_c is the cross-sectional area through which flow takes place, η is the viscosity of the permeating oil, and P_r is the pressure drop over the distance L. The crystal network particle diameter, solid volume fraction, and box-counting fractal dimension are shown as a, ϕ, and D, respectively.

$$B = \frac{a^2}{\tau} \phi^{2/(D-3)}. \tag{8.12}$$

FACTORS AFFECTING OIL MIGRATION IN FATS

As previously mentioned, a number of contributing factors including oil type and chemical composition, viscosity, particle size, ratio of the two lipid phases, mobility of the liquid oil phase, interactions between the solid/liquid phases (contact area), presence of nonfat solid particles, and most importantly, the structural characteristics of the fat crystal network are used to quantify oil diffusivity in fats.

Studies reported the significant influences of fats' chemical composition, such as the level of saturated/unsaturated TAGs, diacylglycerols, or free FAs on either the induction times of crystallization or on the crystallization rates (Chaiseri and Dimick, 1995; Sato and Koyano, 2001; Foubert et al., 2004). Fats with different crystallization rate may display different SFCs and polymorphic behavior, which may affect their diffusivity values (Maleky and Marangoni, 2011a; Sonwai and Rousseau, 2010; Marty et al., 2009a). For example, Marty et al. (2009b) reported that the SFC and, consequently, the liquid oil transport in cocoa butter vary depending on their geographical origin. Oil migration may be further exacerbated when the formulation of the solid fat and/or filing is changed. Two examples of the effects of samples' composition on oil migration are shown in Fig. 8.3. Fig. 8.3A shows the effects of milk fat content on chocolate oil migration (McCarthy and McCarthy, 2008) and cocoa butter emulsifier level (see

(A)   **(B)**

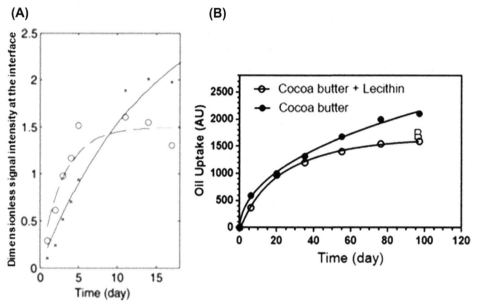

Figure 8.3 (A) Data and curve fitting for days 1–17 at the chocolate–peanut butter filling interface at $T = 30°C$. Data for chocolate with 3.57% and 10% AMF are designated by (x) and by (○), respectively. The model is given as a *solid line* for the low level and dashed for the high level. (B) Oil uptake data and curve fitting for days 1–97 at the cocoa butter–*cream* interface at $T = 20°C$. Data for cocoa butter with and without lecithin are shown by (○) and (•), respectively. Curves for exponential rise of *the uptake* values are also shown in this figure. *((A) Adapted from McCarthy, K.L., McCarthy, M.J., 2008. Oil migration in chocolate-peanut butter paste confectionery as a function of chocolate formulation. J. Food Sci. 73(6), E266–E273.)*

Fig. 8.3B) on oil movement in four two-layer systems. Using MRI and a Fickian-based diffusion model, the authors documented a higher diffusion coefficient in chocolates with higher AMF content (10% w/w) compared with the chocolate with 3.57% AMF. Fig. 8.3B shows the rate of oil movement in cocoa butter with and without emulsifier (Wang et al., 2016) and documents the reduction of oil uptake in cocoa butter samples that contain 0.3% lecithin. The effect of chemical formulation on oil migration could be more pronounced by considering the direct correlation between the lipids' composition and their solid phase structural properties. This has been done by a number of researchers who investigated cocoa butter's oil diffusivity after demonstrating the impacts of the cocoa butter composition on its crystallization kinetics and structural properties (Marty and Marangoni, 2009; Wang et al., 2016; Wang and Maleky, 2018). At first, these studies showed how the average crystalline domain size and the number of the particles in the solid matrix are affected by cocoa butter composition. Then they illustrated the influences of these structural properties on the samples' oil migration rate. Higher permeability coefficients are reported in materials associated with higher amounts of unsaturated fats, larger average particle size, and smaller crystal-line domain size.

Although oil migration rate is strongly dependent on the product composition processing conditions, the way that liquid fat is crystallized, thermal treatments, and storage conditions, have shown to influence the migration process significantly. In fact, the crystal size, the uniformity in crystal distribution, and the interaction between them are induced by the tempering procedure and result in a varying resistance rates to oil migration. For example, Wang and Maleky (2018) conclude that although crystal-lization conditions cannot completely eliminate the effects of minor differences in the chemical composition of cocoa butter on its D_{eff} and overall migration rate (OMR) values, the diffusivity of the foreign oil through the cocoa butter networks can be reduced, more effectively, by the way that they are processed (Wang and Maleky, 2018).

A number of researchers showed that poor control of the crystallization conditions (e.g., temperature, time, shearing, and seeding) ostensibly leads to the formation of undesirable mixtures of crystalline polymorphs and insufficient number of stable crystals, which can promote the liquid oil movement in the crystallized solid lipids (Maleky and Marangoni, 2011a; Marty and Marangoni, 2009b; Lonchampt and Hartel, 2004; Kinta and Hatta, 2007). They also documented that improper tempering of lipids results in a decrease in the final SFC of the product. Fat materials with a lower SFC are more prone to oil diffusivity, and anything that alters solid fat fractions can affect oil migration.

For example, the fluctuation of temperature during storage (postprocessing) can lead to the cycles of melting and subsequent recrystallization of lipid crystals. The continual melting and recrystallization of the solid fraction generates a "pumping" or "peristaltic" action on the liquid fraction, which affects oil migration dramatically (Ali et al., 2001; Smith et al., 2007; Ziegleder et al., 1996a,b). Overall, the higher the storage tempera-tures, the higher the amount of liquid oil phase and the faster the transformation and oil migration. Detailed analysis of the effects of storage temperature on oil transport was carried out by Ziegleder et al. (1996a,b). They monitored the extent of oil migration at a range of temperatures by measuring the triolein content in milk chocolate filled with nougat (as shown in Fig. 8.4). Based on this study, at 20°C, triolein migration into the chocolate developed fat bloom after 100 days of storage. Increasing the storage temper-ature by only 3°C promoted bloom formation occurrence after 20 days. In another study, Ali et al. (2001) report a reduction in a filled dark chocolate hardness from 3.5 to 0.5 kg force after 8 weeks storage at 30°C. However, the hardness had decreased to about 2.8 kg force after 8 weeks of storage at 18°C. These researchers relate their observation at high storage temperatures to the high amount of filling migrated into chocolate and to the eutectic effects of such migration. Ziegler (2009) recommends that storage at a constant temperature below 18°C may extend the shelf life of most confectionary formulations. Interestingly, the effects of the postprocessing and storage temperature on oil migration are not observed when the solid phase is formed under

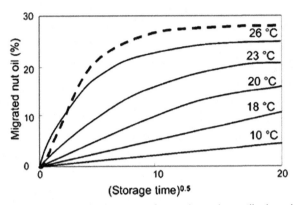

Figure 8.4 Kinetics of oil migration as a function of time through a milk chocolate layer. *Dotted line represents typical data for blooming. (Adapted from Ziegleder, G., Moser, C., GeierGreguska, J.. 1996b. Kinetics of fat migration within chocolate products .2. Influence of storage temperature, diffusion coefficient, solid fat content. Fett-Lipid 98, 253–256.)*

proper crystallization conditions. Miquel et al. (2001) quantified the migration of hazelnut oil into dark chocolate at 28°C and reported more migration of the foreign oil in the undertempered chocolate compared with the normal tempered sample.

To further understand the relationship between the solid lipids' structure and their crystal packing on oil migration rate, several studies monitored and measured mass transfer in systems with differences only in the structural properties of their crystalline networks (Dibildox et al., 2004; Maleky and Marangoni, 2011a,b; Wang et al., 2016; Paluri et al., 2017; Marty et al., 2009b; Choi et al., 2005). Dibildox et al. (2004) crystallized a mixture of peanut oil and chemically interesterified hydrogenated palm oil at different cooling rates and reported that smaller crystals resulting from crystallization at higher cooling rates retard the rate of migration by reducing the permeability coefficient of the network. This observation was confirmed with Choi et al. (2005), who showed a higher diffusion coefficient in chocolate with larger cocoa solid particle sizes (60 μm) compared with the sample with smaller particle sizes (45 μm) (Choi et al., 2005; McCarthy and McCarthy, 2008). These findings could imply that a densely packed network consisting of well-dispersed small crystals results in smaller interparticle voids and perhaps a higher tortuosity. Moreover, studies in various disciplines have shown that the porosity and pore connectivity are both a function of particle size and greatly influence the oil migration rate. For instance, researchers who examined mass transfer in sand and clay showed that porous clay deposits comprised of parallel sheets of clay minerals have a lower permeability to fluid than an unconsolidated deposit of soil with larger pores (Athy, 1930; McDowell-Boyer et al., 1986). This low permeability is due to the size of clay particles that is much finer than soil particles, and thus, the pores between them are very small. Moreover, the voids in clay materials are not interconnected and therefore cannot transmit the fluid or gas across the breadth of the material.

To elucidate the influence of crystalline size in addition to the crystals' alignment, distribution, pores ratio, and connectivity on oil diffusion through fats, a series of studies have performed detailed analysis. Some researchers analyzed the effects of external fields on crystal habit and structural properties of fat crystal networks at different length scales and linked these factors to tortuosity and liquid oil movement through the crystalline matrix (Maleky and Marangoni, 2011a; Sonwai and Rousseau, 2010; Maleky et al., 2012). For instance, Maleky and Marangoni used static (no shear crystallization) conditions and two types of shear: a random shear and an "orienting" shear (in the order of $340 s^{-1}$) to cocoa butter samples' crystallization and compared their oil migration rate. To better highlight their research hypothesis, the authors generated networks with similar SFC ($\sim$78%) and polymorphic form (β_V) but different crystalline packing. (They analyzed the samples' crystal size at micro- and nanoscale and showed a more heterogeneous microstructure network for the static and sheared samples compared with the orienting shear samples (see Fig. 8.5 and Table 8.1). Despite the similarity in the sample crystals' cluster size at microscale, the difference in platelet size between the samples is reported at the nanoscale. Larger particles were observed in the static sample, and much smaller particles were evident in the shear-crystallized samples. After illustrating the crystalline arrangements and the orientation directions of the lamellar layers with respect to the shearing flow in the samples (Fig. 8.6), they demonstrated clearly the effects of crystal orientations and nanoplatelet size on oil migration (Maleky and Marangoni, 2008; Maleky et al., 2011).

Using the flatbed scanning technique, Maleky and Marangoni set up different experimental models and monitored oil transportation from a cream (stained with Nile red dye) to the crystallized cocoa butter samples in different arrangements relative to the shear plane. Overall, the authors demonstrated a rapid movement of the dye front in the static

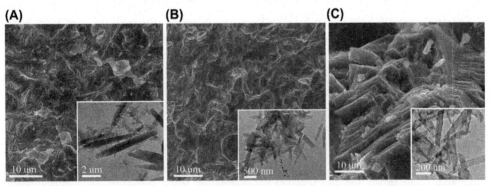

Figure 8.5 Cryo-SEM and cryo-TEM micrographs of the samples: static (A), sheared (B), oriented (C), displaying different crystalline arrangements, nanoplatelet morphology, and size. *(Adapted from Maleky, F., Marangoni, A.G., 2011b. Ultrasonic technique for determination of the shear elastic modulus of polycrystalline soft materials. Cryst. Growth Des. 11(4), 941–944.)*

Table 8.1 Cryogenic Transmission electron microscopy analysis reporting the dimensions of plateletlike structure at the nanoscale and migration rates from floats of 10% value versus time or square root of time for all the samples in model systems A and B

	Crystal properties			Oil migration data			
				Model system A		Model system B	
Sample	Length (nm)	Width (nm)	Aspect ratio	mm/day$^{1/2}$	D_{eff} (m^{2}s^{-1})	mm/day$^{1/2}$	D_{eff} (m^{2}s^{-1})
Static (ST)	2085 ± 128[a]	164 ± 6.2[a]	12.6	0.67[a]	7.57 E-12[a]	5.22[a]	8.13 E-12[a]
Sheared (SH)	386 ± 12.6[b]	132 ± 4.0[b]	2.9	0.56[b]	3.03 E-12[b]	1.69[b]	1.23 E-12[b]
Oriented (OR)	310 ± 7.3[b]	138 ± 5.3[b]	2.3	0.51[b]	1.97 E-12[c]	1.36[b]	2.68 E-13[c]

Superscript letters represents statistically significant differences within each column.
Adapted from Maleky, F., Smith, S.A., Marangoni, A.G., 2011. Laminar shear effects on crystalline alignments and nanostructure of a triacylglycerol crystal network. Cryst. Growth Des. 11(6), 2335−2345 and Maleky, F., Marangoni, A.G., 2011b. Ultrasonic technique for determination of the shear elastic modulus of polycrystalline soft materials. Cryst. Growth Des. 11(4), 941−944, Reproduced by permission of The Royal Society of Chemistry.

samples and showed that such migration rate decreased as the applied shear increased (Table 8.1). By comparing the static and random sheared samples, this finding illustrates the effects of crystal size and aspect ratio on the migration process. However, the results are more interesting when the two sheared samples are compared (the sheared and the oriented samples). When the crystallized cocoa butter was in contact with the cream on only one side (system A in Table 8.1), oil could migrate in a direction parallel to the crystalline alignments and no difference between the migration rates of sheared and oriented samples is recorded. The oil migration rate remained comparable between the two samples when a rectangular piece of cocoa butter was in contact with the cream via five faces (system B in Table 8.1), and oil could migrate through the specimen in all these five directions, parallel and/or perpendicular to the orientation directions. Interestingly, the similarity between the samples' migration rates was not seen when the networks' oil diffusivities were compared. The liquid oil diffusivity decreased significantly when the liquid oil traveled parallel to the direction of crystal orientation in the oriented sample. The reduction was even larger when the oil migrated normal to the orientation planes. Considering the similarity between the average particle sizes (at both nano- and microscales) in samples crystallized under random and orientation shear, this finding highlights the strong effect of crystalline orientation on oil diffusivity. In fact, the preferential alignment of the platelet surfaces parallel to the shear planes of the external shear field may influence other structural properties such as plate spacing, pore size, and pore distributions, all of which impact oil migration (Maleky and Marangoni, 2011a,b). Maleky and Marangoni (2011a,b) also conclude that in comparison with oil migration

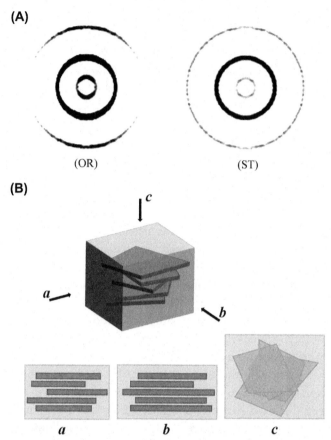

Figure 8.6 (A) The characteristic small-angle Debye diffraction ring of the 002 reflection of both oriented and nonoriented samples. An example of X-ray diffraction pattern of all samples in SAXS: (OR) oriented, (ST) sheared/static. Patterns display a clear orientation of the crystals in the oriented sample, no orientation in the static and sheared samples and (B) schematic illustration of orientation in different planes, a, orientation parallel to the shearing surface; b, orientation, parallel to the flow direction; and c, orientation, perpendicular to the shearing surface. *(Reproduced by permission of The Royal Society of Chemistry.)*

rate, D_{eff} is a more reliable factor to be used for differentiating the effects of the fat structure on oil migration. This conclusion is in agreement with the finding of a recent publication on cocoa butter's chemical composition effects on oil migration (Wang and Maleky, 2018). The authors documented that processing conditions can reduce the diffusivity of a foreign oil through fats (D_{eff}) more effectively than its OMR. This observation about the effect of cocoa butter structure on oil diffusivity could be better quantified by considering other parameters including the fat micro/nanostructure, crystalline alignments, porosity, and the pores' connectivity. In theory, porosity is the volume fraction of the void space which, in a fat network, is the nonsolid portion. As explained before, the connectivity of the pores helps to estimate the matrix tortuosity that influences the liquid oil diffusion through the porous media over time (Eq. 8.10). On the other hand, the value of tortuosity varies based on the definition of porosity for a crystallized fat system. A network porosity (ε) is a nondimensional value that can be defined as volume ratios of either volume filled by oil to total volume (assuming no void fraction) or the volume effectively available for diffusion to total volume (Marangoni et al., 2012). Owing to the experimental limitation for porosity measurement, its calculation has been either ignored ($\varepsilon = 1$) or speculated in some studies. Most of the studies on oil migration present the fat crystalline network porosity by the network void fraction, defined with $\varepsilon = 1 - \text{SFC}/100$ (Barrer, 1968; Cussler, 1990; Tozzi and Lavenson, 2011). However, this definition of ε may not be helpful for understanding the role of the solid phase structure in the diffusion process because it does not take the fat structural factors into consideration. Hence, the best alternative method for calculating the solid fat crystal porosity is its measurements via analyzing the networks microstructure (Paluri et al., 2017; Maleky and Marangoni, 2011a,b). Table 8.2 shows three different tortuosity values calculated based on different definitions of porosity for different cocoa butter samples. Although the calculated tortuosity factors obtained by different methods display similar trends, they are certainly different. Smaller tortuosity values are reported when the network's porosity (the ratio of volume of pores to the total volume) is considered in the calculation of the tortuosity factor. Moreover, their value below 1 indicates the presence of convective flow in the samples. In fact, the decrease in the samples' average platelet size (from static samples to sheared samples) could provide a more densely packed network with well-dispersed particles, smaller interparticle voids, and longer path of diffusivity. Assuming that the solid part was impermeable to the liquid oil and that diffusion took place only through the tortuous pores, these results showed conclusively the effect of crystalline orientation on oil diffusivity. An increase in the tortuosity of the path in oriented samples relative to random shear conditions might account for the decrease in apparent oil migration rate in the oriented samples. The increase in tortuosity with aligned layers of crystals is illustrated in Fig. 8.7 (see Maleky and Marangoni, 2011a,b for more details).

Table 8.2 The values of tortuosity and permeability coefficient of crystallized cocoa butter samples

Sample	Tortuosity values in system A		
	$\tau = \frac{D_0}{D_{eff}}$	$\tau = (1 - \phi)\frac{D_0}{D_{eff}}$	$\tau = \varepsilon \frac{D_0}{D_{eff}}$
Static (ST)	1.09	0.24	0.30
Sheared (SH)	2.67	0.59	0.34
Oriented (OR)	3.99	0.88	0.53

Reproduced by permission of The Royal Society of Chemistry.

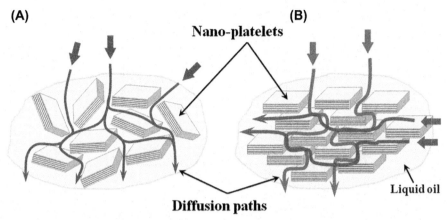

(A)   **(B)**

Nano-platelets

Liquid oil

Diffusion paths

Figure 8.7 Schematic illustration of the tortuosity of the diffusive path in cocoa butter crystallized under various shearing conditions. (A) Shear-mixed cocoa butter crystals showing a random arrangement of crystals. (B) Oriented cocoa butter with fat crystals layers oriented in a direction parallel to shear field. *(Reproduced by permission of The Royal Society of Chemistry.)*

Another important structural attribute of lipid networks with respect to liquid oil diffusivity is the network fractal dimension (D). Fat crystals' fractal dimension measures the self-similarity and degree of mass occupancy in the network and is very sensitive to crystal shape, size, and area fraction (Tang and Marangoni, 2006). Hence, showing the effects of a solid phase fractal dimension on its diffusivity could be a good indicator for presenting the relationship between a solid fat structure and its permeability.

Several studies of oil migrations measured the box-counting fractal dimension of the crystal network and calculated the networks permeability coefficient (*B*) by means of Darcy's law (Eq. 8.12) (Marty et al., 2009b; Wang and Maleky, 2018; Maleky and Marangoni, 2011a). For example, in the study of crystallized cocoa butter, Maleky and Marangoni reported lower fractal dimensions, smaller tortuosity factors, and a higher permeability coefficient for samples with larger average nanoplatelet sizes. This result (illustrated in Fig. 8.8) strongly supports the hypothesis for the interplay between two related, but distinct, phenomena on liquid oil movement, including tortuous path and obstruction. The cocoa butter network's permeability was reduced by about a factor of 100 by performing the crystallization under laminar shear, mainly due to the changes in particle size, shape, orientation, and spatial distribution, at equivalent solid contents. Similar observation is reported in cocoa butter samples with different formulations (Wang and Maleky, 2018). Using MRI analysis for investigating the effects of cocoa butter TAGs and minor compound composition on its oil migration, they reported different fractal dimension and permeability in samples with an equal amount of solid material.

Although these studies clearly highlighted the effects of fat crystal's network structure and processing on oil migration mechanism, it is important to mention that the oil

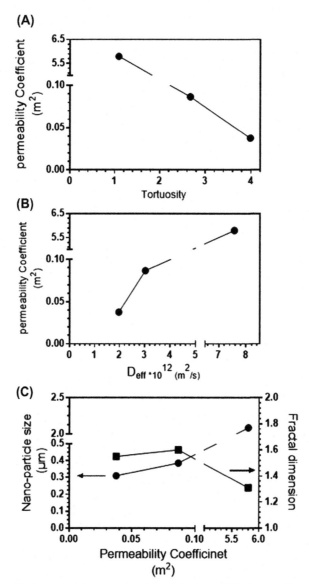

Figure 8.8 Relationship between the permeability coefficient, diffusion coefficient, tortuosity, box-counting fractal dimension and nanoparticle's size in crystallized cocoa butter samples in contact with an oil source. *(Adapted from Maleky, F., Marangoni, A.G., 2011a. Nanoscale effects on oil migration through triacylglycerol polycrystalline colloidal networks, Soft Matter 7, 6012–6024, Reproduced by permission of The Royal Society of Chemistry.)*

migration mechanism and its driving force do not stay constant over time. Analyzing oil migration in a two-layer system (cocoa butter in contact with cream), Maleky et al. (2012) noted two sequential diffusion regimes in the samples (one at storage times less than 35 days and one at storage times greater than 35 days). By examining MR images

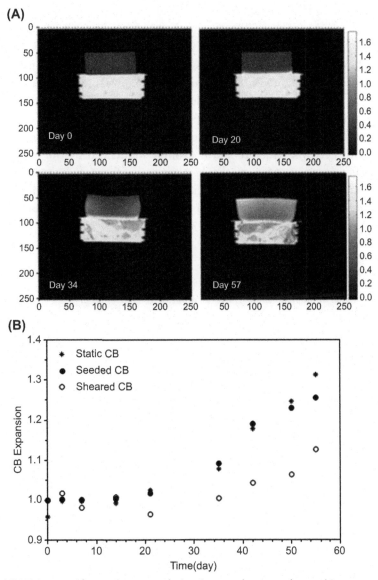

Figure 8.9 (A) MR images of a two-layer sample (static cocoa butter and cream) in storage time of day 0, day 20, day 34, and day 57. (B) Expansion of the cocoa butter region due to oil uptake for the three sample types of cocoa butter. *(Adapted from Maleky, F., McCarthy, K.L., McCarthy, M.J., Marangoni, A.G., 2012. Effect of cocoa butter structure on oil migration. J. Food Sci. 77(3), 74–79.)*

of the samples, they noted that although initially the layer's intensity was uniform and the cross-section of the cocoa butter reflected a rectangular cross-section, the cocoa butter's shape changed and its volume increased after over time (Fig. 8.9A). The quantification of the samples' expansions documented the samples' morphology change during the oil uptake and supported the possibility of the two diffusion regimes (Fig. 8.9B). This result can clearly be related to the possible structural and chemical changes of both solid and liquid phases of the crystallized fat during and after the diffusion process. Interestingly, when the authors compared samples processed under different crystallization conditions, the lowest expansion rate is reported for the sheared samples (Maleky et al., 2012).

In summary, oil migration is not only dependent on crystal habit and the amount of total solid material present but also on factors such as the geometry of the dispersed phase, platelet dimension and shape, and interactions between them. The permeability coefficient varies with network structure and is strongly influenced by the matrices' nano-/microstructural properties. One of the best strategies for reducing diffusivity of oil through a crystalline fat matrix is by affecting the network pore size and tortuosity by inducing a homogenous crystalline distribution and alignments through the media, which the liquid oil must move.

LIST OF SYMBOLS

A Cross-sectional area of the material in which the diffusion is happening, m^2

a Particle diameter, m

B Darcy's law permeability coefficient, m^2

d Characteristic long spacing between two planes producing the X-ray diffraction peak from the nanocrystalline platelets of a given solid phase, in nm

D Fractal dimension of the solid network, nondimensional

D_0 Molecular diffusivity in the absence of the solid matrix, m^2/s

D_{eff} Effective or apparent diffusivity in the presence of the solid matrix, $m^2 \, s^{-1}$

e Porosity, either ratio of volume filled with oil to total volume, or ratio of volume effectively available for diffusion to total volume, nondimensional

K Characteristic constant for the two phases in contact, nondimensional

m_s Mass of material diffused into the matrix at saturation, kg

m_t Mass of material diffused into the matrix at time t, kg

t Time, s

V Total volume of the material in which the diffusion is happening, m^3

τ Tortuosity of the network, nondimensional

ϕ Volume fraction of solids in the material, nondimensional

REFERENCES

Aguilera, J.M., Michel, M., Mayor, G., 2004. Fat migration in chocolate: diffusion or capillary flow in a particulate solid? A hypothesis paper. J. Food Sci. 69, R167−R174.

Alfrey Jr., T., Gurnee, E.F., Lloyd, W.G., 1966. Diffusion in glass polymers. J. Poly. Sci. Part C: Poly. 12, 249−261.

Ali, A., Selamat, J., Man, Y.B.C., Suria, A.M., 2001. Effect of storage temperature on texture, polymorphic structure, bloom formation and sensory attributes of filled dark chocolate. Food Chem. 72, 491−497.

Altimiras, P., Pyle, L., Bouchon, P., 2007. Structure−fat migration relationships during storage of cocoa butter model bars: bloom development and possible mechanisms. J. Food Eng. 80, 600−610.

Altan, A., Lavenson, D.M., McCarthy, M.J., McCarthy, K.L., 2011. Oil migration in chocolate and almond product confectionery systems. J. Food Sci. 76 (6), E489–E494.

Athy, L.F., 1930. Density, porosity, and compaction of sedimentary rocks. AAPG Bull. 14, 1–24.

Barrer, R.M., 1968. In: Crank, J., Park, G.S. (Eds.), Diffusion in Polymers. Academic, New York, pp. 165–217.

Bigalli, G., 1981. Usefulness and Limitations of Fatty Acid Determination in the Confectionery Industry. The Manufacturing Confectioner, pp. 41–47.

Bremer, L.G.B., van Vliet, T., Walstra, P., 1989. Theoretical and experimental study of the fractal nature of the structure of casein gels. J. Chem. Soc., Faraday Trans. 1 85, 3359–3372.

Briones, V., Aguilera, J.M., Brown, C., 2006. Effect of surface topography on color and gloss of chocolate samples. J. Food Eng. 77, 776–783.

Bricknell, J., Hartel, R.W., 1998. Relation of fat bloom in chocolate to polymorphic transition of cocoa butter. J. Am. Oil Chem. Soc. 75, 1609–1615.

Carbonell, S., Hey, M.J., Mitchell, J.R., Roberts, C.J., Hipkiss, J., Vercauteren, J., 2004. Capillary flow and rheology measurements on chocolate crumb/sunflower oil mixtures. J. Food Sci. 69 (9), E465–E470.

Carman, P.C., 1956. Flow of Gases through Porous Media. Butterworths Scientific Publications, London.

Chaiseri, S., Dimick, P., 1995. Dynamic crystallization of cocoa butter II. Morphological, thermal, and chemical characteristics during crystal growth. J. Am. Oil Chem. Soc. 72, 1497–1504.

Choi, Y.J., McCarthy, K.L., McCarthy, M.J., 2005. Oil migration in a chocolate confectionery system evaluated by magnetic resonance imaging. J. Food Sci. 70, E312–E317.

Choi, Y.J., McCarthy, K.L., McCarthy, M.J., Kim, M.H., 2007. Oil migration in chocolate. Appl. Magn. Reson. 32, 205–220.

Crank, J., 1975. Mathematics of Diffusion. Claredon Press, Oxford, England.

Cussler, E.L., 2009. Diffusion on dilute solutions. In: Diffusion: Mass Transfer in Fluid Systems, 3rd. Cambridge Univ Press, pp. 13–50.

Cussler, E.L., 1990. Membranes containing selective flakes. J. Membr. Sci. 52, 275–288.

Deka, K., MacMillan, B., Ziegler, G.R., Marangoni, A.G., Newling, B., Balcom, B., 2006. Spatial mapping of solid and liquid lipid in confectionery products using a 1D centric SPRITE MRI technique. Food Res. Int. 39, 365–371.

Dibildox-Alvarado, E., Rodrigues, J.N., Gioielli, L.A., Toro-Vazquez, J.F., Marangoni, A.G., 2004. Effects of crystalline microstructure on oil migration in a semisolid fat matrix. Cryst. Growth Des. 4, 731–736.

Enscore, D.J., Hopfenberg, H.B., Stannett, V.T., 1980. Diffusion, swelling, and consolidation in glassy polystyrene microspheres. Polym. Eng. Sci. 20, 102–107.

Epstein, N., 1998. On tortuosity and tortuosity factor in flow and diffusion through porous media. Chem. Eng. Sci. 3, 777–779.

Fick, A., 1855. Ueber diffusion. Ann. Phys. 170 (1), 59–86.

Foubert, I., Vanrolleghem, P.A., Thas, O., Dewettinck, K., 2004. Influence of chemical composition on the isothermal cocoa butter crystallization. J. Food Sci. 69, 478–487.

Galdámez, J.R., Szlachetka, K., Duda, J.L., Ziegler, G.R., 2009. Oil migration in chocolate: a case on non-Fickian diffusion. J. Food Eng. 92, 261–268.

Ghosh, V., Duda, J.L., Ziegler, G.R., Anantheswaran, R.C., 2004. Diffusion of moisture through chocolate-flavoured confectionery coatings. Food Bioprod. Process. 82 (C1), 35–43.

Ghosh, V., Ziegler, G.R., Anantheswaran, R.C., 2002. Fat, moisture and ethanol migration through chocolates and confectionery coatings. Crit. Rev. Food Sci. Nutr. 42, 583–626.

Grathwohl, P., 1998. Diffusion in Natural Porous Media: Contaminant Transport, Sorption/desorption and Dissolution Kinetics. Kluwer Academic Publishers, Boston.

Guiheneuf, T.M., Couzens, P.J., Wille, H.J., Hall, L.D., 1997. Visualisation of liquid triacylglycerol migration in chocolate by magnetic resonance imaging. J. Sci. Food Agric. 73, 265–273.

Haghshenas, N., Smith, P., Bergenstahl, B., 2001. The exchange rate between dissolved tripalmitin and tripalmitin crystals. Colloids Surfaces B Biointerfaces 21, 239–243.

Hughes, N., 2008. Potential Applications of Edible Oil Organogels (Master thesis). Food Science Department. University of Guelph.

Kappel, C., Eils, R., 2004. Fluorescence Recovery after Photobleaching with the Leica TCS SP2. Leica Confocal Application Letter August, 2—12.

Karbowiak, T., Gougeon, R.D., Rigolet, S., Delmotte, L., Debeaufort, F., Voilley, A., 2008. Diffusion of small molecules in edible films: effect of water and interactions between diffusant and biopolymer. Food Chem. 106, 1340—1349.

Khan, R.S., Rousseau, D., 2006. Hazelnut oil migration in dark chocolate - kinetic, thermodynamic and structural considerations. Eur. J. Lipid Sci. Technol. 108, 434—443.

Kinta, Y., Hatta, T., 2007. Composition, structure, and color of fat bloom due to the partial liquefaction of fat in dark chocolate. J. Am. Oil Chem. Soc. 84, 107—115.

Korsmeyer, R.W., Gurny, R., Doelker, E., Buri, P., Peppas, N.A., 1983. Mechanisms of solute release from porous hydrophilic polymers. Int. J. Pharm. 15, 25—35.

Kulkarni, S.B., Kariduraganavar, Y.M.Y., Aminabhavi, T.M., 2003a. Molecular migration of aromatic liquids into a commercial fluoroelastomeric membrane at 30, 40, and 50°C. J. Appl. Polym. Sci. 90, 3100—3106.

Kulkarni, S.B., Kariduraganavar, Y.M.Y., Aminabhavi, T.M., 2003b. Sorption, diffusion, and permeation of esters, aldehydes, ketones, and aromatic liquids into tetrafluoroethylene/propylene at 30, 40, and 50°C. J. Appl. Polym. Sci. 89, 3201—3209.

Lee, W.L., McCarthy, M.J., McCarthy, K.L., 2010. Oil migration in 2-component confectionery systems. J. Food Sci. 75 (1), E83—E89.

Lofborg, N., Smith, P.R., Furo, I., Bergenstahl, B., 2003. Molecular exchange in thermal equilibrium between dissolved and crystalline tripalmitin by NMR. J. Am. Oil Chem. Soc. 80, 1187—1192.

Lonchampt, P., Hartel, R.W., 2004. Fat bloom in chocolate and compound coatings. Eur. J. Lipid Sci. Technol. 106, 241—274.

Maleky, F., Marangoni, A.G., 2011a. Nanoscale effects on oil migration through triacylglycerol polycrystalline colloidal networks. Soft Matter 7, 6012—6024.

Maleky, F., Marangoni, A.G., 2011b. Ultrasonic technique for determination of the shear elastic modulus of polycrystalline soft materials. Cryst. Growth Des. 11 (4), 941—944.

Maleky, F., Smith, S.A., Marangoni, A.G., 2011. Laminar shear effects on crystalline alignments and nano-structure of a triacylglycerol crystal network. Cryst. Growth Des. 11 (6), 2335—2345.

Maleky, F., Marangoni, A.G., 2008. Process development for continuous crystallization of fat under laminar shear. J. Food Eng. 89 (4), 399—407.

Maleky, F., McCarthy, K.L., McCarthy, M.J., Marangoni, A.G., 2012. Effect of cocoa butter structure on oil migration. J. Food Sci. 77 (3), 74—79.

Marangoni, A.G., Acevedo, N., Maleky, F., Co, E., Peyronel, F., Mazzanti, G., Quinn, B., Pink, D., 2012. Structure and functionality of edible fats. Soft Matter 8, 1275—1300.

Marty, S., Baker, K.W., Marangoni, A.G., 2009a. Optimization of a scanner imaging technique to accurately study oil migration kinetics. Food Res. Int. 42, 368—373.

Marty, S., Marangoni, A.G., 2009. Effects of cocoa butter origin, tempering procedure, and structure on oil migration kinetics. Cryst. Growth Des. 9 (10), 4415—4423.

Marty, S., Schroeder, M., Baker, K.W., Mazzanti, G., Marangoni, A.G., 2009b. Small-molecule diffusion through polycrystalline triglyceride networks quantified using fluorescence recovery after photobleaching. Langmuir 25 (15), 8780—8785.

Marty, S., Baker, K., Dibildox-Alvarado, E., Rodrigues, J.N., Marangoni, A.G., 2005. Monitoring and quantifying of oil migration in cocoa butter using a flatbed scanner and fluorescence light microscopy. Food Res. Int. 38, 1189—1197.

McDowell-Boyer, L.M., Hunt, J.R., Sitar, N., 1986. Particle transport through porous media. Water Resour. Res. 22 (13), 1901—1921.

McCarthy, K.L., McCarthy, M.J., 2008. Oil migration in chocolate-peanut butter paste confectionery as a function of chocolate formulation. J. Food Sci. 73 (6), E266—E273.

Meyvis, T.K.L., DeSmedt, S.C., Van Oostveldt, P., Demeester, J., 1999. Fluorescence recovery after photobleaching: a versatile tool for mobility and interaction measurements in pharmaceutical research. Pharmaceut. Res. 16, 1153—1162.

Miquel, M.E., Carli, S., Couzens, P.J., Wille, H.J., Hall, L.D., 2001. Kinetics of the migration of lipids in composite chocolate measured by magnetic resonance imaging. Food Res. Int. 34, 773–781.

Myung, H., Lohman, M.H., Hartel, R.W., 1994. Effect of milk fat fractions on fat bloom in dark chocolate. J. Am. Oil Chem. Soc. 71, 267–276.

Paluri, S., Shavezipur, M., Heldman, D.R., Maleky, F., 2015. Analysis of moisture diffusion mechanism in structured lipids using magnetic resonance imaging. Royal Soc. Chem. Adv. 5, 76904–76911.

Paluri, S., Heldman, D.R., Maleky, F., 2017. Effects of structural attributes and phase ratio on moisture diffusion in crystalized lipids. Cryst. Growth Des. 17 (9), 4661–4669.

Peppas, N.A., Brannon-Peppas, L., 1994. Water diffusion and sorption in amorphous macromolecular systems and foods. J. Food Eng. 22, 189–210.

Peppas, N., Sahlin, J., 1989. A simple equation for the description of solute release. III. Coupling of diffusion and relaxation. Int. J. Pharm. 57, 169–172.

Pinte, J., Joly, C., Pie, K., Dole, P., Feigenbaum, A., 2008. Proposal of a set of model polymer additives designed for confocal FRAP diffusion experiments. J. Agric. Food Chem. 56, 10003–10011.

Ritger, P.L., Peppas, N.A., 1987. A simple equation for description of solute release I. Fickian and Non-Fickian release from non-swellable devices in the form of slabs, spheres, cylinders, or discs. J. Contr. Release 5, 23–26.

Rousseau, D., Smith, P., 2008. Microstructure of fat bloom development in plain and filled chocolate confections. Soft Matter 4, 1706–1712.

Sato, K., Koyano, T., 2001. Crystallization properties of cocoa butter. In: Garti, N., Sato, K. (Eds.), Crystallization Process in Fats and Lipid Systems. Marcel Dekker, Inc., New York, pp. 429–456.

Schaefer, C.E., Arands, R.R., van der Sloot, H.A., Kosson, D.S., 1995. Prediction and experimental validation of liquid-phase diffusion resistance in unsaturated soils. J. Contam. Hydrol. 20, 145–166.

Smith, K.W., Cain, F.W., Talbot, G., 2007. Effect of nut oil migration on polymorphic transformation in a model system. Food Chem. 102, 656–663.

Smith, P., Haghshenas, N., Furo, I., Bergenstahl, B., 2009. Development of a Novel NMR Technique for Measurement of Exchange between Liquid and Solid Triglycerides. http://www.food.leeds.ac.uk/mp/LipidConference/PaulSmithAbstract.html.

Sonwai, S., Rousseau, D., 2010. Controlling fat bloom formation in chocolate—impact of milk fat on microstructure and fat phase crystallization. Food Chem. 119, 286–297.

Sonwai, S., Rousseau, D., 2006. Structure evolution and bloom formation in tempered cocoa butter during long-term storage. Eur. J. Lipid Sci. Technol. 108, 735–745.

Svanberg, L., Ahrne, L., Loren, N., Windhub, E., 2011. A method to assess changes in mechanical properties of chocolate confectionery systems subjected to moisture and fat migration during storage. J. Texture Stud. https://doi.org/10.1111/j.1745-4603.2011.00320.x.

Tabouret, T., 1987. Technical note: detection of fat migration in a confectionery product. Int. J. Food Sci. Technol. 22, 163–167.

Talbot, G., October 1989. Fat migration in confectionery products. Confect. Prod. 655–656.

Talbot, G., 1990. Fat migration in biscuits and confectionery systems. Confect. Prod. 56, 265–272.

Tang, D., Marangoni, A.G., 2006. Quantitative study on the microstructure of colloidal fat crystal network and fractional dimensions. Adv. Colloid Interface Sci. 128–130, 257–265.

Taub, I.A., Singh, R.P., 1998. Food Storage Stability. CRC Press, Inc., Boca Raton, Florida.

Thusyanthan, N.I., Madabhushi, S.P.G., 2003. Scaling of Seepage Flow Velocity in Centrifuge Models. CUED/D-SOILS/TR326. http://wwwciv.eng.cam.ac.uk/geotech new/publications/TR/TR326.pdf.

Timms, R.E., 2002. Oil and fat interactions. Manuf. Confect. (MC) 82, 43–57.

Timms, R.E., 2003. Interactions between fats, bloom and rancidity. In: Barnes, P.J.&.A. (Ed.), Confectionery Fats Handbook. Properties, Production and Application. The Oily Press, Bridgwater, pp. 255–294.

Tozzi, E.J., Lavenson, D.M., McCarthy, M.J., Powell, R.L., 2011. Magnetic resonance imaging to measure concentration profiles of solutes diffusing in stagnant beds of cellulosic fibers. Am. Ins. Chem. Eng. J. https://doi.org/10.1002/aic.12578.

Walter, P., Cornillon, P., 2002. Lipid migration in two-phase chocolate systems investigated by NMR and DSC. Food Res. Int. 35, 761–767.

Wootton, M., Weeden, D., Munk, N., 1970. Mechanism of fat migration in chocolate enrobed goods. Chem. Ind. 1052–1053.

Wang, H., Shi, X., Paluri, S.F., Maleky, F., 2016. Effects of processing and added ingredients on oil diffusion through cocoa butter using magnetic resonance imaging. Royal Soc. Chem. Adv. 6 (91), 88498–88507.

Wang, H., Maleky, F., 2018. Effects of cocoa butter triacylglycerides and minor compounds on oil migration. Food Res. Int. https://doi.org/10.1016/j.foodres.2017.12.057 (In press).

Ziegleder, G., Moser, C., GeierGreguska, J., 1996a. Kinetics of fat migration within chocolate products. 1. Principles and analytics. Fett-Lipid 98, 196–199.

Ziegleder, G., Moser, C., GeierGreguska, J., 1996b. Kinetics of fat migration within chocolate products .2. Influence of storage temperature, diffusion coefficient, solid fat content. Fett-Lipid 98, 253–256.

Ziegleder, G., Petz, A.M., 2001. Fat migration in filled chocolates: the dominant influences. Zucker-und Susswaren Wirtschaft 55 (12), 23–25.

Ziegleder, G., Schwingshandl, I., 1998. Kinetics of fat migration within chocolate products. Part III: fat bloom. Fett-Lipid 100, 411–415.

Ziegleder, G., 1985. Improved crystallization behavior of cocoa butter under shearing. Int. Z. Lebensm. Technol. Verfahrenstech. 36 (6), 412–416.

Ziegleder, G., 2006. Understanding bloom from fat migration. In: Proceedings Chocolate Technology, Cologne, Germany, December 12–14.

Ziegler, G.R., 2009. Product design and shelf-life issues: oil migration and fat bloom. In: Talbot, G. (Ed.), Science and Technology of Enrobed and Filled Chocolate, Confectionery and Bakery Products. CRC Press, Boca Raton, FL, pp. 185–210.

CHAPTER 9

Ultra-Small Angle X-ray Scattering: A Technique to Study Soft Materials

Fernanda Peyronel[1], David A. Pink[2]
[1]University of Guelph, Guelph, ON, Canada; [2]St. Francis Xavier University, Antigonish, NS, Canada

INTRODUCTION

The focus of this chapter is on the use of X-rays to understand the solid (crystalline) structures in edible fats and oils. The main advantage of using X-ray techniques is that the measurements are carried out in situ. This means minimum sample manipulation, allowing for the observation of the structure in its native form.

The term "oil" applies to a triglyceride or triglyceride mixture that is liquid at room temperature, whereas the term "fat" applies to such material that is solid at room temperature. Changes in temperature can easily convert oil into fat or vice versa. Edible fat is a term typically used to refer to a soft material that contains both fats and oils. Edible fats, by definition, do not contain water. Moreover, fats are the main structuring agent in high-fat products, such as chocolate, butter, and shortening, i.e., their solid structure comes from crystal networks composed of higher melting point triacylglycerols (TAGs).

Edible fats are composed of more than 95% TAG molecules with minor components, such as phospholipids, free fatty acids, tocopherols, and phytosterols, waxes among others (O'Brien et al., 1964). The backbone of a TAG molecule is a glycerol molecule to which three fatty acids (FAs) are esterified. These long hydrocarbon chains can adopt different conformations depending on the saturation, the length of their hydrocarbon chain, and the presence of trans- or cis-bonds (Small, 1986).

Edible fats are food products typically manufactured by first blending different molten fat and oil stocks at high temperatures and, second, cooling them down under specific conditions (cooling rate, shear, undercooling temperature) to achieve a product that will perform in a particular way. Industries and researchers have found that, by carrying out particular physical analyses, one can predict if a specific fat product will have the right functionality when used.

Physical analyses can be carried out by, for example, differential scanning calorimetry to characterize the melting behavior, low-resolution pulsed nuclear magnetic resonator to determine the amount of solids, or rheology to characterize solid-state structure.

The formation of structure within an edible fat starts with the self-assembly of TAG molecules as the temperature gets lower from the melt. When the temperature decreases

Structure-Function Analysis of Edible Fats, Second Edition
ISBN 978-0-12-814041-3, https://doi.org/10.1016/B978-0-12-814041-3.00009-5

below the melting point of the highest melting TAGs in the mixture, TAG molecules nucleate, or self assemble, to form bilayers (also referred to as lamellae) with the methyl group at the end of each hydrocarbon chain, directed outward from the bilayer. These bilayers stack epitaxially bound by van der Waals interactions, to form anisotropic crystalline nanoplatelets (CNPs).

The CNPs were observed between 1970 and 1987 by various authors such as Jewell and Meara (1970), Poot et al. (1975), and Heertje and Leunis (1987). Since 2010, CNPs had been systematically characterized using Cryo-TEM (Ramel et al., 2016; Acevedo and Marangoni, 2010a, 2010b, Chapter 1 of this book). CNPs were also studied using models and computer simulations for the edible fat systems that contain less than 50%—60% solids (Pink et al., 2013; Quinn et al., 2014). For those edible fat systems containing more than 60% of solids, the concept of grains were used in the simulations (Peyronel et al., 2014a, 2015), and a different interpretation was given to the experimental observations.

For those systems where the amount of solids is below 50%—60%, the solid crystal network is composed of the aggregation of CNPs stabilized by van der Waals forces. Some liquid oil might be attached to the CNPs (Quinn et al., 2014; Peyronel et al., 2014b) with the remaining oil trapped in between the solids that form the crystal network.

Research efforts in the last 50 years have focused on establishing a relationship between the crystalline structure and the functional aspects of edible fats. The crystalline structure comprises many different length scales, ranging from angstroms to nanometers, micrometers, and millimeters. In particular, much work has been done on edible fats using X-rays. X-ray diffraction and small-angle X-ray scattering (SAXS) are routinely used to reveal the crystalline structure at length scales from ~ 2 Å to ~ 90 Å (Clarkson and Malkin, 1934; Garti and Sato, 1988).

However, the study of larger structures via X-ray scattering, those obtained by the supramolecular arrangement of CNPs, began only in 2013 (Pink et al., 2013 and Peyronel et al., 2013). Using the technique of ultra-small angle X-ray scattering (USAXS), our group showed that structures in the region ~ 400 nm to 6 μm could be predicted using a CNP as the basic scattering unit. CNPs' sizes were characterized in some particular systems by Acevedo and Marangoni (2010a; 2010b, 2014), Acevedo et al. (2012a, 2012b). CNPs' sizes fall in the region of ~ 90 to ~ 1000 nm, and these, in turn, form aggregates the sizes of which lie in a region that USAXS can detect (Peyronel et at., 2013).

The aim of this chapter is to show the basic physical concepts of USAXS and present a brief summary of the results obtained between 2013 and 2017. As with X-ray diffraction, one needs a model to be able to explain the USAXS results. This model was developed by our team and used in computer simulations to make predictions about the outcome of USAXS experiments. This chapter reports on two aspects: (1) USAXS experiments and data fitting and (2) data interpretation by using mathematical models of CNPs and by performing computer simulations of CNP aggregation.

X-RAY SCATTERING

The physical phenomenon of X-rays interacting with matter is called scattering. X-rays can be described either as a wave or as a particle. The electromagnetic spectrum covers different wavelengths, with those possessing values between 0.1 and 200 Å corresponding to X-rays. "Photon" is the name given to the X-rays when they are described as particles and as such, the energy they possess is what characterizes them. A simple equation relates the photon's wavelength (λ) to its energy (E):

$$E = \frac{h\,c}{\lambda},\tag{9.1}$$

where h is Planck constant and c is the speed of light.

X-ray scattering is the process in which an X-ray beam, characterized by an essentially definite linear momentum, is scattered from its original trajectory by the electrons comprising an object placed in its path. Typically, one describes the changes in direction and magnitude of the scattered X-ray beam in relation to the incoming X-ray beam. By detecting the location of the scattered photons, one can infer the position of the electrons in the sample; hence, find information about the internal structure of the material in terms of atoms, molecules, or larger structures.

The first consideration is whether one wishes to perform elastic or inelastic X-ray scattering. In the first case, zero energy is transferred between the object and the X-ray beam, and one is limited to investigating the static structure of the object. In the second case, energy transfer can occur and, as a result, one can excite/deexcite the object and so investigate the dynamics of the object. Here, only elastic scattering is considered.

The X-ray techniques used when studying edible fats are as follows: (1) diffraction or wide-angle X-ray scattering (WAXS), (2) small-angle X-ray scattering (SAXS), and (c) USAXS. Each technique makes use of characteristic optics and detectors, but they all have in common that the desired information is obtained from the scattering angle or the scattering vector. In a simple way, the different techniques can be described by the angle formed between the scattered direction and the "zero direction" of the incoming X-ray beam. Fig. 9.1 illustrates one particular scattered angle direction for each of the three different techniques.

Diffraction experiments focus on large angles, the WAXS region and part of the SAXS region. SAXS experiments cover intermediate angles, whereas USAXS looks at angles that are smaller than 1 degree, hence the prefix "ultra." When an incoming X-ray beam strikes a sample, X-rays are scattered into any direction in the space around the sample. Preferences for specific direction reflect properties of the sample, such as atomic spacing, molecular orientation, and aggregation structures. The scattered directions depend on where the sample's electrons are positioned, on the average, in relation to the direction of the incoming X-ray. It is here that one decides whether elastic or inelastic scattering is being investigated by various means, as the photons travel from the scattering

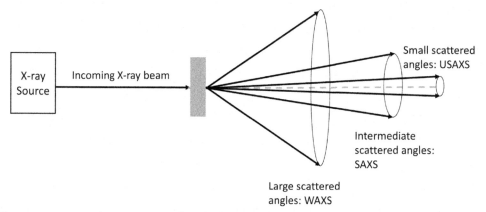

Figure 9.1 Cartoon showing scattered angle directions for each of the three X-ray techniques. USAXS looks at the smallest angles, whereas WAXS looks at the larger ones. *SAXS*, small angle X-ray scattering; *USAXS*, ultra-small angle X-ray scattering; *WAXS*, wide-angle X-ray scattering.

sample to the X-ray detector. There are two kinds of detectors: point ones and 2D ones. Point detectors occupy a position in space, pointing back toward the scattering sample and collect data (intensity of scattered X-rays) for only that single position in space. For that reason, they have to be moved to different positions to collect all the scattered X-ray data. This is not so for 2D detectors (Ponpon, 2005). These detectors present a wall of microscopic essentially point detectors, which accept intensities from a range of scattering directions simultaneously. The reader can imagine a 2D detector as being larger than the cone base shown in Fig. 9.1. This way, the image that the 2D collects is a circle. However, some detectors are designed to only collect the data inside a "vertical slit," rather than the whole circle. Fig. 9.1 does not imply that USAXS detectors should be positioned farther way from the sample than WAXS detectors; it only illustrates the angles.

Among all the X-ray techniques, diffraction is the most widely used. Benchtop equipment that take advantage of the reflection geometry are relatively cheap and do not take too much space. A diffraction experiment gives rise to Bragg peaks from which the distances between atoms or molecules are computed. Diffraction experiment in the WAXS region for edible fat systems cover scattered angles in the range of 5—20 degrees, which gives information about the atomic structure of the material. From this information, one can make conclusions about the ordering of the hydrocarbon chains in TAGs and hence the polymorphism. The part of the SAXS region that is seen in diffraction for edible fat systems covers the scattered angles in the range of 1.5—4 degrees, which gives information about the molecular structure of the material.

Diffraction experiment results are understood on the basis of a simple model: the *unit cell model*. The unit cell is made by the minimum number of atoms forming a particular structure, for example, an atom positioned on each corner of a cube will give rise to a cubic unit cell. The unit cell model assumes that the solid can be completely described by the stacking of *unit cells*, giving rise to an ordered solid. It is this order what gives rise to the X-ray diffraction. An X-ray diffraction shows Bragg peaks, which are the results of constructive interference of the scattered X-rays. Interested readers can learn more by going to Chapter 3.

The SAXS technique looks at smaller angles compared with the diffraction one. The SAXS region can be studied using an X-ray diffractometer, or it can be studied using a dedicated SAXS instrument. The latest one has the optics and a detector set up to collect information only in the SAXS region. There is no precise angle that separates the WAXS from the SAXS region. Many researchers simply called SAXS the region that looks at smaller scattered angles compared with the WAXS region. A SAXS pattern might display some Bragg peaks, but it will also display a region toward the smaller angles in which the scattering intensity can increase as the angle gets smaller. Edible fats have been studied by many in the SAXS region by focusing only on the Bragg peaks. The information obtained from these peaks is linked to the stacking of the molecules, for example, the length of the lamella. The thickness of the crystal can also be computed from the information in those Bragg peaks. To do this, it is necessary to use the Scherrer equation (see Chapter 11).

The USAXS technique looks at the smallest detectable scattered angles. This requires specialized optics and detectors: (1) the optics has to be such that allows the detection of scattered X-rays close to the forward direction of the X-ray incoming beam (primary beam) and (2) careful removal of instrumental scattering is necessary not to confuse the scattered data with that of the primary beam.

These two points are overcome by using a Bonse—Hart instrument and specialized detectors. Before describing the Bonse—Hart camera, it is appropriate to introduce some definitions.

PRINCIPLES OF X-RAY SCATTERING

From the quantum point of view, an X-ray beam is a collection of photons, which can be described by the wave vector k, which indicates the direction of propagation and the wavelength of the photons. In practice, this is the average wave vector because one cannot, in practice, create a beam with a unique wave vector. Assuming that a photon interacting with the sample has a momentum $\hbar k_i$ and the photon leaving the sample has a momentum $\hbar k_s$, then the scattering vector q is given by

$$q = k_s - k_i. \tag{9.2}$$

This section is concerned with elastic X-ray scattering. Elastic scattering is characterized by zero energy transfer, which indicates that the magnitude of incident and scattered wave vectors are the same:

$$|k_i| = |k_s| = 2\pi/\lambda. \tag{9.3}$$

The magnitude of the scattering q vector is defined as

$$q = |k_s - k_i| = \frac{4\pi \sin \theta}{\lambda}. \tag{9.4}$$

It can be seen in Eq. (9.4) the relation between the scattering angle θ and the scattering q vector. This definition is valid for diffraction, SAXS, and USAXS.

The scattering q vector has units of length^{-1}, and q^{-1} represents the length scale of interest, L, of the system under study, where

$$L = \frac{2\pi}{q}. \tag{9.5}$$

Eq. (9.5) shows that the smaller the q vector, the larger the length scale to be detected.

Table 9.1 shows the length scales for USAXS, SAXS, and WAXS. Eq. (9.5) has been used to convert the q-value into length scale.

In this chapter, the name *intensity* and $I(q)$ are used to symbolize absolute intensity as shown in Fig. 9.2. Absolute intensities are obtained after processing the data when the experiment is carried out using a Bonse—Hart instrument.

The question was how can the q dependence of the intensity observed in the USAXS region be explained? The USAXS region does not displayed any Bragg peaks, and it covers length scales beyond the molecular structures. All that one can say is that the solid structure observed in the USAXS region contains larger structures than a simple bilayer of molecules. It was decided that a *model* was needed to explain the q dependence of the scattering intensity obtained. A mathematical model of CNPs was made, and those model CNPs were used as the basic scattering units. Predictions for USAXS experiments were made by allowing the model CNPs, interacting via van der Waals forces, to adopt structures if they aggregate at a particular temperature. The scattered intensity measured is, by definition, proportional to the square of the product of the structure function and the form factor according to the electron distribution in the sample (Als-Nielsen and McMorrow, 2001). From the modeling and computer simulation of the aggregation of the CNPs, the structure function was obtained and compared with the measured intensity $I(q)$ from the experiments. Conclusions were drawn.

Table 9.1 Length scales that are covered by the different X-ray techniques

X-ray technique	Spatial scale	Length scale [Å]	q range [Å^{-1}]
Diffraction/WAXS	Atomic	1—10	6—0.6
SAXS	Molecular	10—100	0.6—0.06
SAXS/USAXS	Nano	100—5000	0.06—0.007
USAXS	Micro	1—20 μm	6×10^{-4}—3×10^{-5}

SAXS, small angle X-ray scattering; *USAXS*, ultra-small angle X-ray scattering; *WAXS*, wide-angle X-ray scattering

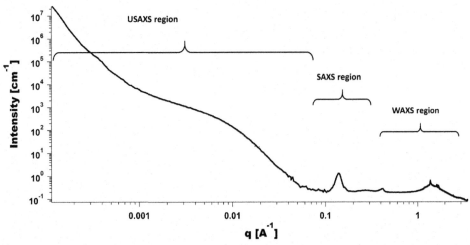

Figure 9.2 Intensity versus scattered q vector shown in a log–log plot, for an edible fat system obtained using three different detectors to cover the WAXS, SAXS, and USAXS region. *SAXS*, small angle X-ray scattering; *USAXS*, ultra-small angle X-ray scattering; *WAXS*, wide-angle X-ray scattering. *(Adapted from Peyronel, F., Ilavsky, J., Pink, D.A., Marangoni, A.G., 2014c. Quantification of the physical structure of fats in 20 minutes: implications for formulations. Lipid Technol. 26 (10), 223–226.)*

MODELING AND COMPUTER SIMULATION PREDICTIONS FOR EDIBLE FATS

All experimental data are interpreted using conceptual or mathematical models. To obtain useful output from the models, one must perform additional mathematical procedures that calculate how the model behaves in prescribed circumstances. If, using these mathematical procedures, the models actually predict something, these predictions can be helpful in understanding experimental data.

Here we distinguish between the laboratory edible fat sample which is to be studied using USAXS, and the model fat system which is to be studied using a mathematical procedure. We assume, initially, that a mixture of lipids of the laboratory sample has been cooled to a final temperature, T, at which there is a fixed concentration of solid CNPs in a homogeneous liquid oil phase. The experimental CNPs were modeled assuming that all "model CNPs" are the same size, their shape does not change as time goes by, and their solubility in the liquid oil is, for all practical purposes, zero. The interactions between model CNPs are then specified. The additional mathematical procedures necessary to obtain useful predictions for USAXS experiments from this model must then be chosen.

To model a CNP, it is convenient and efficient to represent it as a rigid array of spheres, a strategy that lends itself to using computer simulation to study their statics and dynamics. These spheres represent aggregates of TAG molecules, which make up a portion of a model CNP. Each CNP is represented by a 10×10 planar array of hard (they cannot penetrate each other) spheres closely packed in a triangular lattice. The size of the spheres in SI units is easily determined: if each model CNP is 500 nm on a side (Acevedo and Marangoni, 2010a), then the diameter of each of the 10 spheres is 50 nm. The attractive van der Waals interaction between any two spheres is well known (Parsegian, 2005). The models described here assume that the total interaction between two such model CNPs is the sum of the interactions between every pair of spheres each belonging to one of the model CNPs. The van der Waals interaction between spheres on the same model CNP can be ignored because those are essentially rigidly held in place so that their interactions with each other do not change. Accordingly, only the interactions between spheres belonging to two *different* model CNPs are considered.

The attractive van der Waals interaction between two model CNPs depends not only on the TAGs, which comprise these CNPs, but also on the TAG molecules comprising the liquid oil in which the model CNPs are held. This dependence is manifested via the Hamaker coefficient, and its value can be estimated.

Spheres belonging to different model CNPs exhibit an additional interaction: a short-range repulsive interaction to allow for the fact that a pair of fats cannot occupy (much of) the same space simultaneously. Interested readers can find more information in this book, Chapter 7 and in Pink et al., (2013).

The complexity of this model, which comprises the model CNPs together with all their interactions, requires the use of computer simulation to obtain predictions concerning edible oils in thermal equilibrium at some fixed temperature, T.

A proven well-understood technique to use is the Metropolis Monte Carlo (MMC) method (Metropolis et al., 1953; Binder and Heermann, 1997; Landau and Binder, 2000), which proceeds via sequential Monte Carlo steps (MCSs). These involve attempting to move each model CNP and each cluster of model CNPs once at each MCS. In carrying out these procedures, the interactions between spheres belonging to different model CNPs must be taken into account because they can affect the outcomes. The attempted movement of each model CNP in a cluster permits clusters to change their shape and size, so their structure might change. It also enables clusters to form and break up.

To relate computation output to USAXS experiments, the isotropic structure function, $S(q)$ (Pink et al., 2013 and Chapter 7 of this book) was computed. This is done by making use of the knowledge of the positions of all the spheres, which make up the model CNPs. $S(q)$ is obtained from the structure function for oriented samples, $S(\vec{q})$, (which is not relevant to the USAXS experiments described here) by averaging over all directions of $\vec{q}$ keeping its magnitude, $q = |\vec{q}|$, fixed. $S(q)$ is the average Fourier transform of a pair correlation function. One then uses the result that, if the system possesses a fractal dimension, D, then, for some range of q, $\log S(q) \propto -D \log q$. This implies that one is looking for a sufficiently large region of q-space (reciprocal space) in which the experimental USAXS intensity, $I(q)$, exhibits the property that $\log I(q)$ is proportional to log q. Initially, tristearin CNPs in triolein oil (OOO) was studied (Peyronel et al., 2013). One result of computer simulation of the models has been that model CNPs were found to aggregate via their (010) surfaces to form TAGwoods, (one-dimensional) columns of model CNPs adhering to each other via their (010) surfaces. Recent models, driven in part by USAXS experiments on increasingly complex systems, have arisen from the first simple models. Model CNPs, modified to allow for selective adsorption of minority-phase molecules onto some, or all, of their surfaces, have been studied (Quinn et al., 2014). In the case where the (100) and (001) surfaces of the model CNPs are coated by adsorbed molecules, which prevent aggregation at these surfaces, the TAGwoods formed will have their sides coated whereby they can aggregate only at their ends, thereby forming one-dimensional "TAGwood snakes" (Peyronel et al., 2016).

USAXS INSTRUMENTATION AT ADVANCED PHOTON SOURCE

Although Bonse—Hart instruments can be assembled in labs, those setup implies low counts of X-rays coming from vacuum-sealed X-ray tubes. This problem is overcome by performing the experiments at a synchrotron facility.

All experiments described in this chapter were carried out at the Advanced Photon Source (APS), Argonne National Laboratory, USA. If the research is nonpropietary, then National Laboratories in the United States are the places to go, as they are institutions open to researchers (user) free of charge. Different facilities provide different equipment and experts. Users have many options as to where to go based on what they want to study. At Argonne National Laboratory, one has the option to choose between ~ 45 different beam lines and multiple instruments/techniques. The work presented here was done with Dr. Ilavsky and his team taking advantage of the Bonse—Hart USAXS instrument (Ilavsky et al., 2009, 2012).

This Bonse—Hart instrument uses two sets of crystal pairs located in the X-ray beam path. The first pair is the collimating pair and is placed before the sample. The second pair is the analyzer pair and is located immediately after the sample. Crystals are typically made from silicon, Si. Precision grooves are made on each crystal. The incoming X-ray beam bounces in between the pair of crystals that are separated by millimeters. The distances between the grooves and the gap between the crystals are specifically designed to achieve particular resolution. The instrument can be set to use the Si 220 crystal pair, which gives a $1 \times 10^{-4} \text{ Å}^{-1}$ resolution or a Si 440 crystal pair, which gives a resolution of $3 \times 10^{-5} \text{ Å}^{-1}$. The resolution in USAXS identifies the minimum q-value obtainable. The Si 220 crystal pair needs an X-ray beam with an energy of ~ 18 keV, whereas the 440 crystal pair needs an X-ray beam of ~ 24 keV.

It is to note, that this instrument does not make use of a beam stopper, as some SAXS instruments do. Damage to the detector is avoided by a short exposure or, when necessary, by using attenuators or filters on the primary beam itself. The USAXS equipment at

APS uses a photodiode detector with a dynamic range of more than eight decades, which permits measurement of the scattering intensity through the direct beam. By rotating the analyzer crystal pair using small steps, the scattered intensity at various angles can be measured with an extremely high angular precision or resolution. Specific details of the operation of the instrument can be found elsewhere (Ilavsky et al., 2012, 2013).

The work carried out here made use of the one-dimensional geometry arrangement (Long et al., 1991), which enables a q-resolution that is very high in the vertical direction but poor in the horizontal one. This is why the data require a numerical desmearing procedure (Lake, 1967) to recover the differential scattering cross-section or scattering intensity $I(q)$. Desmearing the data is carried out after removing the instrumental scattering. A "blank" or empty sample is used to remove the instrumental scattering. Recording of the beam profile is done every 2 h or whenever the sample holder is moved or changed. The process of removing the instrumentation scattering is called data reduction. The data obtained after reduction and desmearing are the "absolute data."

The Bonse—Hart system permits the collection of a sufficient number of data points in only a few minutes to cover the spatial length scale ~ 100 nm $< L < \sim 10$ μm. The setup at the APS allows to sequentially position the USAXS, Pin SAXS, and WAXS detectors to be able to cover even larger values in the q-region, which go from 1×10^{-4} Å^{-1} or 3×10^{-5} Å^{-1} to ~ 7 Å^{-1}. This is an impressive achievement and is all done automatically by repositioning the detectors without moving the sample.

SAMPLE PREPARATION

Samples can be deposited in capillary tubes or in Grace Bio-Labs silicon isolator (Grace Bio-Labs, Bend, Oregon, USA). The work carried out here was performed using the silicon isolators, which are easy to fill avoiding much sample manipulation. These Grace Bio-Labs isolators are thin squares of silicon with a round or square cavity in the middle that resist temperatures up to 140°C. Isolators can be naked or covered with an adhesive on one or both surfaces. Fig. 9.3A shows a particular isolator with eight cavities. The sample is deposited inside the cavity and "sandwiched" between two microscope glass cover slip attached at the bottom and at the top (Fig. 9.3B).

A melted sample can be directly poured in the cavity, or a semisolid material can be introduced using a spatula. In the latest case, care has to be taken in avoiding any preferential orientation or compression on the samples because the results of the USAXS experiments will detect these changes, and one can then misinterpret the results (Peyronel et al., AOCS conference 2016).

When performing temperature-controlled experiments, the isolator sample holder can be mounted in a Linkam heating/cooling system. The Linkam is positioned in a way that the X-ray beam goes through its center hole and, hence, through the sample.

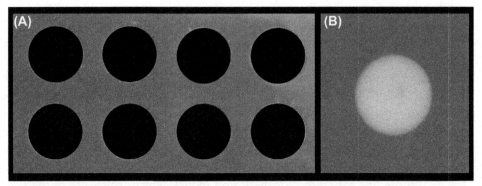

Figure 9.3 Grace Bio-Labs isolators (A, B) used to make the sample holders for edible fat samples (B).

The protocol followed for the temperature cycle on the sample was as follows: (1) the sample was melted in the isolator sample holder at 80°C (2) the temperature was held at 80°C for 15 min, and (3) the isolator sample holder was cooled at either 1°C/min or 30°C/min to achieve either a slow or fast cooled sample.

EXPERIMENTAL DATA FITTING

The analytical models typically used to obtain some parameters from the data collected in USAXS are as follows: the Unified Fit model (Beaucage, 1995, 1996) and the Guinier–Porod model (Hammouda, 2010).

Eq. (9.6) shows the Unified Fit model expression.

$$I(q) = Ge^{-\left(\frac{q^2 R_{gi}^2}{3}\right)} + \frac{B}{q^P}e^{-\left(\frac{q^2 R_{gi-1}^2}{3}\right)}\left[erf\left(\frac{-qR_{gi}}{6^{1/2}}\right)\right]^{3P}, \tag{9.6}$$

where P is the Porod exponent that yields information on the internal structure of the scatterer (Schaefer et al., 2012), R_g is the radius of gyration that gives information on the average size of the scatterer, and $erf(x)$ represents the error function of x. Here, scatterer could be understood as either the basic unit that scatter or a larger structure that is made by the aggregation of some of those basic units. The models allow to link different structural levels. A structural level is considered a section in the q-region that defines one spatial length scale for the material. The subscript i in Eq. (9.6) indicates a particular structural level, whereas -1 indicates the previous structural. The software allow to have these levels link or independent from each other. The exponent $\exp\left(-\frac{q^2 R_{gi}^2}{3}\right)$ shows that the Unified Fit model represents all scattering objects as approximate spheres. The first term in Eq. (9.6) describes the Guinier region, valid for

$q \leq q_1$, and the second term, the Porod power law valid for $q \geq q_1$. The requirement is that the Guinier and Porod terms, together with their derivatives must be continuous at q_1. G and B are the Guinier and Porod scale factors, respectively, with G related to the volume of the scatterers and B containing specific scattered surface area information.

When a fractal interpretation is given to the data, the P value gives information about the fractal dimension of the scatterers. A value of $|P| = 4$ indicates a scatterer or particle with a smooth surface and a fractal dimension $D_s = 2$. If $3 \leq |P| < 4$, then $6-|P|$ is the surface fractal dimension ($2 < Ds \leq 3$). When $1 \leq |P| < 3$, then the value of $|P|$ is the fractal dimension, $D = |P|$. But if $|P| \cdot 4$, all that is known is that there is diffuse interface (Beaucage, 1995 and references therein).

Eq. (9.7) shows the expression for the Porod model, which makes use of the same parameters described for the Unified Fit, but with the difference that this model does not work with spheres.

$$I(q) = \frac{G_2}{q^{s_2}} e^{\left(\frac{-q^2 R_{g2}^2}{3 - s_2}\right)} \qquad q \leq q_2$$

$$I(q) = \frac{G_1}{q^{s_1}} e^{\left(\frac{-q^2 R_{g1}^2}{3 - s_1}\right)} \qquad q_2 \leq q \leq q_1 \qquad (9.7)$$

$$I(q) = \frac{B}{q^P} \qquad q \geq q_1$$

The value of the "s" parameter is used to identify the particular shape of the scatterer. Platelets are present when $s = 2$, cylinders when $s = 1$ and spheres when $s = 0$. Similar to the Unified Fit model, parameters such as the radii of gyration and the power law exponents are obtained.

Both models use nonlinear regression analysis to find the best parameters. The analysis proceeds by identifying a number of "levels" at different q-regions. Each level is viewed as a spatial structural level. A structural level is described by a Guinier function (Guinier and Fournet, 1955) comprising a "knee" and the associated power law regime.

Data fitting requires not only a good fit for all the data points, but it also needs a sound physical explanation. One of the challenges of fitting data is the accuracy of the fit. _Irena_ (Ilavsky and Jemian, 2009) is the software written at APS and used to analyze the data collected there. _Irena_ minimizes χ^2 when finding the parameters for Unified Fit or the Porod model. The large number of parameters (4 per level) makes it difficult to choose the best fit. We decided that the best fit was the one that gave the smallest error for all of the parameters of interest.

As mentioned in the previous section, our interest was to find slopes in order to be compared with the simulation's predictions. These slopes also needed to comprise a large enough q-region to be described as fractal. The following section shows the slope values obtained for different systems and their interpretation based on the simulations. The same as with the slopes, R_g values are parameters obtained from the fitting. We do not place too much importance on this number, as it is an average of the structures giving rise to the R_g. Of course, smaller values are obtained for the nanocrystals than for the microcrystals. But it is hard to put a number on the length, width, and thickness, of either the nano- or the microcrystal when the results are the average of these three dimensions.

APPLICATIONS TO EDIBLE FATS

The experiments involving edible fats carried out using USAXS can be divided into two types: (1) samples that were first melted and crystallized following a particular protocol and (2) commercial materials that were either spooned into the sample holder or cut and gently deposited into the sample holder.

Two types of experiments were carried out in the first protocol: (1) crystallization under static conditions and (2) crystallization under laminar shear. Crystallization under static conditions means that no perturbation was introduced while the temperature was lowered from the melt. Crystallization under shear was limited to crystallizing under a controlled cooling rate while applying a laminar shear stress.

The variables that were changed were cooling rate, solid/liquid ratio, chemical composition, and postcooling storage time.

Fig. 9.4 shows the scattering intensities obtained for a system made of 20% tripalmitin (PPP) in triolein (OOO) measured at 20°C after it was statically crystallized from the melt at 1° per min, which was considered a slow cooling rate.

The data showed in Fig. 9.4 are the absolute data, as explained in the previous section and the best fit (gray line). Hierarchical levels were identified using the Unified Fit model. The data points that define each level are not easy to identify, and many trials were needed before the three levels showed a fit that converged. The hierarchical structure was achieved by linking the levels in the *Irena* software and requiring global convergence.

What follows is the interpretation given to the data using predictions of the models (Pink et al., 2013; Quinn et al., 2014).

Level 1 contains the smaller structures. Level 2 reflects aggregation of the structures identified in Level 1. Level 3 reflects the aggregation of the structures obtained in Level 2. The two parameters of interest obtained from the Unified Fit model were the slopes and the radius of gyration R_g. As mentioned before, the focus of this research has been on the slopes, rather than the R_g. Experience with the fits had shown that the number for R_g that comes out of the fitting varies substantially from fit to fit, which is not the case for the slopes. Nonetheless, the R_g found in Level 1 was associated with the CNPs average nanocrystal size, whereas the R_g found in Level 2 was associated with the size of an average microcrystal.

Slope values were used to describe the surface or mass fractality of the system at each particular spatial length scales. The parameter P (Eqs. 9.6 and 9.7) is the one related to the

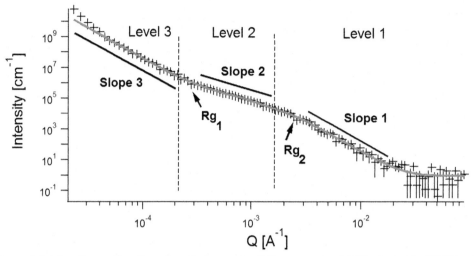

Figure 9.4 Intensity as a function of scattering vector Q for a sample of 20% tripalmitin (PPP) in tri-olein (OOO) crystallized under slow conditions. Shown are the three hierarchical levels identified using the Unified Fit model from Irena. *(Figure adapted from Peyronel, F., Pink, D.A., Marangoni, A.G., 2014d. Triglyceride nanocrystal aggregation into polycrystalline colloidal networks: ultra-small angle x-ray scattering and computer simulation of models. Curr. Opin. Colloid Interface Sci. 19, 459–470.)*

slope (Eqs. 9.6 and 9.7) via $P = -$ slope. The fractal dimension, D, can be obtained from this P parameter, as explained above.

The value of D can then be compared with the results of the models. We used a fractal interpretation when the log $I(q)$ versus log(q) graphs showed a straight line that encompassed more than half a decade of data points.

A value of $s \cong 1$ was obtained when using the Guinier–Porod model for the data shown in Fig. 9.4, as explained by Eq. (9.20). This value indicated that rods originated the scattering. In accord with predictions of models (Pink et al., 2013), it was proposed that the rods were made by the stacking of CNPs on their larger (001) surfaces. These rods were named TAGwoods.

Table 9.2 shows possible P values and its conversion to a fractal dimension (using the explanation given before Eq. (9.7)), D and their interpretation.

One needs to be careful when interpreting the results. The computer simulations were performed using CNPs as the fundamental unit. This means that it is the unit generating the scattering. Hence if the material studied does not have this fundamental unit, then the use of the interpretation given in Table 9.2 might be wrong. In addition, these results were simulated for low to intermediate concentrations of solids. When the amount of solids is larger than 50%, the interpretation changes, as shown by Peyronel et al. (2014a, 2015).

Many edible fat systems had been studied over the last 5 years using USAXS. The scattering intensity was studied using the Unified Fit model to discover the hierarchical

Table 9.2 *P* parameter values obtained after the data was fitted using the Unified Fit model and its conversion to the fractal dimension *D*

P Parameter	D	Interpretation of fractal structure
1	1	Unaggregated rods[1]
1.7 to 1.9	1.7 to 1.9	Diffusion-limited cluster—cluster aggregation[1]
2.0—2.1	2.0—2.1	Reaction-limited cluster—cluster aggregation[1]
3	3	Uniform distribution of structures in space[1]
4	2	Smooth surface
>4	N/A	The surface is not well defined; it is "diffused"
$3 < P < 4$	$2 < D < 3$	Surface fractal dimension, indicating roughness

[1]The interpretation of these slopes came from modeling and computer simulations (Pink et al., 2013).

levels of aggregation. The Porod—Guinier model was used to elucidate the morphology of the aggregates. We had seen that some edible fat systems contain two hierarchical levels, whereas some others show three levels. One needs to be careful again regarding the number of levels and the q-region in which each level appears. Table 9.3 lists the *P* value parameter obtained during past studies.

Table 9.4 shows the parameters obtained for some commercial products.

Table 9.3 Different edible fat systems and the number of hierarchical levels identified using Unified Fit and the corresponding *P* and *S* values

System	Number of levels	Values of P	S Value	Reference
5, 10, 15, 20% SSS in OOO 22 days storage	3	$P_1 = 4$ $P_2 = 1.8$ to 2.1 $P_3 = 3$	~1	Peyronel et al. (2013)
20% SSS in OOO 0.5 deg/min (slow cooling)	3	$P_1 = 4$ $P_2 = 1.9$ $P_3 = 3.8$	1.2	Peyronel et al. (2013)
20% SSS in OOO 30 deg/min (fast cooling)	3	$P_1 = 4$ $P_2 = 1.2$ $P_3 = 3.8$	1.2	Peyronel et al. (2013)
20% SSS in OOO 2 years of storage	2	$P_1 = 4$ $P_2 = 2$	1	Peyronel et al. (2014b)
20% PHCO in OOO 48 h storage	3	$P_1 = 4$ $P_2 = 1.5$ $P_3 = 4.1$	1.5	Peyronel et al. (2014b)
20% SSS, 16% Shea, 64% HOSO Storage: 48 h and 8 months	3	$P_1 = 4$ $P_2 = 1.1$ $P_3 = 4.1$	1.2	Peyronel et al. (2014b)
SSS, cotton seed, OOO Various concentrations Storage: 6 days and 8 months	3	$P_1 = 4.1$ to 4.3 $P_2 = 1.1$ $P_3 = 4.1$	~1.0	Peyronel et al. (2014b)

HOSO, high oleic sunflower oil; *OOO*, triolein; *PHCO*, partially hydrogenated canola oil; *SSS*: tristearin.

Table 9.4 Commercial products and the number of hierarchical levels identified using Unified Fit and the corresponding P values, S values, and solid fat content (SFC) at a particular temperature stated in brackets

System	Number of levels	Values of P	S Value	SFC	Reference
Milk fat Storage: 6 months	2	$P_1 = 3.9$ $P_2 = 1.8$	>2	11% (23°C)	Peyronel et al. (2015)
Milk fat Storage: 2 months	3	$P_1 = 4.0$ $P_2 = 2.1$ $P_3 = 4.3$	2		Ramel et al. (2016)
Palm olein Storage: 1 year	2	$P_1 = 4.0$ $P_2 = 2.1$	>2	9% (23°C)	Peyronel et al. (2015)
Cocoa butter Storage: 1 year	3	$P_1 = 3.7$ $P_2 = 2.7$ $P_3 = 4.4$	>2	56% (23°C)	Peyronel et al. (2015)
Hydrogenated PKO-based fat Storage: 27 months	3	$P_1 = 3.6$ $P_2 = 2.4$ $P_3 = 4.2$	1.5	52% (23°C)	Peyronel et al. (2015)
Laminated rolling shortening	3	$P_1 = 4.0$ $P_2 = 1.5, 1.8, 2.1$ $P_3 = 4.3.6 \; or \; 2.8$	~1.5	22%–26% (20°C)	Macias- Rodriguez et al. (2017)
Cake shortening	2	$P_1 = 4.0$ $P_2 = 4.2$	~1	20% (20°C)	Macias- Rodriguez et al. (2017)

SUMMARY

We have described how to use the technique of USAXS, identifying all the steps necessary for its successful implementation. We have also described experiments for simple and complex systems. We have pointed out that to understand the scattering intensity output, it is necessary to develop a mathematical model that can be used to predict some characteristics of the structure function.

A mathematical model of CNPs was devised and used in computer simulations to predict the USAXS outcome. By comparing the experimental results with the model predictions, conclusions about the aggregation of the CNPs were drawn.

The models have shown that the structures obtained by the aggregation of CNPs were dependent on (1) the solid-to-liquid ratio, (2) the TAGs present in the edible fat, and (3) the processing conditions. The computer simulations explained the mechanisms of formation of the solid structure with spatial length scales from hundreds of nanometers to a few micrometers.

The experimental data were fitted to a maximum of three hierarchical levels, which can be summarized as follows:

Structural level 1 identified the primary scatterers, the CNPs, which appeared at the largest q-values in the USAXS regime.

Structural level 2 was observed in a q-region for smaller values than those in structural level 1. Structural level 2 was associated with either the aggregation of CNPs into TAGwoods, or the aggregation of those TAGwoods.

Structural level 3 was associated with the spatial distribution of the clusters observed in structural level 2. In some cases when the system came to equilibrium, the clusters were found to be uniformly distributed in space on average. In other cases, however, the information obtained for this level indicated that the interface of the cluster with its surroundings was not well defined.

ACKNOWLEDGMENTS

The authors would like to thank Dr. Bonnie Quinn for making the model CNPs and carrying out the simulations. The authors also want to thank Dr. Jan Ilavsky and his team at beamline 9ID for their invaluable support and help through the data collection and analysis. This research used resources of the Advanced Photon Source, a US Department of Energy (DOE) Office of Science User Facility operated for the DOE Office of Science by Argonne National Laboratory under Contract No. DE-AC02-06CH11357.
This work was supported by the Natural Sciences and Engineering Research Council of Canada and by ACENET.

REFERENCES

Als-Nielsen, J., McMorrow, D., 2001. Elements of Modern X-ray Physics, second ed. John Wiley and Sons, Ltd., West Sussex, England.

Acevedo, N.C., Marangoni, A.G., 2010a. Characterization of the nanoscale in triacylglycerol crystal networks. Cryst. Growth Des. 10 (8), 3327–3333.

Acevedo, N.C., Marangoni, A.G., 2010b. Toward nanoscale engineering of triacylglycerol crystal networks. Cryst. Growth Des. 10 (8), 3334–3339.

Acevedo, N.C., Block, J.M., Marangoni, A.G., 2012a. Critical laminar shear-temperature effects on the nano- and mesoscale structure of a model fat and its relationship to oil binding and rheological properties. Faraday Discuss 158, 171–194.

Acevedo, N.C., Block, J.M., Marangoni, A.G., 2012b. Unsaturated emulsifier-mediated modification of the mechanical strength and oil binding capacity of a model edible fat crystallized under shear. Langmuir 28, 16207–16217.

Acevedo, N.C., Marangoni, A.G., 2014. Functionalization of non-interesterified mixtures of fully hydrogenated fats using shear processing. Food Bioprocess Technol. 7, 575–587.

Beaucage, G., 1995. Approximations leading to a unified exponential/power-law approach to small-angle scattering. J. Appl. Crystallogr. 28, 717–728.

Beaucage, G., 1996. Small-angle scattering from polymeric mass fractals of arbitrary mass-fractal dimension. J. Appl. Crystallogr. 29, 134–146.

Binder, K., Heermann, D., 1997. Monte Carlo Simulation in Statistical Physics: An Introduction, vol. 3rd. Springer.

Clarkson, C.E., Malkin, T., 1934. Alternation in long-chain compounds. Part I-l. An x-ray and thermal investigation of the triglycerides. J. Chem. Soc. 666–671.

Garti, N., Sato, K., 1988. Crystallization of fats and fatty acids. In: Crystallization and Polymorphism of Fats and Fatty Acids. Dekker, New York, pp. 254—259.

Guinier, A., Fournet, G., 1955. In: Fournet, G. (Ed.), Small-angle Scattering of X-rays. Wiley, New York.

Hammouda, B., 2010. A new guinier—porod model. J. Appl. Crystallogr. 43, 716—719.

Heertje, I., Leunis, M., 1997. Measurement of shape and size of fat crystals by electron microscopy. LWT - Food Sci. Technol. 30 (2), 141—146.

Ilavsky, J., Jemian, P.R., 2009. Irena: tool suite for modeling and analysis of small-angle scattering. J. Appl. Crystallogr. 42 (2), 347—353.

Ilavsky, J., Jemian, P.R., Allen, A.J., Zhang, F., Levine, L.E., Long, G.G., 2009. Ultra-small-angle X-ray scattering at the advanced photon source. J. Appl. Crystallogr. 42, 469—479.

Ilavsky, J., Allen, A.J., Levine, L.E., Zhang, F., Jemian, P.R., Long, G.G., 2012. High-energy ultra-small-angle X-ray scattering instrument at the advanced photon source. J. Appl. Crystallogr. 45, 1318—1320.

Ilavsky, J., Jemian, P.R., Allen, A.J., Zhang, F., Levine, L.E., Long, G.G., 2013. Ultra-small-angle x-ray scattering instrument at the advanced photon source: history, recent development, and current status. Metall. Mater. Trans. A 44, 68—76.

Jewell, G.G., Meara, M.L., 1970. A new and rapid method for the electron microscopic examination of fats. J. Am. Oil Chem. Soc. 47, 535—538.

Lake, J.A., 1967. An iterative method of slit-correcting small angle x-ray data. Acta Crystallogr. 23, 191—194.

Landau, D.P., Binder, K., 2000. A Guide to Monte Carlo Simulations in Statistical Physics, Vol. 2nd. UK. Cambridge University Press, Cambridge.

Long, G.G., Jemian, P.R., Weertman, J.R., Black, D.R., Burdette, H.E., Spal, R., 1991. High-resolution small-angle x-ray scattering camera for anomalous scattering. J. Appl. Crystallogr. 24, 30—37.

Macias-Rodriguez, B.A., Peyronel, F., Marangoni, A.G., 2017. The role of nonlinear viscoelasticity on the functionality of laminating shortenings. J. Food Eng. 212, 87—96.

Metropolis, N., Rosenbluth, A.W., Rosenbluth, M.N., Teller, A.H., Telle, E., 1953. Equation of state calculations by fast computing machines. J. Chem. Phys. 21, 6.

O'Brien, R.D., Jones, L.A., King, C.C., Wakelyn, P.J., Wan, P.J., 1964. Edible oil and fat products. In: Shahidi, F. (Ed.), Bailey's Industrial Oil and Fat Products, vol. 2. Wiley -Interscience Publishers, New York, p. 192.

Parsegian, V.A., 2005. Van der Waals Forces. A Handbook for Biologists, Chemists, Engineers, and Physicists. Cambridge University Press, Cambridge.

Peyronel, F., Ilavsky, J., Mazzanti, G., Marangoni, A., Pink, D., 2013. Edible oil structures at low and intermediate concentrations: II. Ultra-small angle x-ray scattering of in situ tristearin solids in triolein. J. Appl. Phys. 114, 234902.

Peyronel, F., Quinn, B., Marangoni, A.G., Pink, D.A., 2014a. Ultra small angle x-ray scattering for pure tristearin and tripalmitin: model predictions and experimental results. Food Biophys. 9 (4), 304—313.

Peyronel, F., Quinn, B., Marangoni, A.G., Pink, D.A., 2015. Edible fat structures at high solid fat concentrations: evidence for the existence of oil-filled nanospaces. Appl. Phys. Lett. 106 (2), 023109.

Peyronel, F., Quinn, B., Marangoni, A.G., Pink, D.A., 2014b. Ultra small angle x-ray scattering in complex mixtures of triacylglycerols. J. Phys. Condens. Matter 26 (46), 464110.

Peyronel, F., Ilavsky, J., Pink, D.A., Marangoni, A.G., 2014c. Quantification of the physical structure of fats in 20 minutes: implications for formulations. Lipid Technol. 26 (10), 223—226.

Peyronel, F., Pink, D.A., Marangoni, A.G., 2014d. Triglyceride nanocrystal aggregation into polycrystalline colloidal networks: ultra-small angle x-ray scattering and computer simulation of models. Curr. Opin. Colloid Interface Sci. 19, 459—470.

Peyronel, M.F., 2015. Ultra Small Angle X-Ray Scattering Studies of Triacylglycerol Crystal Networks (Ph D thesis). University of Guelph, Canada.

Peyronel, F., Bonnie, Q., Pink, D.A., Marangoni, A.G., 2016a. Side-coated TAGwppds and the Formation of Dense Aggregated of Intertwined TAGwoods Strings. Advances in X-ray Analysis. V60. International Center for Diffraction Data. www.icdd.com.

Peyronel, F., Pink, D.A., Marangoni, A.G., 2016b. Which length scale is affected on sheared edible fat systems?. In: Annual Meeting of the American Oil Chemists' Society, Salt Lake City, Utah, USA, May 1—4.

Pink, D.A., Quinn, B., Peyronel, F., Marangoni, A.G., 2013. Edible oil structures at low and intermediate concentrations: I. Modelling, computer simulation and predictions for x-ray scattering. J. Appl. Phys. 114, 234901.

Ponpon, J.P., 2005. Semiconductor Detectors for 2D X-ray Imaging (docslide.net/documents).

Poot, C., Dijkshoorn, W., Haighton, A.J., Verburg, C.C., 1975. Laboratory separation of crystals from plastic fats using detergent solution. J. Am. Oil Chem. Soc. 52, 69–72.

Quinn, B., Peyronel, F., Gordon, T., Marangoni, A., Hanna, C.B., Pink, D.A., 2014. Aggregation in complex triacylglycerol oils: coarse-grained models, nanophase separation, and predicted x-ray intensities. J. Phys. Condens. Matter 26, 464108.

Ramel, P., Peyronel, F., Marangoni, A.G., 2016. Preliminary studies on the nanoscale structure of milk fat. Food Chem. 23, 224–230.

Schaefer, D.W., Kohls, D., Feinblum, E., 2012. Morphology of highly dispersing precipitated silica: impact of drying and sonication. J. Inorg. Organomet. Polym. Mater. 22, 617–623.

Small, D.M., 1986. The Physical Chemistry of Lipids: From Alkanes to Phospholipids. Plenum Press, New York.

CHAPTER 10

Fat Crystallization and Structure in Bakery, Meat, and Cheese Systems

Kristin D. Mattice, Alejandro G. Marangoni
University of Guelph, Guelph, ON, Canada

OVERVIEW

Fat is a major component in many types of food products, existing in either solid or liquid state. As an ingredient in food systems, fat contributes mouthfeel, functionality, lubrication, and flavor. Solid fats are semisolid in nature due to the presence of a large proportion of high-melting triacylglycerols (TAGs), which form crystalline structures at room temperature, resulting in a network that confines the lower-melting TAGs. Certain foods require solid fat for functionality, including high-fat bakery, meat, and cheese products. For consumers, it has been shown that the properties and content of fat contained is more integral to the texture and overall acceptance in solid or semisolid foods (Bolhuis et al., 2017; Le Calvé et al., 2015). This, in general, encompasses foods that contain primarily solid fat, emphasizing the need to understand the physical properties of solid fats within these food systems.

Different foods have requirements of solid fat specific to their application, and the properties that achieve these requirements have been studied extensively. However, the conditions that exist when testing these fats in bulk do not match the conditions that exist once a fat has been incorporated into a food system. Depending on the processing conditions of the food, fats can be subjected to high temperatures and shear and introduced to many other ingredients. In many cases, the temperatures required are greater than the melting temperature of the fats contained, erasing any crystal memory and creating opportunity for changes in fat crystallization and structure on recooling. In addition, the nonfat ingredients within each of these products also serve as potential entities for interaction with the fat, with possibility of interactions directly with components, or as an indirect result of the food's structure. Thus, the properties of fats analyzed before incorporation into food systems may not still exist. Determining how solid fats behave within the matrix of well-known food products is therefore of interest because these would be the properties that exist at the time of consumption. It would also lead to a better understanding of the interactive role of fats within specific food systems.

Structure-Function Analysis of Edible Fats, Second Edition
ISBN 978-0-12-814041-3, https://doi.org/10.1016/B978-0-12-814041-3.00010-1

FUNCTIONALITY OF FAT IN FOOD SYSTEMS

In general, fat provides not only lubricity, texture, and mouthfeel to food products but also application-specific functionality. The role of fat in bakery, meat, and cheese products will be detailed in the following sections.

Fat in Bakery Systems

Most bakery products contain very little to no solid fat; however, pastries or laminated products are an exception. Laminated dough products include puff pastry, Danishes, and croissants, which contain approximately 30%—40% fat by weight. Each characteristically contains numerous alternating layers of fat and dough created by repeated rolling and folding but will differ in the softness of the dough, the layering process used, and the content of yeast. By creating these thin layers of solid fat within the dough, the fat is then able to act as a moisture barrier during baking. Moisture contained in the dough turns to vapor at baking temperatures and expands; however, many hydrophobic layers of fat prevent this vapor from simply escaping, and as a result, the layers separate and bake individually (Baldwin et al., 1972; Renzetti et al., 2016; Rogers, 2004). Laminated doughs possess the capability of rising 80%—600% their initial height (Deligny and Lucas, 2015). For context, oven-rising bread dough will only rise to 20%—100% of the initial height. The fats used in lamination have earned the name "roll-in shortenings" because of the extensive rolling and folding that takes place.

Optimum functionality of roll-in fats is linked to specific crystallization properties. The fats must be soft and spreadable to facilitate layering without tearing the dough layers but not so soft that they leak out under pressure. A certain solid fat content (SFC) is required to exist in solid state, as well as for the entrapment and stabilization of small bubbles of air. The layering of fat within the wheat dough results in a horizontal distribution of nonuniform air bubbles (Deligny and Lucas, 2015). This aids in the separation of the dough layers creating the visible rise and yielding an airy, flaky appearance. Another factor involved in the plasticity and functionality of roll-in fats is polymorphism. Polymorphism is often used as an indicator for fat functionality in the food industry, and in the case of laminated doughs, optimum plasticity has been correlated to the presence of β' crystals. As such, most roll-in shortenings are manufactured using conditions that preferentially form β' polymorphs, including specific cooling rates and shear during cooling. These crystals tend to be smaller in size with needlelike shapes, creating strong, plastic shortenings (Macias-Rodriguez and Marangoni, 2016). It has also been suggested that fats stable in the β' form have a greater ability to retain separate layers during rolling and manufacturing (Garcia-Macias et al., 2011). The concern is that conversion to β polymorphs, the most stable crystal form, can occur over time. These crystals are often larger and associated with imparting a hard, brittle texture and an undesirable waxy mouthfeel (Baldwin et al., 1972; DeMan et al., 1991). It cannot be assumed that the properties of a roll-in shortening before incorporation remain after baking, thus there has been work done to investigate the potential changes in crystallization behavior and structure of solid fats contained within high-fat baked products, specifically croissants (Mattice and Marangoni, 2017, 2018). In addition, roll-in shortenings often contain a significant amount of saturated and even *trans* fatty acids. Current recommendations point to limiting amounts of saturated fatty acids and the removal of *trans* fatty acids from the diet. Understanding the interactive role of fats within baked laminated dough products can lead to the eventual development of acceptable alternatives that match functionality but with improved fatty acid profiles.

Fat in Processed Meat Systems

Meat from animals naturally contains a considerable amount of fat, and the composition and properties of animal fats, including pork fat (lard), beef fat (tallow), and poultry fat have been well established (Alm, 2013). The melting points of tallow, lard, and chicken fat are around 43–47°C, 38–44°C, and 31–37°C, respectively. This melting behavior ensures that the fat is in liquid state on consumption of freshly cooked meat, a desired characteristic. The melting temperatures also reflect the amount of saturated fatty acids in each type of fat, where a greater amount leads to a higher-melting fat. The average saturated fatty acid content of common meat fats is approximately 55%–60% for tallow, 42%–44% for lard, and only 30% for chicken fat. However, this content will vary based on where the fat was sourced within the animal (Feiner, 2006c). Tiensa et al. (2017) determined the polymorphic behavior of animal fats contained within commercially prepared pâtés, each containing only β or a combination of β' and β crystals.

With processed meat products, fat is added separately to increase the juiciness and improve texture. This classification includes frankfurters and paté, also known as liver sausage. These products contain finely ground meat and fat, where the structure of the meat tissue has been completely destroyed and the material appears homogenous (Lawrie, 2006). Frankfurters will contain from 25% to 30% fat, whereas pâté can contain between 25% and 40% fat (Feiner, 2006a,b). For frankfurters, the meat mixture is prepared and then stuffed into casings and then cooked. In the case of pâté, the meat is first cooked and then finely ground with liver and solid fat (Feiner, 2006b). The meat mixture is stuffed into casings or filled into a mould and cooked. The fat in processed meats provides flavor and juiciness, but because the protein structure has been destroyed, the fat also plays a significant role in the texture of these products by acting to solubilize the protein gel network. Additionally, fat will act as a filler, preventing shrinking on cooking (Feiner, 2006a). Traditionally, it would be animal fat that is added to these products; however, there are health concerns regarding their high saturated fatty acid content. Now, attempts to reduce the saturated and overall fat content in meat products by substituting with less-saturated, nonanimal sources is becoming more common (Delgado-Pando et al., 2011;

Tiensa, 2017; Youssef and Barbut, 2011; Zetzl et al., 2012). This brings on a need to understand the properties of fat contained in processed meats when considering the replacement of animal fats with other solid fat sources not traditionally used in this application. Processed meats also contain a considerable amount of fat, making the properties of the fat quite relevant to understanding product quality.

Fat in Cheese Systems

As in the case of most fat-containing food products, fat in cheese products provides many desirable sensory properties, including mouthfeel, texture, and flavor. However, the different types of cheese products that exist can be drastically different from one another, making the importance of fat vary among the different types of cheeses. Cheese will contain approximately 20%—35% fat, with reduced fat cheeses containing anywhere from 1% to 15% fat (Health Canada, 2008). Variations in fat content between cheese types is caused by a combination of factors, including the production process involved, the protein content as well as the moisture content (Guinee and McSweeney, 2006). In cheese, fat is also critical to the microstructure, where fat can be thought of as existing as dispersed globules within a protein gel (Martini and Marangoni, 2007; Ramel and Marangoni, 2017). By occupying this space within the protein matrix, it has been said that the globules also hinder any further aggregation of the protein (Guinee and McSweeney, 2006). Increasing or decreasing the amount of fat will therefore affect the casein matrix, either by disturbing extensive network formation through formation of larger, aggregated fat globules, or by allowing for a continuous protein matrix with uniform fat globule dispersion, respectively.

Cheese customarily contains milk fat, which is composed of a unique blend of TAGs and minor components (including free fatty acids, mono- and diacylglycerols) that give it physical and crystallization properties unlike other solid fat (Wright et al., 2000). The complexity of milk fat has been extensively researched, but like many products, there is limited work studying the crystallization properties milk fat contained within a food matrix. Ramel and Marangoni (2017) determined that despite the fact that bulk milk fat tends to crystallize only into β' crystals (resultant of its complex composition), β crystals were formed when milk fat was contained within a cheese or processed cheese matrix. The authors cited the possible causes either as the result of the physical constraints from the food matrix itself or as the result of interactions with other crystalline components within the food matrix. Regardless, this study demonstrated the importance of studying milk fat as a component of food systems.

Fat reduction in cheese has been a target of the food industry for some time. However, there are significant quality sacrifices made when fat content is reduced, mainly due to inherent changes in rheology, texture, and flavor (Bryant et al., 1995; Drake et al., 2010; Rogers et al., 2009; Tunick et al., 1993). Despite concerns, many low-fat cheese products are available; however, research into producing highly acceptable, low-fat cheese is ongoing. To assist in this endeavor, investigation into the fundamentals of fat behavior within cheese matrices can provide further insight and lead to great advancement.

ANALYSIS OF FAT CONTAINED WITHIN FOOD MATRICES

There are many established methods for fat analysis with respect to polymorphism, SFC, microstructure, melting behavior and more. However, the introduction of matrix components presents a challenge, as these components can interact with fat, changing the properties that exist when the fat is in bulk, and interfering with the analysis. Yet, the analysis of entire food matrices using the same analytical methods has the ability to reveal more about the interactive relationship of fat and other components, only determinable when the matrix is intact. Therefore, this type of analysis is beneficial, but considerations must be made to account for any effects or influence from other ingredients.

Crystal Polymorphism by Powder X-ray Diffraction

Polymorphism classification is achieved by powder X-ray diffraction (XRD), using equipment such as that from Rigaku, Japan. Polymorphic forms are identified from the resulting spectra in the wide-angle X-ray scattering (WAXS) region. MDI's Jade 6.5 software (Rigaku, Japan) is used to examine the WAXS spectra, where the presence of characteristic peaks identifies the presence of different polymorphic forms. In contrast, XRD in the small-angle X-ray scattering region allows for determination of the size or thickness of TAG lamellae. Three principal polymorphic forms can exist in fats: α, β', and β, where α crystals are the least stable and β are the most. Characteristic Bragg peaks in an XRD spectrum are used to identify the polymorphic form(s) present (Table 10.1). Peaks corresponding to d-spacing values of 4.3 Å are not always included when discussing the presence of β' polymorphs; however, they have been associated with this polymorphic form previously and have appeared in the case of certain roll-in shortenings so have, therefore, been included in this table.

Table 10.1 D-spacing corresponding to the characteristic Bragg peaks in the wide-angle (short spacings) X-ray scattering regions for the three polymorphs existing in solid fats

Polymorphic form	d-spacing (Å)
α	4.1
β'	4.2, 4.3, and 3.8
β	4.6, 3.8, and 3.7

Identification of the polymorphic forms present in fat contained within a food matrix is largely similar to the procedure for the same fats in bulk. Comparison of results from the same fats in bulk and in the matrix allows identification of the impact of the food matrix on polymorphism. However, the matrix components in baked goods, meat, and cheese products may interact and influence polymorphic conversion or stability. There may also be the appearance of peaks unrelated to fat crystals caused by other crystalline components. These may confuse the classification of polymorphic forms or contribute to overall noise and reduced resolution in the XRD spectra.

Based on the characteristic Bragg peaks that appear in the XRD spectra from different roll-in shortenings and butter, as well as croissants made with the respective fats (Fig. 10.1), the polymorphic form of the shortenings can be easily identified, even

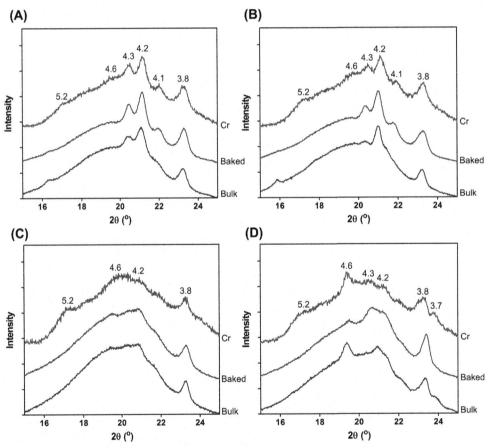

Figure 10.1 Wide-angle X-ray spectra of roll-in fats in bulk, after baking, and baked into a croissant (Cr): (A) hydrogenated shortening 1, (B) hydrogenated shortening 2, (C) butter, and (D) nonhydrogenated shortening. *(Reproduced from Mattice, K.D., Marangoni, A.G., 2017. Matrix effects on the crystallization behaviour of butter and roll-in shortening in laminated bakery products. Food Res. Int. 96, 54—63.)*

when the fat is contained within a croissant. The major concern regarding shortening polymorphism is whether the shortenings remain in β′ form, or converts to β, the most stable form. Recent work identified that nonhydrogenated shortenings will completely convert to β polymorphs over time, but hydrogenated shortenings largely remain stable in the β′ form (Mattice and Marangoni, 2017, 2018). Monitoring the formation of β polymorphs is observed by the development of a peak corresponding to a d-spacing value of 4.6 Å, as well as the shrinking of peaks corresponding to d-spacings of 4.3 Å and 4.2 Å. Additionally, the appearance of a peak at 3.7 Å, separate from that at 3.8 Å, indicates that β polymorphs are predominating (Marangoni and Wesdorp, 2013a). Comparison of fats contained within croissants with fats that have been baked using the same conditions as for croissants allows for determination of whether notable changes in polymorphism result simply by the melting and recooling of fats or rather the presence of the food matrix (Fig. 10.1).

When croissants are analyzed after some storage time, different peaks appear, unique to those commonly observed in fats (Fig. 10.2). This is caused in croissants by the fact that wheat starch is gelatinized during baking; however, the starch will undergo retrogradation over storage time. Retrograded starch causes the appearance of peaks characteristic to crystalline starch in the XRD spectrum. Native wheat starch exists in a semicrystalline form and will produce characteristic peaks in an XRD spectrum. These peaks are not produced once the starch has been gelatinized, when granules swell and lose their crystallinity on hydration and heating (Fig. 10.3A and B). This gelatinized state is associated with physical changes including increased viscosity and gel formation (Ratnayake and Jackson, 2007; Wang et al., 2015). After gelatinization, amylose and amylopectin chains begin to precipitate from their gelled state and reassociate into ordered structures over time, in a process known as retrogradation. Retrogradation has also been described as the return of starch components to their granular or crystalline state, evident by the return of a weak pattern over time (Bayer et al., 2006; Miles et al., 1985; Morris, 1990; Roulet et al., 1988). Therefore, XRD can be used to monitor starch retrogradation. In a comparison of the characteristic Bragg peaks corresponding to polymorphic crystals of fat and starch, most of the peaks appear at unique locations, making it easy to discern which peaks result from each source. However, it is apparent that both native crystalline and retrograded starch will cause the appearance of a peak at 5.2 Å (Fig. 10.3A and C). This peak is easily mistaken as being associated with β crystals because of established associations with this peak and this polymorphic form (Marangoni and Wesdorp, 2013a). Therefore, one must use peaks at 4.6 Å as the true indicator of the presence of β crystals.

The polymorphism of fat crystals contained within a pâté or liver sausage matrix has been determined by XRD, and the results, compared with the same fats contained after extraction from the matrix (Tiensa et al., 2017). According to characteristic Bragg peaks, pâté will either contain only β or some combination of β and β′ crystals (Fig. 10.4). Comparison of samples established that polymorphism of a fat within pâté largely mirrors

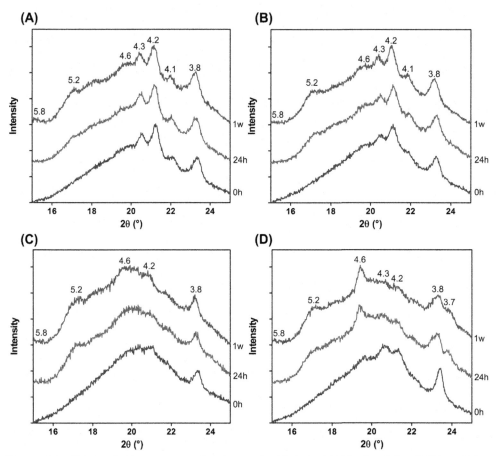

Figure 10.2 Wide-angle X-ray spectra of croissants freshly baked (0 h), after 1 day (24 h) and after 1 week of storage prepared from (A) hydrogenated shortening A, (B) hydrogenated shortening B, (C) butter, and (D) nonhydrogenated shortening. *(Reproduced from Mattice, K.D., Marangoni, A.G., 2017. Matrix effects on the crystallization behaviour of butter and roll-in shortening in laminated bakery products. Food Res. Int. 96, 54–63.)*

that of the same fats after extraction from the matrix with few exceptions, allowing considerable certainty that the major peaks appearing at 4.6 Å, 4.2 Å, and 3.8 Å are indeed caused by fat crystals. This comes despite the fact that the spectra of pâté samples display many more peaks in between 23 and 25 degrees region compared with fat analyzed alone and had considerably increased noise overall (Fig. 10.4). It is understandable that peaks corresponding to polymorphic crystals would be the most prominent and identifiable despite the presence of matrix components, as the other major component in pâté is protein, which does not exist in crystalline state and therefore does not prompt any peaks in an XRD spectrum.

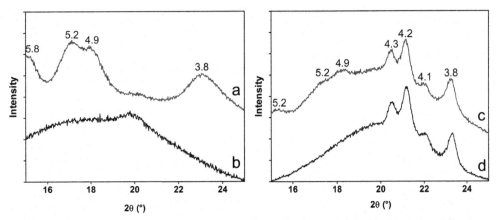

Figure 10.3 Wide-angle X-ray spectra of (A) crystalline wheat starch (obtained in isolated form at a local supermarket), (B) gelatinized wheat starch, (C) roll-in fat baked with crystalline wheat starch, and (D) roll-in fat baked with gelatinized wheat starch. *(Adapted from Mattice, K.D., Marangoni, A.G., 2017. Matrix effects on the crystallization behaviour of butter and roll-in shortening in laminated bakery products. Food Res. Int. 96, 54–63; Mattice, K.D., Marangoni, A.G., 2018. Gelatinized wheat starch influences crystallization behaviour and structure of roll-in shortenings in laminated bakery products. Food Chem. 243, 396–402.)*

Samples of processed cheese loaf, processed cheese slices, and cream cheese prepared by different brands and in different countries have also been analyzed for polymorphism using XRD (Ramel and Marangoni, 2017). For cream cheese, distinct peaks corresponding to polymorphic crystals were easily observable (Fig. 10.5C). The fact that these peaks were not a result of nonfat matrix components was confirmed by comparison with the XRD spectra of anhydrous milk fat, which displayed similar peak size and distribution (Fig. 10.5A). However, numerous nonfat peaks were obtained for processed cheese loaf and slices (Fig. 10.5B and D). Despite the fact that the composition of these two products is largely the same (differing mainly in the size and format in which the product is packaged), these additional peaks are much more prominent and sharp in processed cheese slices, indicating that there are many factors involved extending beyond the presence of matrix components. Identification of each of these additional peaks requires the systematic analysis of each matrix component, and thus, with the complexity of cheese composition, has not yet been carried out to completion. However, Ramel and Marangoni (2017) did determine that lactose, another crystalline component, is likely not the cause of these unspecified, prominent peaks (Fig. 10.5D). But even with the presence of these unexplained peaks, the largest and most significant peaks were still those associated with fat crystal polymorphism. Confirmation that these peaks were indeed caused by fat polymorphic crystals was carried out by XRD analysis of samples at 45°C, above the melting temperature of milk fat. Here, the prominent, sharp peaks assumed related to polymorphic crystals disappeared, meaning that polymorphism identification in cheese samples can be carried out with substantial certainty (Fig. 10.6).

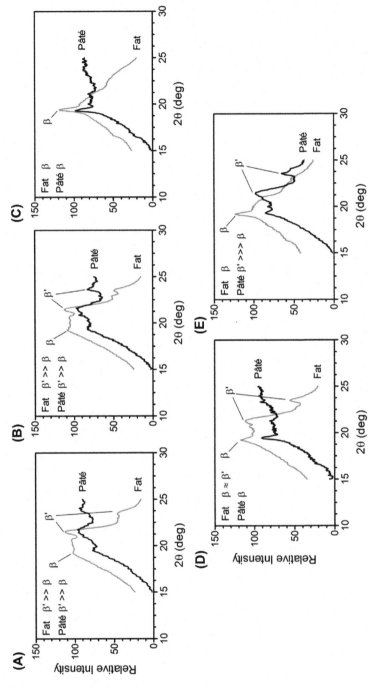

Figure 10.4 Wide-angle X-ray diffraction patterns for five different commercial brand pâtés (A–E) and their extracted fats. *(Reproduced from Tiensa, B.E., Barbut, S., Marangoni, A.G., 2017. Influence of fat structure on the mechanical properties of commercial pate products. Food Res. Int. 100, 558–565.)*

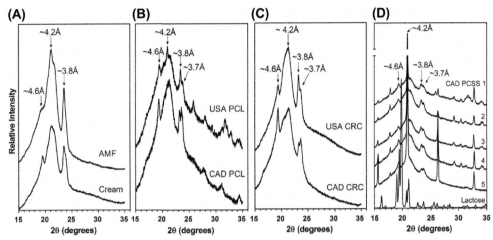

Figure 10.5 Wide-angle X-ray diffraction patterns of (A) anhydrous milk fat (AMF), cream; (B) processed cheese loaf (PCL) produced in United States and Canada (CAD); (C) cream cheese original brick (CRC) produced United States and CAD; and (D) different brands of processed cheese single slices (PCSS)—(1) kraft (CAD), (2) black diamond (CAD), (3) compliments (CAD), (4) no name (CAD), and (5) kraft (USA), and lactose after storage at 0°C. Lines indicate the peaks corresponding to d-spacings, characteristic of β′ (4.2, 3.8, and 3.7 Å) and β (4.6 Å) polymorphs. *(Adapted from Ramel, P.R., Marangoni, A.G., 2017. Characterization of the polymorphism of milk fat within processed cheese products. Food Struct. 12, 15—25.)*

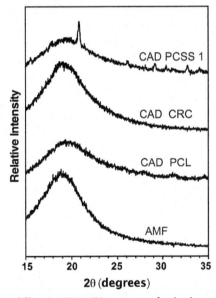

Figure 10.6 Wide-angle X-ray diffraction (WAXD) patterns of anhydrous milk fat (AMF), cream cheese original brick (CRC), processed cheese loaf (PCL), and processed cheese single slices—kraft brand (PCSS 1) produced in Canada (CAD) after heating at 45°C for 30 min. Lines indicate the peaks corresponding to d-spacings characteristic of the β′ polymorph of fats (4.2, 3.8, and 3.7 Å) and β (4.6 Å) polymorph of fat. *(Reproduced from Ramel, P.R., Marangoni, A.G., 2017. Characterization of the polymorphism of milk fat within processed cheese products. Food Struct. 12, 15—25.)*

Thermal Behavior by Differential Scanning Calorimetry

In general, fat is the only component of a food product that will melt in the range of 0–60°C. This means that differential scanning calorimeter (DSC) measurements of entire food matrices can be carried out without special considerations for peaks from nonfat ingredients. However, bakery, meat, and cheese samples still only contain a fraction of fat, unlike the majority of samples traditionally analyzed that are 100% fat. Therefore, any measurements of the enthalpy of melting must be multiplied accordingly to be able to directly compare it with other samples of differing fat contents. Measurement of the thermal behavior of fats and fat-containing products can be done using a TA Mettler Toledo DSC (Mettler Toledo, Mississauga, ON, Canada). Samples are weighed into aluminum crucibles and subjected to specific heating and cooling conditions. The peak temperatures of melting and crystallization (T_m or T_c) are determined using STARe software (Mettler Toledo), observable in the melting endotherm and crystallization exotherm, respectively. Observations of location of peak maximum, as well as peak height and width, can be used to interpret thermal information about the samples. Sharp peaks indicate homogenous crystal materials such that melting of the crystals occurs all at one temperature. In contrast, broad peaks suggest that the TAG crystals present are more heterogeneous, existing in many different sizes, such that melting of the different crystals occurs over a range of consecutive temperatures. This can be thought of as many smaller peaks that overlap very closely such that individual peaks cannot be distinguished and only one wider peak is visible. The presence of multiple peaks indicates the melting or crystallization of different crystalline species or different polymorphic forms. In certain systems containing complex solid fats, fractionation can occur after baking and cooling, identified when two smaller peaks take the place of a former larger peak on remelting. Here, the slow cooling of the fats causes higher and lower melting fractions to fractionate, resulting in two distinct melting points when heated.

Studies investigating the thermal behavior in cheese systems have been carried for some time (Gliguem et al., 2009; Lopez et al., 2006; Ramel and Marangoni, 2017; Tunick et al., 1989), with studies on bakery and meat systems occurring only recently (Mattice and Marangoni, 2017; Tiensa et al., 2017). In each of these many investigations, the food matrix did not cause the appearance of any unidentifiable peaks nor did it prevent the melting of fat contained, allowing for evaluation of the overall thermal behavior, identification of melting temperatures, and observations of changes in thermal behavior caused by matrix components. Measurement of the melting behavior determined that fat contained within croissants will crystallize heterogeneously, based on peak width (Fig. 10.7) (Mattice and Marangoni, 2017). In pâté, the first melting peak of fat contained in a pâté matrix occurs at a different temperature compared with extracted fat analyzed separately (Fig. 10.8) (Tiensa et al., 2017). The magnitude and direction of the difference depend on the type of pâté, where no universal trend between pâtés and their extracted

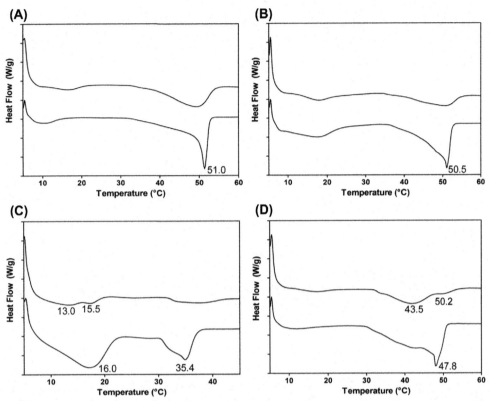

Figure 10.7 Differential scanning calorimeter melting endotherms of fats in bulk (bottom) and after baking within a croissant (top): (A) hydrogenated shortening 1, (B) hydrogenated shortening 2, (C) butter, and (D) nonhydrogenated shortenings. *(Adapted from Mattice, K.D., Marangoni, A.G., 2017. Matrix effects on the crystallization behaviour of butter and roll-in shortening in laminated bakery products. Food Res. Int. 96, 54–63.)*

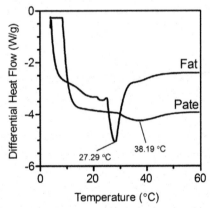

Figure 10.8 Differential scanning calorimetric (DSC) traces for one type of commercial brand pâté and its extracted fat-melting peaks. *(Adapted from Tiensa, B.E., Barbut, S., Marangoni, A.G., 2017. Influence of fat structure on the mechanical properties of commercial pate products. Food Res. Int. 100, 558–565.)*

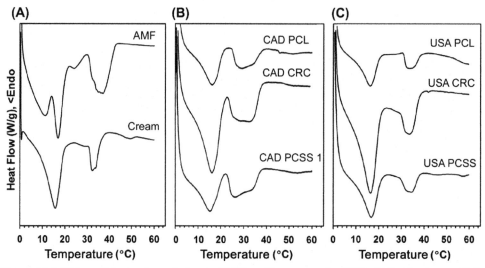

Figure 10.9 Differential scanning calorimetry (DSC) melting curves (0–60°C) of (A) anhydrous milk fat (AMF) and cream; (B) processed cheese products produced in Canada (CAD)—processed cheese loaf (PCL), cream cheese original brick (CRC), processed cheese single slices—kraft (PCSS 1); and (C) processed cheese products produced in United States. *(Adapted from Ramel, P.R., Marangoni, A.G., 2017. Characterization of the polymorphism of milk fat within processed cheese products. Food Struct. 12, 15–25.)*

fats have been detected. Although melting peak maximums (T_m) were identified for all pâté samples, the behavior of different pâtés on melting and remelting is so variable that more research is required to identify the relationship of the protein matrix and the fats' thermal behavior. Milk fat contained in cheese has been shown to have three distinct melting fractions, corresponding to low-, mid-, and high-melting fractions, visible as three distinct peaks each with its own peak maximum (T_m) (Fig. 10.9A). However, the melting endotherm of some processed cheeses only display two melting peaks, resulting from two of these fractions melting together (Fig. 10.9B and C) (Ramel and Marangoni, 2017).

Solid Fat Content by Pulsed Nuclear Magnetic Resonance

The ratio of crystalline to liquid TAGs at a given temperature is expressed by the percent of SFC. The traditional and fastest method used for fats is the direct SFC measurement according to AOCS Official Method Cd 16b-93. Measurements for all samples can be carried out using a Bruker PC/20 Series Minispec pulsed nuclear magnetic resonance (NMR) analyzer (Bruker Optics Ltd., Milton, ON). Glass NMR tubes are filled with the fat samples and equilibrated at each temperature for 30 min before measurement. Using this method allows for fast analysis, where the equipment will directly determine the SFC as a percentage.

Many additional factors must be considered when measuring the SFC of fat contained within a food matrix, as the direct method of SFC measurement will find the amount of total solids in the sample, fat, or otherwise. Accounting for this additional signal can be done in a few different ways. First, when possible, fat-free samples of the food product can be prepared. The signals achieved here must then be multiplied by the fraction of the food product that the nonfat ingredients make up. These values can then be subtracted from the values obtained for the samples of the entire food product at each respective temperature. Second, the indirect measurement method can be used, according to the AOCS Official Method Cd 1681. In this standard method, the signal of each sample is measured at both the desired temperature, as well as at 60°C, when fat is completely melted. This accounts for the nonfat ingredients because these components will not melt, allowing them to be subtracted. This method takes more time; therefore, to speed the determination of a general trend, it is also possible to measure using the direct method, with the addition of a measurement at 60°C. When all measurements have been taken, manual subtraction of the value at 60°C from each other value can give a good estimate of the SFC (Fig. 10.10). However, the signal obtained from nonfat ingredients does change slightly with changes in temperature; therefore, this faster method can only provide an estimate of the general trend, not precise SFC values.

Another barrier is the fact that many food systems contain some water. Water is also in a liquid state and therefore adds to the signal of fat in the liquid state. Ensuring that

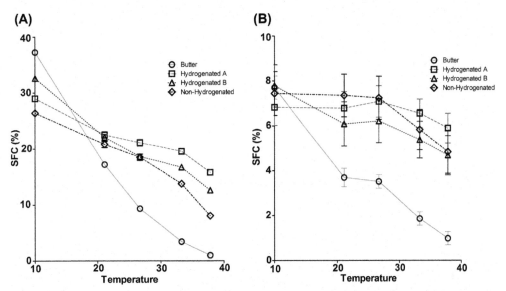

Figure 10.10 Solid fat content curves as a function of temperature for (A) roll-in fats and (B) croissants. *(Reproduced from Mattice, K.D., Marangoni, A.G., 2017. Matrix effects on the crystallization behaviour of butter and roll-in shortening in laminated bakery products. Food Res. Int. 96, 54–63.)*

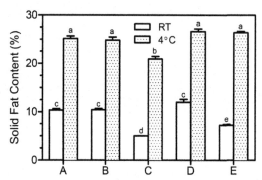

Figure 10.11 Solid fat contents of commercial pâtés (products A—E) tested at room temperature (RT) and at 4°C. Bars indicate standard error of the mean. n = 3 per sample. *(Reproduced from Tiensa, B.E., Barbut, S., Marangoni, A.G., 2017. Influence of fat structure on the mechanical properties of commercial pate products. Food Res. Int. 100, 558—565.)*

samples do not differ in water content can allow for the assumption that while each of the SFC measurements are not precise to only fat, each measurement should be modified by approximately the same magnitude. This, again, only allows for determination of a general trend rather than precise values.

One can also return to the use of DSC where these additional solid components were not a barrier. Because fat is the only component that melts, the enthalpy of melting (obtained from integration of the melting endotherm) is proportional to the amount of solid fat contained in the system. Therefore, direct comparison of the enthalpy of melting obtained for an entire food matrix with that of a solid fat on its own would provide indication of the degree of difference between the SFC in each sample. Again, the food product sample does not contain 100% fat; therefore, the measured enthalpy must be multiplied accordingly before comparison.

Crystallization Kinetics

The rate of fat crystallization on cooling is affected by the presence of impurities in the system (Rousset, 2002). In comparison with a system of pure fat, a food product contains a large amount of what would elsewhere be considered "impurities," including components that are crystalline themselves. Essentially, the presence of matrix components in a food product could act to speed crystallization by providing a surface on which fat can crystallize. The Avrami model is commonly used to fit fat crystallization data, as the variables in this model are used to identify and quantify the nature of crystal growth in a system (Marangoni, 1998; Sharples, 1996; Wright et al., 2000). The equation is in the form:

$$\frac{\text{SFC}(t)}{\text{SFC}(\infty)} = 1 - e^{-k(t)^n},$$

where k is the Avrami constant that represents the crystallization rate constant and n is the Avrami exponent or index of crystallization. The Avrami exponent n pertains to the geometry of crystal growth or the crystal growth mechanism (Marangoni and Wesdorp, 2013b). The Avrami constant, k, takes into account both nucleation and crystal growth, where a larger value indicates a faster crystallization rate and vice versa. This model is applied to SFC measurements taken at fixed time intervals, as samples are cooled under isothermal conditions until the maximum SFC at that given temperature is reached (Fig. 10.12). Beginning at temperatures above the melting point of the lipid material in the sample allows for certainty that no fat crystals will exist from the starting point, in addition to eliminating any thermal history. Even if samples contain nonfat ingredients, normalization of the data suffices, as the parameter of interest is only the time to reach the maximum SFC for that specific sample. A sample that crystallizes at a faster rate will display larger SFC values in a shorter amount of time, and a corresponding k value can be calculated. The nonlinear regression can be carried out using software such as GraphPad Prism (Graphpad Software, San Diego, CA, USA).

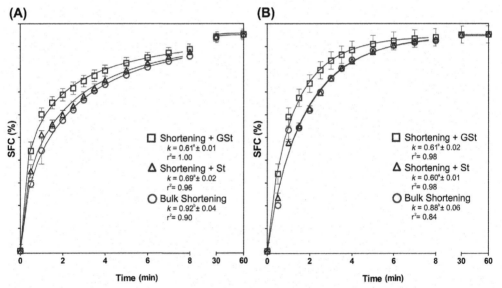

Figure 10.12 Changes in SFC as a function of time during the isothermal cooling (10°C) of (A) hydrogenated shortening and (B) nonhydrogenated shortening, each in bulk, and after baking with gelatinized wheat starch (+GSt) or crystalline wheat starch (+St). Solid lines correspond to the Avrami fit of the model to the data. Values with the same superscript letter are not statistically different ($P < .05$). *(Adapted from Mattice, K.D., Marangoni, A.G., 2018. Gelatinized wheat starch influences crystallization behaviour and structure of roll-in shortenings in laminated bakery products. Food Chem. 243, 396–402.)*

MICROSTRUCTURE

In a multicomponent food product, microscopy allows for visualization of how the different components in a food system are physically structured together. Because of the complexity of samples of an entire food system, specific stains are used to allow easier identification of the different components.

One technique is light microscopy. Light microscopes, such as Model BX60 (Olympus Optical Ltd., Tokyo, Japan), can be used, and images are captured using a computerized image analysis system (such as Image-Pro Plus, Version 5.1, Media Cybernetics Inc., Silver Spring, MD, USA). Here, thin slices of samples are cut, fixed to slides, and stained. The selection of stain depends on the sample composition. For meat samples, staining can be done with Masson's stain, which stains muscle fibers red and connective tissue blue. The fixing process used for meat systems can be a complex procedure but results in the removal of fat globules from a fixed protein matrix allowing for easy distinction between the fat and protein matrix (Youssef and Barbut, 2009). In analysis of pâté samples, Tiensa et al. (2017) obtained a two-dimensional view of the microstructure, with the resultant images displaying the protein matrix as a pink background, and white sections where the globules were once present (Fig. 10.13).

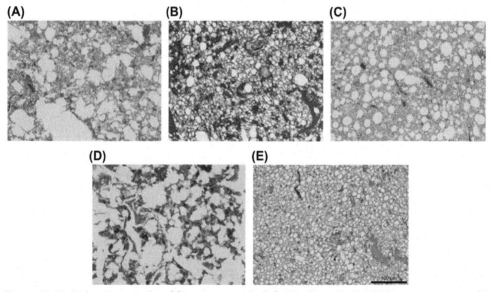

Figure 10.13 Light micrographs of five commercial pâtés (products A–E). White areas represent fat globules that were removed during sample preparation for microscopy. Scale bar = 100 μm. *(Reproduced from Tiensa, B.E., Barbut, S., Marangoni, A.G., 2017. Influence of fat structure on the mechanical properties of commercial pate products. Food Res. Int. 100, 558–565.)*

Another option is confocal laser scanning microscopy (CLSM), which images only light emitted from the fluorescence of dyes that have been taken up by certain components within the sample. This allows for the visualization of separate components without requiring actual physical separation. Microstructural analyses can be made using a confocal Leica TCS SP2 inverted microscope (Leica Microsystems, Heidelberg, Germany). Images can then be processed using Leica LAS AF Lite version (Leica Microsystems Inc., Wetzlar, Germany). Again, thin slices of sample are cut and stained. Ramel and Marangoni mounted thin slices of processed cheese and used Fast Green FCF to stain protein and Nile red to stain fat (Ramel and Marangoni, 2017). The lasers were excited according to the stains, at 488 nm for Nile Red and 633 nm for Fast Green with a triple dichroic filter (488/543/633 nm wavelength). This also allowed for a two-dimensional view of the microstructure of only the protein and fat within processed cheese. Their images displayed red globules of milk fat embedded in a green continuous protein network (Fig. 10.14).

For a better view of the surface of samples, a three-dimensional image can be obtained by cryogenic scanning electron microscopy (cryo-SEM), with equipment such as a Hitachi S-570 SEM unit (Tokyo, Japan). Images can be captured digitally using the Quartz PCI imaging package (Quartz Imaging Corp., Vancouver, BC, Canada). In this technique, samples are flash-frozen in liquid nitrogen (approximately −207°C). The frozen samples are then fractured to expose the structure within. Samples are sputter coated with 30 nm of gold to allow visualization. No stains are used in this technique; therefore, there are no colors to visually distinguish the different components. However, different components can be distinguished because of the dimensionality in the resultant images. Analysis by Ramel and Marangoni (2017) of processed cheeses allowed for visualization of the spherical fat globules seen as surrounded by the cheese matrix (Fig. 10.15).

Texture and Rheology

Crystallization characteristics of fats contained within whole food systems, including polymorphism, SFC, and thermal behavior, can be used as indicators for functionality and sensory properties. However, texture analysis and rheometry can uncover the true relationships between structure and crystallization properties with the physical properties experienced on eating.

For croissants and other baked goods, texture analysis can provide an indication of firmness and the potential sensorial consequences of using different roll-in fats. A Model TA.XT2 (Stable Micro Systems, Texture Technologies Corp., USA) can be affixed with a TA-42 knife blade attachment to measure the force required to cut a croissant. The maximum force during cutting occurs at the time just before the bulk of the croissant

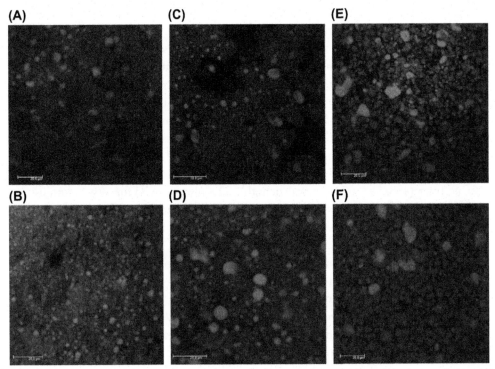

Figure 10.14 Confocal laser scanning micrographs (CLSM) of various processed cheese products—(A) USA processed cheese single slices (PCSS); (B) CAD processed cheese single slices—kraft brand (PCSS 1); (C) USA processed cheese loaf (PCL); (D) CAD processed cheese loaf (PCL); (E) USA cream cheese (CRC); and (F) CAD cream cheese (CRC). Fat globules can be distinguished as red (light gray in print versions) globular or irregular structures while the protein matrix is stained green (dark gray in print versions). Scale bars correspond to 20 μm. The depth of optical sectioning from the surface of samples was varied depending on the optical density of the samples. *(Reproduced from Ramel, P.R., Marangoni, A.G., 2017. Characterization of the polymorphism of milk fat within processed cheese products. Food Struct. 12, 15—25.)*

yields to the knife attachment and two separate pieces form (Fig. 10.16) (Mattice and Marangoni, 2017). Determining differences in the firmness of samples containing unlike fats (previously shown to have varying crystallization properties) can indicate whether crystallization behavior can influence texture at the eating level. The challenge when it comes to baked goods is that the process of starch retrogradation causes significant changes in firmness; therefore, it can be difficult to know the cause of any observed variations over croissant shelf life. Therefore, comparison must be made between baked goods that have been stored for the same amount of time. In a previous study, croissants made with nonhydrogenated roll-in shortenings that experience polymorphic instability also experienced significant changes in firmness over a 1-week shelf life, thereby correlating polymorphic conversion to the eating quality over time (Fig. 10.16) (Mattice and Marangoni, 2017).

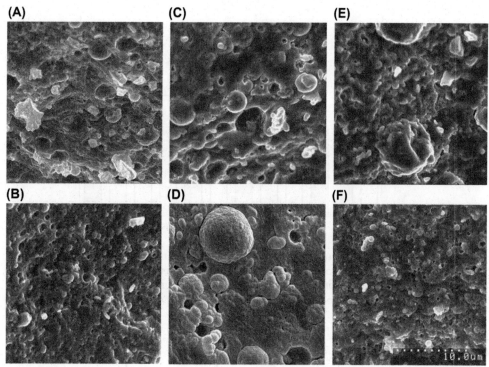

Figure 10.15 Cryogenic scanning electron micrographs (cryo-SEM) of various processed cheese products—(A) USA processed cheese single slices (PCSS); (B) CAD processed cheese single slices—kraft brand (PCSS 1); (C) USA processed cheese loaf (PCL); (D) CAD processed cheese loaf (PCL); (E) USA cream cheese (CRC); and (F) CAD cream cheese (CRC). Scale bar corresponds to 10 μm. *(Reproduced from Ramel, P.R., Marangoni, A.G., 2017. Characterization of the polymorphism of milk fat within processed cheese products. Food Struct. 12, 15–25.)*

Texture analysis by back extrusion can also characterize the hardness of food systems that are only semisolid, such as pâté. Back extrusion involves penetration of a sample that has been stuffed into a test tube with a cylindrical, stainless steel probe with a truncated spherical tip. The rate and depth of penetration are controlled, and the average force over the final 25% of the penetration depth is reported. Tiensa et al. (2017) measured the hardness of different pâtés at different temperatures using this method, and the authors concluded that hardness was proportional to the SFC in the particular pâté sample (Fig. 10.17). This trend was also true for samples of fat on their own; therefore, this study illustrates that the presence of the pâté matrix does not affect this behavior characteristic of fats and emphasizes the importance of fat within a pâté system.

Texture profile analysis (TPA) is a method that can characterize multiple physical parameters of a food sample, including hardness, springiness, cohesiveness, and chewiness. Here, cylindrical pieces of sample are compressed twice to 75% of their original height by

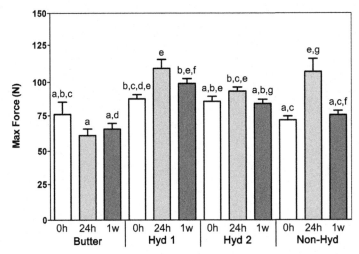

Figure 10.16 Maximum force required to cut croissants comparing croissants prepared with the different roll-in fats at different storage times. Bars with the same letter are not statistically different. *(Adapted from Mattice, K.D., Marangoni, A.G., 2017. Matrix effects on the crystallization behaviour of butter and roll-in shortening in laminated bakery products. Food Res. Int. 96, 54–63.)*

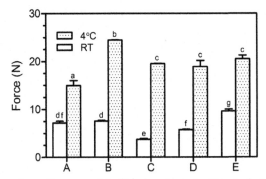

Figure 10.17 Hardness values of commercial pâtés (products A–E) tested at room temperature (RT) and 4°C. Bars indicate standard error of the mean. n = 2–10 per sample. *(Reproduced from Tiensa, B.E., Barbut, S., Marangoni, A.G., 2017. Influence of fat structure on the mechanical properties of commercial pate products. Food Res. Int. 100, 558–565.)*

a texture analyzer. Zetzl et al. (2012) performed TPA on frankfurter samples prepared with different types of fat and correlated hardness and chewiness to the binding of protein and fat at interfaces, which in itself depends on fat globule size.

Small and large deformation rheometry are used to characterize the viscoelastic properties of systems by applying specific stress or strain to the material and measuring the corresponding response. Small deformation rheology probes the linear viscoelastic region, whereas large deformation rheology probes the nonlinear region. These techniques

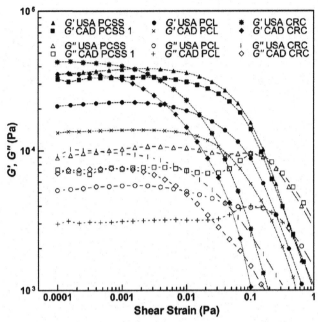

Figure 10.18 Storage and loss moduli (G′ and G″) of processed cheese products obtained from conducting amplitude sweep (increasing shear strain from 10^{-4} to 1 Pa) at a constant angular frequency (ω) of 1 Hz and temperature of 7°C. *(Reproduced from Ramel, P.R., Marangoni, A.G., 2017. Characterization of the polymorphism of milk fat within processed cheese products. Food Struct. 12, 15–25.)*

can be used to determine the storage modulus (G′), which describes the elastic or solid properties, the loss modulus (G″), which describes the viscous or liquid properties. Dynamic oscillatory experiments can be used to understand the system's response to changes in the amount of strain applied. Ramel and Marangoni (2017) used a Physica MCR 302 (Anton Paar, Graz, Austria) rheometer to measure the rheological properties of processed cheeses (Fig. 10.18). This analysis allowed for the connection between the yield strain of samples and their microstructure, where indications of a weaker structure using CLSM and cryo-SEM corresponded to a lower yield strain value.

CONCLUSIONS

The many methods and considerations for the analysis of fat contained in a food matrix outlined in this chapter enable one to overcome barriers brought on by the presence of nonfat ingredients. Because there is a general desire for products containing less-saturated and *trans* fat (highly present in solid fats) and less fat overall, these techniques allow for a better understanding of the role solid fat plays in different food systems, as well as what must be accounted for should fat be reduced, removed, or replaced in a food product.

Overall, analysis of fat contained in bakery, meat, and cheese systems revealed that traditional fat analysis methods are possible with only minor adjustments to procedure and analysis. The studies discussed also determined that the presence of a matrix components in a whole food system does in some way impact or influence the crystallization behavior and structure of the fat contained, highlighting the importance of this type of research.

REFERENCES

Alm, M., 2013. In: Animal Fats. AOCS Lipid Library.

Baldwin, R.R., Baldry, R.P., Johansen, R.G., 1972. Fat systems for bakery products. J. Am. Oil Chem. Soc. 49, 473−477.

Bayer, R.K., Cagiao, M.E., Baltá Calleja, F.J., 2006. Structure development in amorphous starch as revealed by X-ray scattering: influence of the network structure and water content. J. Appl. Polym. Sci. 99, 1880−1886.

Bolhuis, D.P., Costanzo, A., Keast, R.S.J., 2017. Preference and perception of fat in salty and sweet foods. Food Qual. Prefer. 64, 131−137.

Bryant, A., Ustunol, Z., Steffe, J., 1995. Texture of cheddar cheese as influenced by fat reduction. J. Food Sci. 60, 1216−1219.

Delgado-Pando, G., Cofrades, S., Ruiz-Capillas, C., Solas, M.T., Triki, M., Jimenez-Colmenero, F., 2011. Low-fat frankfurters formulated with healthier lipid combinations as functional ingredient: microstructure, lipid oxidation, nitrite content, microbiological changes and biogenic amine formation. Meat Sci. 89, 65−71.

Deligny, C., Lucas, T., 2015. Effect of the number of fat layers on expansion of Danish pastry during proving and baking. J. Food Eng. 158, 113−120.

DeMan, L., DeMan, J.M., Blackman, B., 1991. Physical and textural characteristics of some North American shortenings. J. Am. Oil Chem. Soc. 68, 63−69.

Drake, M.A., Miracle, R.E., McMahon, D.J., 2010. Impact of fat reduction on flavor and flavor chemistry of Cheddar cheeses. J. Dairy Sci. 93, 5069−5081.

Feiner, G., 2006a. Cooked sausages. In: Feiner, G. (Ed.), Meat Products Handbook: Practical Science and Technology, first ed. CRC Press, Boca Raton, FL, pp. 239−286.

Feiner, G., 2006b. Spreadable liver sausage and liver pate. In: Feiner, G. (Ed.), Meat Products Handbook: Practical Science and Technology, first ed. CRC Press, Boca Raton, FL, pp. 451−475.

Feiner, G., 2006c. The protein and fat content of meat. In: Feiner, G. (Ed.), Meat Products Handbook: Practical Science and Technology, first ed. CRC Press, Boca Raton, FL, pp. 3−32.

Garcia-Macias, P., Gordon, M.H., Frazier, R.A., Smith, K., Gambelli, L., 2011. Performance of palm-based fat blends with a low saturated fat content in puff pastry. Eur. J. Lipid Sci. Technol. 113, 1474−1480.

Gliguem, H., Ghorbel, D., Lopez, C., Michon, C., Ollivon, M., Lesieur, P., 2009. Crystallization and polymorphism of triacylglycerols contribute to the rheological properties of processed cheese. J. Agric. Food Chem. 57, 3195−3203.

Guinee, T.P., McSweeney, P.L.H., 2006. Significance of milk fat in cheese. In: Fox, P.F., McSweeney, P.L.H. (Eds.), Advanced Dairy Chemistry, Lipids, vol. 2. Springer International Publishing, Boston, MA, pp. 377−440.

Health Canada, 2008. Nutrient Value of Some Common Foods, pp. 1−68.

Lawrie, R.A., 2006. The eating quality of meat. In: Lawrie, R.A., Ledward, D.A. (Eds.), Lawrie's Meat Science, seventh ed. CRC Press, Boca Raton, FL, pp. 279−341.

Le Calvé, B., Saint-Léger, C., Babas, R., Gelin, J.L., Parker, A., Erni, P., Cayeux, I., 2015. Fat perception: how sensitive are we? J. Texture Stud. 46, 200−211.

Lopez, C., Briard-Bion, V., Camier, B., Gassi, J.-Y., 2006. Milk fat thermal properties and solid fat content in emmental cheese: a differential scanning calorimetry study. J. Dairy Sci. 89, 2894−2910.

Macias-Rodriguez, B.A., Marangoni, A.A., 2016. Physicochemical and rheological characterization of roll-in shortenings. J. Am. Oil Chem. Soc. 93, 575−585.

Marangoni, A.G., 1998. On the use and misuse of the Avrami equation in characterization of the kinetics of fat crystallization. J. Am. Oil Chem. Soc. 75, 1465−1467.

Marangoni, A.G., Wesdorp, L.H., 2013a. Crystallography and polymorphism. In: Marangoni, A.G., Wesdorp, L.H. (Eds.), Structure and Properties of Fat Crystal Networks, second ed. CRC Press, Boca Raton, FL, pp. 1−24.

Marangoni, A.G., Wesdorp, L.H., 2013b. Nucleation and crystalline growth kinetics. In: Marangoni, A.G., Wesdorp, L.H. (Eds.), Structure and Properties of Fat Crystal Networks, second ed. CRC Press, Boca Raton, FL, pp. 27−99.

Martini, S., Marangoni, A.G., 2007. Microstructure of dairy fat products. In: Tamime, A. (Ed.), Structure of Dairy Products. Blackwell Publishing, Ames, IA, pp. 72−103.

Mattice, K.D., Marangoni, A.G., 2017. Matrix effects on the crystallization behaviour of butter and roll-in shortening in laminated bakery products. Food Res. Int. 96, 54−63.

Mattice, K.D., Marangoni, A.G., 2018. Gelatinized wheat starch influences crystallization behaviour and structure of roll-in shortenings in laminated bakery products. Food Chem. 243, 396−402.

Miles, M.J., Morris, V.J., Orford, P.D., Ring, S.G., 1985. The roles of amylose and amylopectin in the gelation and retrogradation of starch. Carbohydr. Res. 135, 271−281.

Morris, V.J., 1990. Starch gelation and retrogradation. Trends Food Sci. Technol. 1, 2−6.

Ramel, P.R., Marangoni, A.G., 2017. Characterization of the polymorphism of milk fat within processed cheese products. Food Struct. 12, 15−25.

Ratnayake, W.S., Jackson, D.S., 2007. A new insight into the gelatinization process of native starches. Carbohydr. Polym. 67, 511−529.

Renzetti, S., de Harder, R., Jurgens, A., 2016. Puff pastry with low saturated fat contents: the role of fat and dough physical interactions in the development of a layered structure. J. Food Eng. 170, 24−32.

Rogers, D., 2004. Functions of fats and oils in bakery products. Food Technol. 15, 572−574.

Rogers, N.R., Drake, M.A., Daubert, C.R., McMahon, D.J., Bletsch, T.K., Foegeding, E.A., 2009. The effect of aging on low-fat, reduced-fat, and full-fat Cheddar cheese texture. J. Dairy Sci. 92, 4756−4772.

Roulet, P., MacInnes, W.M., Würsch, P., Sanchez, R.M., Raemy, A., 1988. A comparative study of the retrogradation kinetics of gelatinized wheat starch in gel and powder form using X-rays, differential scanning calorimetry and dynamic mechanical analysis. Food Hydrocolloids 2, 381−396.

Rousset, P., 2002. Modeling crystallization kinetics of triacylglycerols. In: Marangoni, A.G., Narine, S.S. (Eds.), Physical Properties of Lipids. Marcel Dekker, New York, NY, pp. 11−46.

Sharples, A., 1996. Overall kinetics of crystallization. In: Sharples, A. (Ed.), Introduction to Polymer Crystallization. Edward Arnold, London, UK, pp. 44−59.

Tiensa, B.E., 2017. Fat and Organogel Structure within Pate and Their Influence on Texture and Sensory Attributes (MSc thesis). University of Guelph.

Tiensa, B.E., Barbut, S., Marangoni, A.G., 2017. Influence of fat structure on the mechanical properties of commercial pate products. Food Res. Int. 100, 558−565.

Tunick, M.H., Basch, J.J., Maleeff, B.E., Flanagan, J.F., Holsinger, V.H., 1989. Characterization of natural and imitation Mozzarella cheeses by differential scanning calorimetry. J. Dairy Sci. 72, 1976−1980.

Tunick, M.H., Mackey, K.L., Shieh, J.J., Smith, P.W., Cooke, P., Malin, E.L., 1993. Rheology and microstructure of low-fat Mozzarella cheese. Int. Dairy J. 3, 649−662.

Wang, S., Li, C., Copeland, L., Niu, Q., Wang, S., 2015. Starch retrogradation: a comprehensive review. Compr. Rev. Food Sci. Food Saf. 14, 568−585.

Wright, A.J., Hartel, R.W., Narine, S.S., Marangoni, A.G., 2000. The effect of minor components on milk fat crystallization. J. Am. Oil Chem. Soc. 77, 463−475.

Youssef, M.K., Barbut, S., 2009. Effects of protein level and fat/oil on emulsion stability, texture, microstructure and color of meat batters. Meat Sci. 82, 228−233.

Youssef, M.K., Barbut, S., 2011. Fat reduction in comminuted meat products-effects of beef fat, regular and pre-emulsified canola oil. Meat Sci. 87, 356−360.

Zetzl, A.K., Marangoni, A.G., Barbut, S., 2012. Mechanical properties of ethylcellulose oleogels and their potential for saturated fat reduction in frankfurters. Food Funct. 3, 327−337.

CHAPTER 11

Methods Used in the Study of the Physical Properties of Fats

Peyronel Fernanda
University of Guelph, Guelph, ON, Canada

INTRODUCTION

This chapter is concerned with describing the fundamentals and setup procedures of five techniques that have become indispensable in characterizing food, and in particular, systems of fats and oils: X-ray diffraction (XRD), proton nuclear magnetic resonance (^{1}H-NMR), differential scanning calorimetry (DSC), rheometry and microscopy. The intent of this chapter is to describe the basics of state-of-the-art equipment in a practical way so that, while being easy to comprehend, it is also sufficiently complete.

It is important to understand the molecular structure of fats and oils from the atomic length scale to the large one such as millimeters. XRD focuses on the atomic to molecular length scale when working with powder diffractometers, as shown in Chapter 3. X-ray scattering helps elucidate the nano- to mesolength scale as shown in Chapter 9. ^{1}H-NMR also provides information regarding these larger lengths scales, but it is more suited to give information regarding diffusion than static structures. Consuming food is all about large-scale activities: biting, chewing, appreciating changes in texture, and mouthfeel. Accordingly, understanding food properties on submillimeter to centimeter scales is essential. This is the area of DSC, which is concerned with heat flow and thermal phase transitions and phase changes. Understanding the mechanical characteristics of food is also critical, and it is here where rheometry plays a fundamental role. When wanted to "see" the structure, one moves to microscopy. The area of microscopy, with all the different kinds of microscopes can nowadays cover length scales from few nanometers to many millimeters.

This chapter is divided into five sections, which address the abovementioned techniques: XRD, ^{1}H-NMR, DSC, rheometry, and microscopy. Each section is introduced by a box, which briefly describes the instrument, states the parameters to be controlled while doing the experiment, describes the relevant information that can be obtained and, with the exception of microscopy, it presents a typical graph of the results that can be obtained.

Structure-Function Analysis of Edible Fats, Second Edition
ISBN 978-0-12-814041-3, https://doi.org/10.1016/B978-0-12-814041-3.00011-3

X-RAY POWDER DIFFRACTOMETER

About the instrument

It uses a monochromatic X-ray beam that gets diffracted by a material.

A typical Bragg–Brentano geometry keeps the sample stationary while both the X-ray source and the detector are moved to collect the data.

Relevant information obtained for crystalline fats

Positions and shape of Bragg peaks allow to find the following:

1. Lamellar size
2. Polymorphism
3. Crystalline domain

Parameters to control

The range of Bragg angles to analyze

The rate at which data are collected

In some cases, the temperature of the sample

Typical output for fats

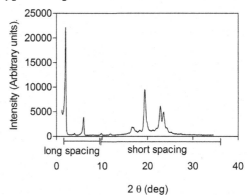

Principles of the Technique

X-rays have been used for about 100 years to help elucidate the structure of matter. They were discovered by the German physicist R. Röntgen in 1895. The first application, to radiography, was made by placing the X-ray source on one side of the object and a photographic film on the opposite side. The less dense matter allowed more X-rays to go through it than did denser matter, and this, in turn, enabled the recognition of, for example, a bone fracture. In 1912, diffraction by crystals was discovered. This application focused on using X-rays to explain the internal structure of matter. X-rays are electromagnetic radiation with energies ranging from ~ 120 eV to ~ 120 Kev, which is much higher than the energy, ~ 10 ev, of visible light. Owing to its high energy and corresponding short wavelength, X-rays penetrate matter, thereby interacting with internal electron densities to show the structure of the material. The wavelengths used in X-ray crystallographic diffraction experiments ranges from ~ 0.5 to ~ 25 Å, which is the length appropriate to reveal molecular and atomic distributions. Fig. 11.1 illustrates the different physical processes that can arise after X-rays interact with matter.

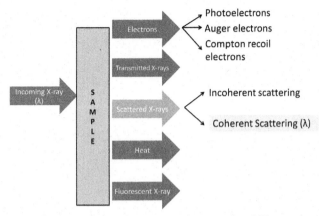

Figure 11.1 Schematic diagram of X-ray scattering, showing the different physical processes that arise after a substance (sample) has been irradiated with an incoming X-rays beam. Powder X-ray diffraction deals only with coherent scattering.

From a material structure point of view, the relevant physical process after X-rays interact with matter is scattering. X-ray scattering takes place when an X-ray photon, a discrete bundle or quantum of electromagnetic radiation, interacts with an electron. The photon can transfer energy to or from the electron, thereby undergoing inelastic scattering, or it can interact without net loss or gain of energy via elastic scattering. This is the only effect, of all those shown in Fig. 11.1 that is used in powder XRD analysis. The elastic scattering implies that the energies of the incoming and scattered photon are equal, which indicates that the wavelengths are necessarily equal.

Diffraction is the "accumulation or amplification" of the scattering of photons in those particular directions in which the amplitudes of the radiation scattered from different atoms are in phase (below). When those amplitudes are in phase, the scattering is said to be coherent. In principle, a coherent scattering pattern can be recorded using the reflective classical geometry setup for XRD shown in Fig. 11.2.

The technique of diffraction has evolved through the years. The first machines used films to record the diffraction patterns, whereas nowadays there is a generation of detectors that acquire data in two dimensions. Fig. 11.2 shows a typical Bragg–Brentano setup (Pecharsky and Zavalij, 2009, page 281), where on the left side of the samples is the X-ray source and on the right side is the detector. The X-ray source is a vacuum-sealed tube and requires continuous cooling. The incident and diffracted optic consists of slits and monochromators used to collimate (direct) the X-ray beam before and after it encounters the sample. Single slits are used to collimate in the vertical direction, whereas Soller slits are used to collimate in the horizontal axis. Fig. 11.2 can be viewed as a case where the

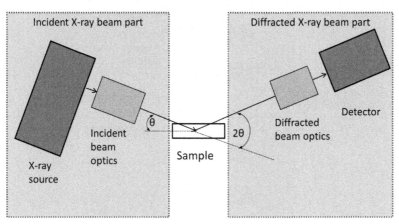

Figure 11.2 Setup for a scattering experiment using X-ray diffraction in a reflection geometry.

sample does not move, but the X-ray source and the detector, mounted on a goniometer, are made to rotate always to keep the relative angle between them constant. Using the reflection geometry, one can study wide angles, wide-angle X-ray scattering (WAXS) and only a section of small angles, small-angle X-ray scattering or SAXS. For angles that are small, the detector gets aligned perpendicular to an axis that is tangential to the surface of the sample, generating two problems: (1) the detector cannot take the direct beam and (2) there is a difference on the area of the sample irradiated at these small angles compare with the area irradiated at larger angles. This difference can manifest as differences in intensity because the data are not normalized per unit sample area. When looking at small angles, it is advisable to work in transmission mode, rather than in reflective mode. In the past 20 years, the technique of ultra-small angle X-ray scattering was developed at synchrotron facilities. The brilliance achieved in a synchrotron is many orders of magnitude greater than that of a conventional X-ray vacuum tube. To be able to see the scattering at very small angles of incident, a Bonse—Hart instrument is used.

XRD by a crystalline structure is best detected when the X-ray wavelength is of the same order of magnitude as the size of the features that one is trying to observe. These could be molecular features such as, for example, bilayer dimensions, primitive unit cell distances, and size of scattering centers, or it could be atomic features, such as distance between hydrocarbon chains.

Diffraction experiments can be separated according to the diffracted angle, θ. In particular, for edible fats, the WAXS region covers the angles between 12 degrees $< 2\theta < 35$ degrees, whereas the SAXS region covers the angles between 1 degree $< 2\theta < 12$ degrees. Each region probes a different length scale inside the material structure, as shown in Table 11.1.

Table 11.1 Information obtained from the two regions of X-ray diffraction used in fats

Region name	Diffracted angle	Characteristic size of scattering object (Å)	Information obtained for fats
WAXS	6 degrees $< 2\theta <$ 18 degrees	2 to 10	Primitive unit cell dimensions. Polymorphism
SAXS	0.5 degree $< 2\theta <$ 6 degrees	30 to 300	Lamella spacing and crystal domain size

The angular relationship between the incident and diffracted beam gives rise to Bragg peaks. Diffraction experiments give three kinds of information (to be discussed in more detail below):

1. Peak position
2. Peak intensity
3. Peak shape

Peak Position

The observed position of maximum of the Bragg peak provides information about the location of the lattice atomic planes in the crystal structure. To understand how these "lattice planes" are arranged, a brief explanation of Bragg theory is provided here. In fats, these lattice planes can be used to identify the polymorphism and lamellae spacing (explained later on). Bragg theory uses the simple notion of mirror reflections, where the diffraction from a crystalline sample can be explained as a reflection of the incoming X-rays by a series of crystallographic planes. A regular three-dimensional array of points can be considered as the basis of a crystal. From these points, "crystallographic planes" can be defined according to the position of the atoms in the crystal, considering that parallel atomic planes are similar in the atom arrangement. The orientation of the planes in the lattice is defined in terms of the notation introduced by the English crystallographer Miller. Three numbers, given in brackets specified the "Miller indices" of an atomic plane (hkl). The advantage of using this system is that parallel to any atomic plane in any lattice defined by the indices (hkl), there is a whole family of parallel atomic planes that are equidistant with a spacing d_{hkl} between them. The Miller indices (hkl) refer to the atomic plane that passes through the origin. In Bragg's approach, each atomic plane in the set (hkl) is considered a scattering object. Bragg proved that diffraction from a set of objects (in this case the atomic planes) separated by an equal distance, is only possible at certain angles. The diffracted beams are found only for special situations, which means, only for certain angles. Only the outgoing diffracted X-ray beams that are in phase will contribute to a constructive interference, all the others will cancel among themselves. To have a diffracted beam, parallel lattice atomic planes of atoms diffract the X-ray via constructive interference. Bragg's law can be geometrically derived with the aid of Fig. 11.3.

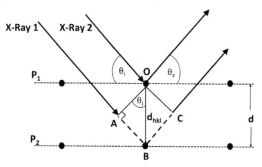

Figure 11.3 Parallel lattice atomic planes P_1 and P_2 showing the constructive interference of a front wave of parallel X-ray beams 1 and 2 (see explanation in text).

The incident X-ray beams 1 and 2 impinge at an angle θ_i onto a set of parallel planes (hkl), P_1 and P_2, in Fig. 11.3. The question is when will coherent scattering be observed via constructive interference? The answer lies with the understanding of wave interference. It is well known that constructive interference happens when the difference in path length between the two waves is an integer multiple of the wavelength. Let us see when this happens for the picture drawn in Fig. 11.3. The two X-ray beams are part of a flat front wave. The difference in path between the two scattered wave directions is given in Eq. (11.1):

$$\text{"path length of beam \#1"} - \text{"path length of beam \#2"} = \overline{AB} + \overline{BC}, \qquad (11.1)$$

where the segment $\overline{BC}$ is the shorter side of the rectangular triangle with vertices O, B, C; having 90 degrees between $\overline{OB}$ and $\overline{CB}$ as shown in Fig. 11.3. The same is valid for the segment $\overline{AB}$, for which the 90 degrees is between segments $\overline{OA}$ and $\overline{AB}$. Coherent or constructive interference must satisfied the relation in Eq. (11.2)

$$\overline{AB} + \overline{BC} = n\lambda \qquad (11.2)$$

with n an integer and

$$\overline{AB} = d_{hkl}\sin(\theta_i); \quad \overline{BC} = d_{hkl}\sin(\theta_s), \qquad (11.3)$$

where θ_i is the angle between the direction of the incident X-ray and the atomic planes P_1 and P_2, θ_s is the angle between the direction of the scattered X-ray and the atomic planes P_1 and P_2, d_{hkl} denotes the distance between the two lattice atomic planes. The subscript "hkl" has been used to alert the reader to think about the Miller index. It is obvious that $\overline{AB} = \overline{BC}$ only when the incident and scattered angle $\theta_i = \theta_s$ are the same. Therefore, under these conditions, one can write

$$2\,\overline{BC} = n\lambda = d_{hkl}\sin(\theta), \qquad (11.4)$$

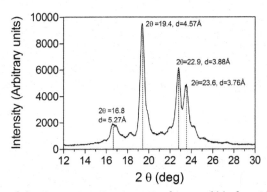

Figure 11.4 Histogram of the X-ray scattering intensity for an edible fat with a high percentage of solid material. The histogram shows intensity versus 2θ in the region 12–20 degrees. d was computed using Eq. (11.6) and a value of $\lambda = 1.54$ Å

where $\theta = \theta_i = \theta_s$. The integer n is called the "order" of the reflection. Its value is always taken as 1 because higher order can be represented as the relation shown in Eq. (11.5)

$$d_{hkl} = n\, d_{nh\ nk\ nl}. \tag{11.5}$$

Eq. (11.4) shows the relation required for parallel atomic planes, separated by a distance d to generate constructive interference. It can be simplified to obtain Eq. (11.6), what is known as the Bragg's law:

$$\lambda = 2d_{hkl} \sin(\theta). \tag{11.6}$$

The results of a diffraction experiment are usually plotted as values of intensities versus 2θ (rather than θ) and the plot is called a "powder diffraction pattern" or "histogram" (Fig. 11.4). Fig. 11.4 shows the WAXS region for an edible fat. The intensity is not an absolute value because it has not been normalized. Fig. 11.4 shows the peak positions as a function of 2θ and the d-values corresponding to those peak positions. This d-value is obtained using Eq. (11.6) and the corresponding wavelength used in the experiment, $\lambda = 1.54$ Å for a cooper tube.

Scattering theory uses the scattering q vector to represent the difference between the direction of the incoming X-ray and the direction of the scattered X-ray. This vector q is related to the scattering angle θ. The relation between q and θ is given by Eq. (11.7)

$$q = \frac{4\pi \sin(\theta)}{\lambda} \tag{11.7}$$

The reader can learn more about the q vector in Chapter 3.

Peak Intensity

In the study of edible fats using a benchtop powder X-ray diffractometer, relative intensities of peaks rather than absolute values are used. Typically, these relative intensities are used to identify the kind of polymorphs present in the edible fat.

The intensity value include factors such as the following: scattering by an electron, thermal scattering, unit cell scattering, as well as geometrical factors due to the measuring equipment.

What follows is a simple explanation for the basis of crystallography and XRD. More detail can be found in Cullity and Stock (2001).

Scattering From an Electron

Thomson (1906) demonstrated that the intensity, I, of the beam scattered elastically by a single electron of charge e (in C) and mass m (in kg) at a distance r (in meter) from the electron is given by Eq. (11.8)

$$I = I_o \left(\frac{\mu_o}{4\pi}\right)^2 \left(\frac{e^4}{m^2 r^2}\right) \sin^2 \alpha, \tag{11.8}$$

where I_o = intensity of the incident beam, α is the scattering angle between the incident X-ray beam and the direction of the scattered X-ray beam and $\mu_o = 4\pi \ 10^{-7}$ m kg/C^2. The reader must have in mind that an X-ray beam can be thought as a radiation wave or as a bundle of particles (the photons). This has tremendous implications because the theory of X-rays can be explained using either quatum mechanics (particle) or the wave theory.

Scattering From an Atom

When an X-ray beam encounters an atom, each electron will scatter following Thompson's equation. In an atom with Z electrons, the scattering is the sum of the waves for each electron, which might involve differences in phases for the scattered waves (Cullity and Stock, 2001, page 128). A quantity f, called the "form factor" or "atomic form factor," describes the efficiency of scattering from a given atom in a given direction and is defined by the ratio of two amplitudes (Cullity and Stock, 2001, page 129) shown in Eq. (11.9)

$$f = \frac{amplitude\_of\_wave\_scattered\_by\_an\_atom}{amplitude\_of\_wave\_scattered\_by\_one\_electron}. \tag{11.9}$$

The actual calculation involves $\sin(\theta)$ rather than θ, so the net effect is that f decreases as $\sin(\theta)/\lambda$ increases.

An expression for the atomic scattering factor, f_o^{ij}, is given by Pecharsky and Zavalij (2009, page 211) Eq. (11.10)

$$f_o^j \frac{\sin\theta}{\lambda} = c_o^j + \sum_{i=1}^{4} a_i^j e^{\left(-b_i^j \frac{\sin\theta}{\lambda}\right)}, \tag{11.10}$$

where the coefficients can be found in the International Tables for Crystallography (1999) and which depend on the atom used. The atomic scattering factor can then be calculated for each atom.

Scattering by a Unit Cell

The atomic scattering factor describes the results of interference effects within the lattice due to one atom. But, for the entire crystal, the coherent scattering coming from all the atoms present in the crystal must be considered. The mere fact that the atoms are arranged in a periodic fashion in space means that the scattered radiation is now limited to certain directions with respect to the direction of the incoming radiation. These directions are determined by Bragg's law, and the challenge is to find the intensity of the radiation diffracted by a crystal as a function of the *atom* position in the *unit cell*. Because the crystal is formed by repeating the fundamental unit cell, it is enough to consider how the atoms arranged in the unit cell affect the diffracted intensity. The result of combining all "atomic form factors" (f) for the atoms present in a unit cell gives rise to a function called "structure factor," F, Eq. (11.11), which can be expressed by (Cullity and Stock, 2001, page 136)

$$F_{hkl} = \sum_{n=1}^{N} f_n e^{2\pi i(hu_n + kv_n + lw_n)}, \tag{11.11}$$

where N identifies the number of atoms contained in the unit cell. F is a complex number and expresses both the amplitude and phase of the resultant wave. The square of its absolute value is proportional to the scattered X-ray intensity. The (hkl) reflection and the set $\{u_n, v_n, w_n\}$ identify the coordinates of the end atoms.

When performing an X-ray experiment, the intensity is recorded. The integrated intensity corresponding to a particular set of planes, or for a particular (hkl), can be computed using Eq. (11.12) (Pecharsky and Zavalij, 2009, page 184)

$$I_{hkl} = K p_{hkl} L_\theta P_\theta A_\theta T_{hkl} E_{hkl} |F_{hkl}|^2, \tag{11.12}$$

where K is a scale factor that comes from the normalization of the integrated intensities with absolute calculated intensities; p_{hkl} takes into account the number of symmetrically equivalent reflections; L_θ is the Lorentz factor, which takes into account two factors: the finite size of the reciprocal lattice points and the variability of the Debye rings (Pecharsky and Zavalij, 2009, page 153); P_θ accounts for a partial polarization of the scattered electromagnetic wave; A_θ accounts for the absorption in both the incident and diffracted beam; T_{hkl} is the preferred orientation factor: it takes into account possible deviation from a complete randomness in the crystallites orientations; E_{hkl} is the extinction coefficient, which takes into account a deviation from the kinematical diffraction models; F_{hkl} is the structure factor. It is defined by the details of the crystal structure of the material.

Peak Shape

Specific information on the crystal structure can be obtained from the shape of the peak. For sufficiently large and strain-free crystallites, the diffraction theory predicts that the

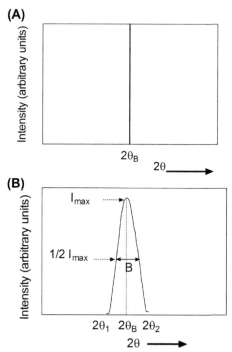

Figure 11.5 Hypothetical Bragg peak (A) and experimental observed Bragg peak (B) showing the effect of the crystallite size.

lines of the powder pattern will be exceedingly sharp (Fig. 11.5A). In actual experiments, lines of such sharpness are never observed because of a combine number of instrumental and physical factors that broaden the "pure" diffraction line profile (Fig. 11.5B). When there is no strain or imperfection in the sample, the Scherrer (1918) Eq. (11.13) can be used to find the thickness of the crystallite,

$$\text{thickness} = \frac{K\lambda}{B\cos(\theta)}, \tag{11.13}$$

where B is the full width at half maximum (FWHM) in radians of the peak under study on the 2θ scale in radians (Fig. 11.5B), λ is the wavelength (in nm), and K is a constant approximately equal to unity and related both to the crystallite shape and to the way that B and the thickness are defined.

Scherrer original derivation (Scherrer, 1918) was based on the assumptions of a Gaussian line profile and small cubic crystal of uniform size. This gave a value of K = 0.94. The derivation calculated by Klug and Alexander (1974, page 689) showed a value of K = 0.89 for no particular crystal. It has been quoted (West, 1984) that this equation is valid only if the crystallites have a thickness smaller than 100 nm. The

Scherrer equation shows that the thickness of the crystal increases as the width of the peak decreases. It is also known that broadening increases with the diffraction angle, and for that reason, this equation has been used in our laboratory for symmetrical peaks that are in the region of $1 < 2\theta < 10$ degrees.

Instrument Description

Our laboratory is equipped with a rigaku automated powder X-ray diffractometer, MultiFlex DXG, theta/2 theta 2 KW, operating at 40 KV and 44 mA, with a Cooper X-ray tube (wavelength of 1.54 Å). The denomination $\theta/2\theta$ indicates that the angle between the sample and the incoming X-ray direction is θ and between the incoming X-ray direction and the direction to the detector is 2θ, as shown in Fig. 11.6.

The sample position is fixed, in contrast with some instruments that make the sample spin on its axis. In addition, this sample holder uses a flat sample holder, rather than a capillary tube. This instrument has a Peltier system for temperature control of the sample holder. A Peltier system is a device that measures relative temperatures by using the Peltier effect for which an electric current or a difference in voltage is used. It requires continuous cooling and is typically controlled by software. Sensors on the housing of the electronic indicates the software if the temperature had been reached or not. The Peltier system requires continuous cooling, which in our case is provided by an external water bath.

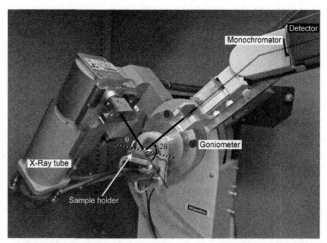

Figure 11.6 A goniometer comprises the two arms shown here and the sample holder. The right arm holds the monochromator and detector, whereas the left one holds the X-ray tube. The sample's position is fixed at the center of the goniometer.

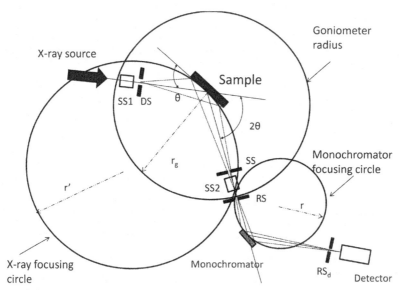

Figure 11.7 Schematic principle of operation of the Rigaku Multiflex using the Bragg–Brentano geometry.

The system has four main parts/components: X-ray tube, sample holder, monochromator, and detector. The X-ray tube and the detector are each attached to one of the two arms of the goniometer. The arms move together but in opposite directions and follow the measuring diffractometer circle (Fig. 11.7) to maintain the relation $\theta/2\theta$ at all times. The distance between the X-ray source and the sample is equal to the distance between the sample and the detector.

Fig. 11.7 is a schematic representation of the system showing the optics that collimates the X-ray beam, both before the sample and after it, as well as showing the circular path that the X-ray tube and the detector followed.

Fig. 11.7 illustrates the following elements:

- DS: goniometer divergence slit.
- SS: soller slit assembly (SS1 on "tube" side, SS2 on detector side), a series of closely spaced parallel plates, parallel to the diffractometer circle (i.e., the plane of the paper), designed to limit the axial divergence of the beam.
- θ: incident angle of the X-ray beam onto the sample.
- 2θ: diffracted angle of the X-ray beam.
- SS: goniometer scatter slit. This system is designed in such a way that both DS and RS are located at the same distance from the sample and should have the same value (either $1/2$ degree or 1 degree)
- RS: monochrometer receiving slit.
- RS_d: detector receiving slit.

- r': Primary focal circle radius
- r$_g$: Goniometer radius
- r: Secondary focal circle.

The DS determines the sample's area to be irradiated. It should be noted that as the incident angle of irradiation changes, the area that is irradiated changes. Both SS and DS must have same aperture or the same number. The monochromator (possessing a graphite crystal with $2d = 6.708$ Å) works such as a filter, allowing only certain wavelengths to reach the detector.

The angle θ can be varied from 1.0 degree to about 80 degrees in this machine. Angles smaller than 1 degree can burn the detector; therefore, they are avoided. The user has to have in mind that in the case of edible fats, the Bragg peaks obtained at small angles are nonsymmetrical and that the position of the first peak with high intensity (commonly referred as (001)) is usually imprecise. In our instrument, we use the (002) or (003) diffraction peaks to estimate the position of the (001) by multiplying by 2 or 3, respectively.

Calibration

No daily check is necessary for the use of this machine. The machine gets serviced and checked once a year by the instrument maker. The user can check the performance by using a standard material, such as silver behenate to ensure that the scattering peaks for the small angles have not shifted.

Experimental Procedure

A countertop diffractometer is simple to operate. The X-ray instrument requires an external refrigeration cooling system to prevent the X-ray tube from getting burnt out. The efficiency of an X-ray tube to generate useful X-rays is only 10%, which means that there is a lot of heat to be dissipated. The refrigeration cooling system uses a water reservoir to prevent the X-ray tube from overheating. The water must circulate whenever the X-ray tube is on. The X-ray tube has to be on at full power before it can be used for measurements. This is achieved by carrying out a procedure called "aging." Aging consists of incrementing the voltage and the current that goes into the X-ray tube in a series of steps until 2 KW, with a voltage of 44 kV and a current of 44 mA is reached. Typically, the whole procedure takes 20 min. Aging of the X-ray tube is necessary; otherwise the X-ray tube life is shortened by taking it to full power in only one step. When the instrument is not in use for more than 24 h, it is completely turned off. For periods of inactivity of less than 12 h, the X-ray tube is left at minimum activity. This is done by using an aging down procedure that brings the voltage from 44 to 20 kV and the current from 40 to 1.7 mA.

The Rigaku X-ray diffractometer is considered a safe machine in the sense that there is no X-ray leakage and that it has safety features in place. For example, the door has a safety mechanism that shuts down the X-ray (no X-ray is produced at all) in case someone tries to open or close it without following the correct procedure. This is to avoid the door being opened while the X-rays are being ejected from the X-ray tube and bouncing off the sample. In addition, if any fast movement is performed on the door while closing it, the X-ray automatically cuts the beam by closing the shutter and making the alarm go off.

The specimen is placed in either a glass or an aluminum holder, which is then placed on the sample holder inside the machine. The sample holder provided by Rigaku is a piece of metal that gets screw (using two small screws) to the goniometer to hold the sample holder. The powder technique requires that no specific orientation is introduced and that the surface of the material is smooth and level with the top of the cavity in the sample holder (Pecharsky and Zavalij, 2009, Chapter 12). The goniometer is software-controlled. The user needs to specify the conditions for the analysis: starting angle, final angle, step size, scan speed, and the DS and SS slit sizes. The slits help maximize the intensity as a function of θ, making sure that the incoming and scattered X-rays are directed (collimated) onto the sample and the detector. A typical value for this instrument is $\frac{1}{2}$-degree slits. The starting angle as well as the final one depends on whether one is using the wide- or small-angle region. The step size or sampling width represents the step that the detector is moved between data collection. It is generally maintained at 0.02 degree. The scan speed determines how long the detector will collect at each position. The faster the detector moves, the less the resolution is achieved. A typical full scan for fats, covering both wide- and small-angle regions, will start at 1 degree and finish at 30 or 35 degrees, with a step size of 0.02 degree and a scan speed of 1 deg/min. The Rigaku instrument allows the following types of scan: (1) continuous, where data are collected continuously as the X-ray tube moves, (2) fixed time, where the X-ray tube moves a specific interval of degrees, stops, and collects data only at this position for the duration of the selected period of time before moving to the next position, and (3) fixed counts, where the detector moves a specific interval of degrees and collects a specified amount of counts.

Sample Preparation

A powder diffraction pattern is achieved when there is a sufficiently large number of small randomly oriented particles. This to make sure that all orientations are taken into account without any preferential one. This random orientation is achieved by ensuring that the samples contain particles with dimensions between 10 and 50 μm and that they are randomly oriented. If the material is not in the form of small particles, particle sizes should be reduced. This is not a problem with edible fats, which are composed of crystals possessing a suitable size and no net orientation, if a high shear was not applied during the manufacture of the product. The orientation of the crystals can be eliminated by properly mounting the sample in the holder. Rigaku sells sample holders made from glass with a cavity that has the dimensions of 20 × 20 mm and 0.3 mm in depth. In addition, any machine shop can make them out of different materials, for example, aluminum. Filling the cavity with a powderlike material is done by using a flat spatula and making sure not to compress the sample in a particular direction. For samples that are pastelike, the same procedure is followed, resisting the temptation to squeeze down on the sample. Hard fat samples cannot be mounted on the cavity of the glass slide because the process of trying to convert them to powder causes friction and melting. When the material needs to be melted before the analysis, then the material is simply poured into the cavity in liquid form. The only precaution is not to overfill or underfill the sample holder. The design of the glass slide and its cavity is such that it takes into account the slits and geometry of the goniometer. The user is not required to do any adjustments.

Data Interpretation for Fats: A Case Study

As mentioned before, the powder diffraction technique is used in edible fats to determine characteristics of atomic plane spacings. As data are being collected, it gets displayed on the screen as a histogram of the signal intensity versus 2θ. Farther processing is done with a fitting analysis program such as Jade 9.0 Plus 1995−2011 (MDI, Livermore, California, USA) or Igor Pro 6.2 (WaveMetrics, Oregon, USA).

Bragg peaks in the WAXS (~ 10 degrees $< 2\theta < 30$ degrees) region give information about the polymorphism of the edible fat sample, whereas Bragg peaks in the SAXS (~ 1 degree $< 2\theta < \sim 8$ degrees) region give information about the lamellae and crystal thickness.

The XRD AOCS (AOCS Method Cj-2- 95) states that if the sample shows a peak at d $= 4.15$ Å, then the polymorphic form is á; however, if two peaks appear at positions 3.8 Å and 4.2 Å, then the polymorphic form is â'; whereas, if the peak appears at position 4.6 Å, then the polymorphic form is â. These d-spacings obtained from the Bragg peaks defines the subcell within the crystal lattice (Small, 1966), which can be hexagonal (á), orthorhombic (â'), or triclinic (â). Larsson (1966) introduced a criterion for the classification of the different crystal forms. He defined the á polymorphic form as the one that has a single peak at 4.15 Å, the β' form as the one with two peaks at positions 3.8 Å and 4.2 Å, and the β form as the one that shows three peaks, at positions 4.6 Å, 3.8 Å, and 3.7 Å.

Fully hydrogenated canola oil (FHCO) was used to generate an X-ray spectrum for each of the three polymorphic forms. The range of angle study is in the region of 1 degree $< 2\theta < 35$ degrees. The scan was performed with a step size of 0.02 and a scan speed of 1 deg/min. The DS and SS slits were chosen to be $^1/_2$ degree. The material was melted in an oven at 80°C and kept at that temperature for 15 min to erase all crystal memory. It was then poured into a heated XRD glass slide and immediately placed into the sample holder in the XRD, where the temperature was set and held for a definite length of time to control and obtain different cooling rates. The α polymorph was obtained when the temperature was set at 20°C and held there until complete crystallization of the fat was achieved (about 1 min). The β' polymorph was obtained by setting the temperature on the sample holder to 55°C and held for 90 min. The cooling rate for these polymorphs was calculated as 0.27°C/min. The β polymorph was obtained by setting the temperature to 62°C for 30 min and then reducing it to 58°C in steps of 1 degree at a time and with a waiting time of 15 min at each temperature which gave a cooling rate of 0.02°C/min.

Long spacings or small angles ($2\theta = 1-5$ degrees) are used to determine the longitudinal packing of the lamella. The d-spacing obtained from the Bragg peak position corresponds to the distances between the lamellae and yields information regarding the unit cell or lamella size. The other information obtained is the lamellar thickness of the sample, which is obtained from the d-value of any of the peak's position in the small-angle region according to the diffracting atomic plane under consideration. The histogram shown in Fig. 11.8 displays the Bragg peak positions in the small-angle region for two peaks. The highest peak is the (001) reflection and the smallest one is the (003). Because the Rigaku machine does not give accurate values for the scattering intensity at small angles, we used the (003) reflection to obtain the correct (001) value. The (003) d-position is multiplied by 3 to obtain the (001) value, which gives the lamellae size: 49.4 Å for α, 45.6 Å for β', and 45.0 Å for β.

The longitudinal packing of the TAG molecules in the lamellae is based on its constituent fatty acid (FA) alignment. This d-distance, obtained from the first-order reflection in the short-angle region, is used to estimate the packing of FA in the lamella. Many

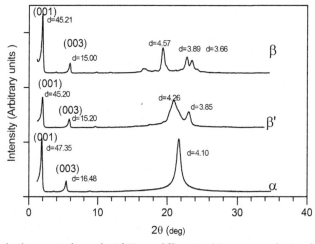

Figure 11.8 Fully hydrogenated canola oil X-ray diffraction histograms obtained at three different cooling rates. The three spectra show the d-spacing value associated with the peak positions for both, the small- and the wide-angle regions.

lamellae stacked together constitute the thickness of a crystal lattice. Typically, if one knows the material under study, one can predict the longitudinal length of the FA. The chair configuration, known as 2L, means that the total length of the longitudinal packing is the length of two FA chains. When longitudinal packing contains three FAs, molecules in the lamellae are in a tuning fork configuration, known as 3L. To find out whether FHCO has crystallized in a 2L or 3L configuration, a simple calculation was performed as follows.

The predominant FA present in FHCO contains 18 carbons. Carbons are joined in a zigzag form maintaining a 120 degrees angle. The distance between three carbons is 2.54 Å (Small, 1966), which means that the total carbon chain length is $2.54 \times 8.5 = 21.54$ Å. One has also to have in mind the presence of the glycerol and the carbon and oxygen attached to each hydrocarbon length. A rough estimate for 18 carbons using the 21.54 Å indicates that a 2L packing will give a value close to 43 Å, whereas a 3L packing will give a value close to 63 Å. Our results are closer to 43 Å than 63 Å, which means that is safe to say that FHCO has crystallized in a 2L conformation. By including the carbon from the glycerol one can also infer if the lamellae contains hydrocarbon chains packed, forming an angle with the atomic plane that causes the scattering or not.

The thickness of the crystal can be obtained from the data in the long spacings. The Scherrer equation (Eq. 11.13) is used in this calculation. As mentioned above, it is valid only for a well-defined, symmetric peak in the small-angle region rather than for peaks from the wide-area region because peak broadening increases with the diffraction angle. Accordingly, peak (003) was chosen for this calculation and $\lambda = 1.54$ Å for copper. The three polymorphs were analyzed to obtain both the crystal domain size and the number of lamellae contained in the crystal domain. The results are shown in Table 11.2.

The powder X-ray technique allows the user to properly identify the polymorphic form present in a fat sample. It also allows the user to get parameters that describes the crystal domain and lamella size. The characterization of the fats is useful for the development and manufacture of new fat-based products. The polymorphism form present in a

Table 11.2 Data obtained from the (003) peak for the three polymorphic forms of fully hydrogenated canola oil (B, θ, and lamella size) and the results of calculating the crystal domain size (using Eq. 8.13) as well as the number of lamellae present in a crystal domain

	α polymorph	β′ polymorph	β polymorph
B measured from the (003) peak	0.00226 rad	0.00261 rad	0.00261 rad
θ (deg)	2.67	2.89	2.97
Crystal domain (using Eq. 11.13)	619.0 Å	536.4 Å	536.4 Å
Lamella size	49.4 Å	45.6 Å	45.0 Å
Number of lamellae in a crystal domain	12.5	11.75	11.9

fat product plays an important role in the sensory attributes of the final product, such as texture and mouthfeel. It is also known that the shelf life of the final product might be limited by a polymorphic transformation. For example, manufacturers make margarine in the β' because it gives the product the much desire functionality. On the contrary, margarine in the β polymorphic form has a dull appearance and a mottled surface, and the texture becomes brittle (deMan and deMan, 2001).

PULSED NUCLEAR MAGNETIC RESONANCE BENCHTOP SPECTROMETER

Technique principles
This technique uses a radio frequency (RF) pulse to perturb the magnetic moment of protons in the hydrogen atoms to study relaxation times while the sample is interacting with an applied constant magnetic field B_o

Relevant information obtained for fats
Solid fat content (SFC) (percent)
Direct method
Indirect method
Spin–spin relaxation time, T_2
Spin–lattice relaxation time, T_1

Parameters to control
Number of scans
Recycle delay (waiting time between scans)
Tau: the waiting time between applied RF pulses
The time that the RF is applied
Sample thermal history

Typical output
Free induction decay (FID) curve

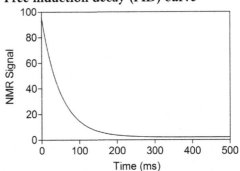

Principles of the Technique

Nuclear magnetic resonance (NMR) has developed into a standard laboratory technique encompassing dipolar and quadrupolar NMR, covering a range of nuclei possessing various spin states. Dipolar systems are those possessing spin magnitude of $\hbar/2$ such as hydrogen protons. It is these nuclei that are analyzed by the instrument described in this section. The technique is named either pNMR (p for proton) or ^{1}H-NMR to indicate that the hydrogen proton is being studied. In particular, this section deals with a small benchtop instrument that applies a RF pulse, hence the technique is also called low-resolution pulsed ^{1}H-NMR.

Pulsed NMR is a nondestructive technique in which the material is subjected to electromagnetic excitation and its response monitored over time.

The pulsed NMR process can be viewed in four steps: (1) the sample is subject to an applied continuous external magnetic field, B_0, in the z-direction, (2) the sample responds to B_0 generating a net magnetization in the equilibrium state, (3) a RF pulse is applied during a limited time causing the net magnetization to change to a none-equilibrium state, and (4) the net magnetization relaxes to its equilibrium state. The relaxation generates a signal, which is monitored over time and which provides information about the system.

During step (1), the sample is inserted into the external permanent magnetic B_0, which induces a net magnetic moment vector M_0, the net magnetization. Step (2) is the appearance of this induced net magnetization, which is not a static one, but it rather "process," it rotates around the Z-direction of the applied external magnetic field. The precession of M_0 takes place at a particular frequency of rotation, called the Larmor frequency, ω_L, and which depends on the magnitude of B_0. Because the hydrogen protons are usually targeted with a benchtop NMR instrument, B_0 is set to 40 T. The external magnetic field B_0 interacts with the sample's natural nuclear magnetic moments. All nuclei are characterized by their magnetic properties determined by the value of their "spin," I. This quantity, I, can possess quantized values, $I = 0, 1/2, 1, 3/2$ in units of $\hbar$. Nuclei with $I = 1/2$ possess a magnetic dipole moment, M_0, and no higher multipoles. Each nuclear dipole in an external magnetic field, B_0, possesses two quantized energy states defined by whether the Z-component of the dipole moment points in the direction of B_0 ("spin up") or in the direction opposite to B_0 ("spin down"). The "spin up" state possesses a lower energy than the "spin down" state so that, at any finite temperature, the "spin up" state will be more populated than the "spin down" state: the magnetic dipole will preferentially be aligned along the Z-axis, in the direction of the external field B_0. The contribution of all these dipoles generate M_0 (Fig. 11.9), the total magnetization moment due to all the protons moments.

Step (3) consists in applying another magnetic field, a RF-pulsed B_{RF} for certain time. To observe NMR, the nuclei moments that make the net magnetization M_0 and which are aligned predominantly along the Z-axis are perturbed out of alignment with the application of a B_{RF}. This pulsed magnetic field will affect the orientation of the spins, in which those spins that were in the "spin up" state now go to a "spin down" state and vice versa. Instead of following individual spins, one follows the state of the net magnetization M_0. For example, a pulse with the right magnitude and the right orientation can cause a "tip over" of M_0 from the original Z-direction to, for example, the Y direction in the (X,Y) plane. This is called a 90-degree pulse. The magnitude of this magnetic field B_{RF} is selected to satisfy Eq. (11.14)

$$\omega_L = \gamma B_{RF}, \tag{11.14}$$

where ω_L is the proton Larmor frequency and γ is the gyromagnetic ratio, the ratio of the classical magnetic moment to the classical angular momentum.

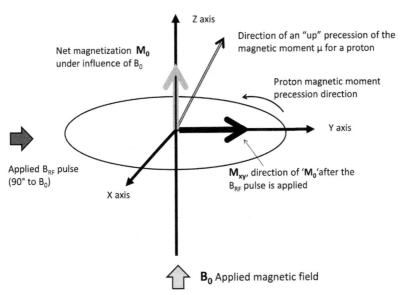

Figure 11.9 Schematic representation of the applied magnetic field B_o (in the Z-direction), the net magnetization moment, M_o the direction of the applied RF pulse, B_{RF} and the subsequently "tipped" of M_o into the (X,Y) plane.

Step (4) in the NMR process stars after the B_{RF} pulse is terminated. The precession of the net magnetization, as it goes back to its Z original position can be detected by a coil, which records the alternating voltage that the relaxation process induces. This coil is placed surrounding the sample, and it is used to generate the B_{RF} and to collect the signal. The voltage (V) is given by the Faraday's law $V = d\Phi/dt$. To maximize the change in the magnetic flux Φ through the coil, the coil axis is oriented perpendicular to B_o, hence only the component of the net magnetization in the XY plane, M_{xy}, is recorded over time.

The physical process is complicated, as there are some nuclei that are in the up state and others in the down state, which indicates that the precession is opposite. It is worth mentioning that the precession of spins in the XY plane lasts only for a finite time. The decay happens due to three effects: (1) B_o is not perfectly uniform. Its consequence is that protons in different parts of the sample precess at slightly different frequencies and get out of phase with each other, thereby decreasing the value of M_o, (2) because protons in a sample belong to different molecular environments, the precession of each of them will be at a different frequency, causing a loss of phase coherence and a decay of M_o, and (3) transition between spin up and spin down happen due to electromagnetic interactions between the protons and the surrounding particles, the lattice, again causing a decay of M_o. The NMR process follows only the net magnetization M_o, which precesses around Z. The M_{xy} signal components decreases over time, whereas the Z-component of M_o, M_z increases. These phenomena give rise to two relaxation processes, T_1 and T_2 to be discussed below.

The "FID" is the name given to the time-domain signal obtained during the relaxation process. In addition, the relaxation process itself is referred as the FID. In this section, we present applications of this technique to the determination the SFC as well as the time constants for two relaxation processes called T_1 and T_2.

Solid Fat Content

Compared with nuclei in liquid phases, nuclei in solid phases interact more strongly with other nuclei in their neighborhood because of the slower dynamics of their atoms compared with atoms in a liquid; therefore, they relax very quickly (<10 μsec) and can be easily distinguished from the slower relaxing nuclei of more mobile atoms or molecules in a liquid phase. This characteristic is exploited to determine SFC on edible fat systems. The SFC measurement requires the application of only one B_{RF} pulse, which is applied for a finite amount of time to rotate 90 degrees the magnetization vector M_o. The instrument monitors the time evolution of the signal, as the sample nuclei relaxes.

It has been established that edible fats can be studied by using a B_{RF} pulse for about 3 μs, followed by a dead time of the receiver of about 7 μs (Van Putte and van den Enden, 1974). The FID obtained after the B_{RF} pulse is removed is detectable for times that range from milliseconds to seconds. The American Oil Chemists' Society (AOCS) suggests two methods to calculate the SFC: the direct method and the indirect method.

Direct Parallel Nuclear Magnetic Resonance Method (AOCS Cd 16b-93 Method, 2000)

This methodology requires the values from the FID at two different times. Protons in solid state relax faster than in liquid state. It has been established that the signal value, S_{S_SL}, 11 μsec after the B_{RF} pulse was terminated is considered to have information about the liquid and solid protons together, whereas the signal value, S_L, 70 μsec after the B_{RF} pulse was terminated is considered to represent only the protons in liquid state. Fig. 11.10 illustrates the FID and the two values that are extracted to compute the SFC which is the percentage of protons in the solid state.

Most commercial instruments possess software that computes the SFC based on the two numbers extracted. Eq. (11.15) is used to find the SFC:

$$SFC = \frac{(S_{S\_SL} - S_L)F}{S_L + (S_{S\_SL} - S_L)F + D},$$

(11.15)

where F is an empirical correction factor (established during calibration) to account for the detector dead time and D is the digital offset factor (also established during calibration).

Indirect Parallel Nuclear Magnetic Resonance Method (AOCS Cd 16-81 Method, 2000)

This method uses the signal from the FID that represents only the liquid protons, S_L (Fig. 11.10). IN this case, the SFC is calculated using two values obtained (1) after the

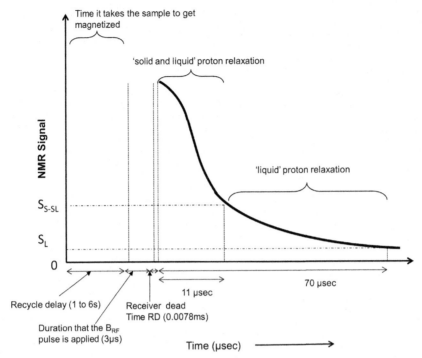

Figure 11.10 Free induction decay (FID) for a fat sample obtained after a radio frequency pulse at 90 degrees was applied. The two relevant signals correspond to the solid—liquid response (S_{S-SL}) and the liquid response (S_L).

material was completely melted and (2) after the material was brought to a particular temperature by following the right tempering protocol. To account for the effect of temperature on instrument sensitivity, it is also necessary to measure a reference or a standard oil, that should be liquid at both temperatures. Eq. (11.16) is used for this case is

$$\%\text{SFC} = 100 - \left(\frac{\text{Oil}_{60} \times \text{Sample}_T}{\text{Sample}_{60} \times \text{Oil}_T}\right) 100. \tag{11.16}$$

Subscripts indicate the temperature at which the FID is obtained. This calculation is based on the assumptions that both the standard oil and the sample are completely liquid at 60°C and that the standard is completely liquid at the final measurement temperature. Although these assumptions are probably justified in many cases, they may not always be correct, particularly when making measurements at low temperatures. In the case of fats that melt above 60°C, the official method gives no provisions, but common sense indicates that the sample and oil temperature at which the reference measurement is performed (Oil_{60}, Sample_{60}) needs to be adjusted accordingly.

Relaxation

The relaxation experiment differs from the SFC in two aspects: the signal is continuously detected over a period of time and more than one B_{RF} pulse are typically applied. Depending on the method used, only particular values of the FID are used to generate a new "envelope" that is later process to extract the desired parameters.

As mentioned in the introduction, the nuclear relaxation can be divided into two separate physical processes that are characterized by two times, T_1 and T_2 (Deleanu and Pare, 1997; Blumich, 2005; Traficante, 2005). The external permanent magnetic field B_0 acting on the sample generates a net magnetization in the Z-direction M_0, which is the result of having some nuclei spins precessing in the up direction (along the +Z-axis) and some in the down Z-direction. When a RF pulse, B_{RF}, is applied during a certain time, the nuclei spins are disturbed. Relaxation refers to the process by which those spins come back to the original M_0 magnitude and direction.

The T_1 process is concerned with spin—lattice of longitudinal relaxation, which appears as soon as M_0 is moved from its equilibrium. The populated spin states are influenced by the environment, the lattice of the other nuclei around it. The spins can exchange energy with electrons and other nuclei in the sample resulting in a fast approach to the equilibrium state.

The T_1 process is characterized by a decay as shown in Eq. (11.17), with T_1 being the characteristic time of the process:

$$M_z = M_o\left(1 - e^{-\frac{t}{T_1}}\right). \qquad (11.17)$$

One of the oldest methods for measuring T_1 is called the inversion recovery method, otherwise known as the "180-τ-90 degrees" method, where τ (tau) is a time interval. In this method, a 180-degree pulse is applied to achieve a population inversion. The net magnetization M_o is flipped from the Z-direction to the −z-direction so that the "spin down" state becomes more favorably occupied than the "spin up" state, and thus the Z-component of the net magnetization is antiparallel to the external field B_o. On removal of B_{RF}, the net magnetization begins to relax to its equilibrium position. An arbitrary time, τ (tau), is allowed to pass before the next pulse in the sequence is applied. This waiting time is called the "inversion time." A 90-degree pulse is then applied to rotate the net magnetization vector onto the Y axis to be able to record the signal because the coil is positioned in a way that can only detect signals in the XY plane. At least two measurements need to be made with different inversion times to have sufficient equations to solve for the many unknowns. Experiments have shown that the value of τ should be $> 5 \times T_1$ for the system to completely recover.

The second relaxation process is the spin—spin process or transverse relaxation. The relaxation of M_o in XY plane is related to the spread of the Larmor frequencies due to the proton's spin moments that are not moving in unison. This is a product of having inhomogeneities in B_o, which causes different moments to precess at different rates. The effect of having spins that "fan out" as they precess is also influenced by random fluctuations of other magnetic fields, such as the dipole field of one proton moving past another. The decay of M_{XY} happens at another time scale compared with the decay in the Z-axis. The characteristic time for this process is called T_2, the *spin–spin or transverse* relaxation time (Forshult, 2001; Capozinni and Cremonini, 2009).

Typically, the transverse relaxation is modeled to take into account both effects. To do this, the exponent in the decay is expressed as $-t/T_2^*$ with $T_2^* = 1/T_2 + 1/T_{inhom}$, which includes both events: the spin–spin and the field inhomogeneities. This indicates that the decay is governed by T_2^*, rather than the desire T_2.

The spin ECHO is a technique used to obtain T_2. A spin ECHO is a phenomenon that cancels the effect of field inhomogeneities and makes measurements of T_2 possible. A measurement of the signal is done when the ECHO takes place. One method of measuring T_2 is by following the Hahn spin-echo sequence (Hahn, 1950) that consists of applying a 90-degree B_{RF}, a delay time, a second B_{RF} pulse at 180 degrees and a second delay time.

Another approach of measuring T_2 is by following the Carr–Purcell–Meiboom–Gill (CPMG) sequence (Carr and Purcell, 1954). The spin-echo sequence of the CPMG method consists of the following sequence of events after the first 90-degree pulse: $[\tau$ -180°- τ - τ -180°- FID Detection$]_n$. The FID for this situation differs from the one shown for SFC. This signal is "constructed" by using the largest value of the signal detected after the ECHO. It consists of several overlapping sinusoidal signals with slightly different frequencies giving the FID a different shape to the one obtained from the SFC measurements. The envelope of this FID resembles a decay, which has a "steepness" that depends on the transverse relaxation time (Forshult, 2001). The T_2 relaxation constant is the time that the spin coherent has been reduced to 37% (1/e) of its original value.

Spin–spin relaxation measurements can be used to study the mobility (time of relaxation) of protons, to elucidate the degree of confinement of the oils in organogels (Laredo et al., 2011; Rogers et al., 2008a,b). This technique has been used in cookie dough to monitor the mobility of water protons when changing the base fat material (Assifau et al., 2006; Goldstein and Seetharaman, 2011).

Instrument Description

Our laboratory is equipped with a Bruker mq20 Series PC 120 NMR analyzer, MiniSpec (Bruker Optics, Milton, ON, Canada), as shown in Fig. 11.11. The data acquisition is performed using the MiniSpec software V2.51 Rev 00/NT by Bruker Biospin Gmbh (2002) operating at 20 MHz and 0.47 T.

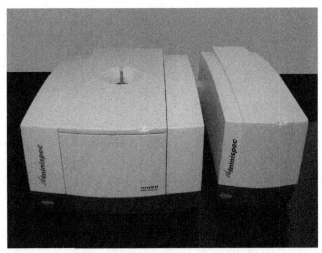

Figure 11.11 The MiniSpec nuclear magnetic resonance (NMR) analyzer used in our laboratory showing an NMR inserted in the measuring cell.

Calibration

A distinction should be made as to whether one is calibrating for measuring the SFC or calibrating for relaxation.

Solid Fat Content Calibration

Every 24 h, the routine "Daily Check" is performed for the measurements of SFC. After a successful execution of this routine, the MiniSpec software will indicate "validated," which means that the instrument is running within specifications and that the necessary parameters to perform a correct measurement were recorded (for example, constant F and D in Eq. 11.15). The Daily Check routine requires three standard calibration samples—polymers that will produce an FID corresponding to specific SFC values—supplied by Bruker. A problem that might arise is that the Daily Check routine does not ask for the three standard calibration samples. This can be solved by enabling the "SFC" option into the "Instrument Settings" in the MiniSpec menu. It is during the daily check that the correction factor (F) and digital offset factor (D) are calculated/determined.

Relaxation Calibration

The calibration for the case of relaxation experiments is performed by running the sample of interest, using "Update Settings" from the Instrument menu. This will allow the instrument to automatically set the values of the gain, the magnetic field and the length of the 90 and 180 pulses. Once the specific sample was run, similar samples can be studied without any farther adjustments.

Sample Preparation

The instrument uses NMR glass tubes that are inserted into the measuring cell (Fig. 11.11) for a determined time period (ranging from seconds to minutes) to perform the measurement. The NMR tubes are 10 mm in diameter and 180 mm in height. The sample is placed inside the NMR tube. Depending on the protocol followed, the sample is deposited in the NMR tube from the melt or from its original semisolid state.

The amount of sample introduced in the NMR tube and the position of the tube in the measuring cell are very important, especially when working with relaxation measurements. A sample that is not completely located within the coil will give an erroneous result. In the MiniSpec/instrument, the coil that generates the B_{RF} extends 1.25 cm to each side of the coil's center. The coil is 1.5 cm in height, which means that the signal extends beyond the coil. If the sample were to be larger than this 4 cm, part of the sample would not interact with a full 90- degree or 180-degree pulse, thus resulting in erroneous measurements. It is thus recommended (Minispec User's Manual, 1989) to have less than 2.5 cm of sample inside the NMR tube. Care must be taken to center the sample properly within the coil height. The user should check with the manufacturer where the center of the coil is in relation to either the bottom or the top of the measuring cell.

Depending on the scope of the experiment, an already crystallized fat or a melted one might need to be deposited inside the NMR tube. Introducing a crystallized fat into the bottom of an NMR tube is not easy, if not impossible. Different strategies have been used, such as filling up another reservoir that gets dropped into the NMR tube. The back of a glass pipette, a clear glass shell vial, and a short Teflon tube have been successfully used for this purpose. Bruker now commercializes a tool that consists of two concentric cylinders of which the inner one can be lifted up and down to pick up the sample and present it to the sample tube. Bruker recommends this tool for samples such as margarine or mayonnaise or any others with similar properties.

A melted fat is easy to pour inside the NMR tube. Specific protocols might need to be followed to follow some standards. The protocol depends on the type of fat under study and the kind of experiment to be performed. Regardless of the protocol or method, the NMR must be dry and clean before it is inserted inside the measuring cell, to prevent hardware damage. Rubber or plastic stoppers are recommended to prevent the entry of dust, solvent vapors, or moisture.

Protocols for the Sample Preparation to Measure Solid Fat Content

We will briefly describe the direct parallel and serial method from AOCS (Cd 16b-93 revised in 2000) and the indirect parallel and serial method from AOCS (Cd 16—81 revised in 2000) and comment on differences with the IUPAC method (IUPAC Method 1.25).

The Direct Method

The "direct" methodology separately measures the NMR signals of the hydrogen protons in the solid and the liquid phases. The SFC is calculated from their ratio as stated in a previous section. The protocol starts by melting the fat using a temperature of 100 degrees $\pm 2°C$ for less than 5 min. No suspended solids should be seen. If necessary, the fat must be passed through a paper filter and then melted again. After thoroughly mixing it, the sample is transferred to the glass NMR tube, filling it to the correct height and keeping it capped. The fat is then tempered following the protocols given in Table 11.3. The tempering is different depending on whether a stabilizing fat is being studied or not. Stabilizing fats are those for which the polymorphic form must be stable before reproducible SFC results can be obtained (i.e., cocoa butter) (Timms, 2005; AOCS Method Cd 16b-93, 2000).

After the tempering is completed, the measurements are started. The selected temperatures depend on the needs and interests of the investigator. Temperatures of 10, 21.1, 26.7, 33.3, and 37.8°C are used in the AOCS Official Method, while the IUPAC Standard Method indicates measurements at 10, 20, 25, 30, 35, and 40°C. A systematic description of this methodology can be found in both the AOCS Official Method Cd 16b-93 and the IUPAC Standard Method 2.150 [IUPAC].

Other suitable temperatures might be to go from 10 to 60°C in increments of 5°C to obtain adequate melting profiles, although, the final temperature might be dictated by the fat and whether it has completely melted or not. To measure a full profile from a single tempering sequence, it is necessary to have a preequilibrated tempering block or a water bath for each temperature under investigation. If a dry tempering block is unavailable, then after removing the sample from the water bath, rapid and complete drying of the tube must be carried out before significant temperature drift has occurred. Tempering blocks can be easily made by drilling holes in a stainless steel block or other suitable

Table 11.3 Tempering Protocols of Stabilizing and Nonstabilizing Fats

Nonstabilizing fats	Stabilizing fats
Melt at 100°C	Melt at 100°C
100°C, 15 min	100°C, 15 min
60°C, 5—15 min	60°C, 5—15 min
0°C, 60 $\pm$ 2 min	0°C, 90 $\pm$ 5 min
Hold for 30—35 min at each measurement Temperature	26°C, 40 $\pm$ 0.5 h
	0°C, 90 $\pm$ 5 min
	Hold for 60—65 min at each measurement
Follow parallel or serial method	Follow parallel or serial method

Adapted from Coupland, J., 2001. Determination of Solid Fat Content by Nuclear Magnetic Resonance, Current Protocols in Food Analytical Chemistry. John Wiley & Sons, Inc.

conductive metal. As they are frequently brought to the required temperature in a water bath, it is worthwhile selecting a metal that is not prone to rust.

A parallel setup is generally recommended for the measurements. In this setup, a different NMR tube is used for each measurement temperature (Table 11.3). However, when the amount of sample is limited, it may be necessary to use a serial setup. In this kind of setup, after measuring a sample at the first lowest temperature, it is transferred to the next warmest tempering block or water bath, held at the measurement temperature for the appropriate incubation time, and then measured. The process is repeated until the entire temperature range has been covered. Note that the SFC of a given sample is a function of thermal history; therefore, serial and parallel measurements may give dissimilar results.

The Indirect Method

This technique is used to measure the protons in the liquid phase of the sample. The indirect method is also reproducible and accurate, but it is not as fast as the direct method. Four measurements must be performed at two temperatures to calculate the percentage of solids. The method starts the same as with the direct method, by melting both the standard oil and the fat at 100 degrees $\pm 2°C$ for less than 5 min. After thoroughly mixing the sample, both sample and oil are poured into glass NMR tubes. They should be filled to the correct height and kept capped. The temperature of both tubes are brought to 60°C and after equilibration, the SFC is measured. This will give two of the values to be used in Eq. (11.16): Oil_{60} and $Sample_{60}$. The next step is to temper both the standard and the fat, following the tempering steps in Table 11.3. After finishing the tempering, both tubes are brought to the desired measuring temperature. After 30 min of equilibration (60 min when working with stabilizing fats), the SFC is measured to give the values of "Oil_T" and "$Sample_T$" to be used in Eq. (11.16). The final step is to use Eq. (11.16) with the measured numbers to obtain the % SFC.

Measuring the Solid Fat Content to Monitor Isothermal Crystallization

This section does not follow an official protocol, but it explains what we follow in our laboratory.

In isothermal crystallization, the sample is melted and then incubated at the desired crystallization temperature. 80°C for 15 min should be enough to melt the sample and erase all crystal memory. The SFC is then measured at particular intervals of time, while keeping the samples inside the water bath at the crystallization temperature. Care should be taken to wipe and dry the NMR tubes before inserting them into the NMR cell. The time intervals chosen for the measurement and duration of the experiment depend on the fat and temperature of crystallization. One common protocol is to measure every 30 s for the first 15 min, every minute for the following 15 min and every 10 min for the remaining 60 min. After this, readings are made at intervals of 1 h.

Protocol for Sample Preparation for Relaxation Measurements

The same protocols as the one used in measuring the SFC might be followed, depending on whether the sample is to be crystallized under particular conditions or analyzed from the original material without any tempering or standardization.

Experimental Procedure

Step 1: Choose the kind of measurement to perform.

If the welcome box is active, then the first thing to do is to choose the kind of experiment to carry out. This can be done by choosing Setup > Customize.

Selecting "SFC Analyser," will allow for SFC measurements. Either Quality Control Analyzer of NMR Analyser will allow for relaxation measurements. We use Quality Control in our lab.

Step 2: Open the correct application, typically sfc_lfc for SFC measurement using the direct method or t2_cp_mb for the T_2 relaxation.

Step 3: Carry out a daily check for SFC or "update setting" when doing relaxation. As mentioned above, SFC requires the use of the three standards provided by Bruker. Relaxation requires the use of one of the samples to be studied.

Step 4: Set up the necessary parameters to make a meaningful measurement. This can be done by inputting the desire parameters in (1) Application Configuration Table or (2) Acquisition Parameter Table.

Instrument setting also shows some of the parameters used during the measurement.

Parameters such as recycle delay, gain, number of scan, and τ, need to be set. Some of them are set when the calibration sample is run, but they can also be changed manually by the user. As mentioned above, this routine will tune the gain, the magnetic field, and the generation of the 90-degree and 180-degree pulses, and should be performed with the sample at room temperature and properly positioned. The recycle delay is the waiting time before a new scan is taken and needs to be 5 to 10 times longer than the anticipated longest T_1. The Application Configuration Table includes parameters that are sample-dependent, such as the recycle delay. Another parameter that can be changed manually is the number of echos not recorded. This will allow collect more data points as the system relaxes. Care must be taken in choosing a meaningful number. The number of echoes recorded needs to be large enough to generate a line, which resembles a decay where the tail needs to become asymptotic. In many experiments, the data are not sampled at each echo, but rather a "dummy echo" is used where no sampling point is collected. Use of an odd number of dummy echoes (e.g., 1, 3, 5…) in the MiniSpec causes the program to sample at every (e.g.,) 2nd, 4th, 6th echo. As only a limited number of data points can be acquired, this provides a way of extending the time scale to ensure the entire relaxation event is recorded. Care is necessary in not setting a high number of dummy echos as vital information for those first points could be lost.

Depending on the kind of measurement to be performed, it might be necessary to refrigerate the measuring cell of the NMR for the sample not to change it structure while it is being measured. This could be a problem when doing T_2 measurements, as the data acquisition is done over a minute or more. The temperature of the cell can be controlled by attaching a water bath to the measuring cell. It is our experience that the magnet temperature should be below 30°C when performing T_2 measurements.

It is common practice to average several FIDs (done by the software) to reduce the contribution of the noise component. As the number of scans increases, the signal component increases relative to the noise component.

To obtain the characteristic time constants T_2, it is necessary to apply an inverse Laplace transform to the data collected by the application. For example, the CONTIN (Provencher, 1982) software application along with MiniSpec software version 2.3 (Bruker, Milton, ON, CA) was designed to carry out this kind of analysis. Other programs can also be used, such as Igor (WaveMetrics, Portland, Oregon, USA).

Case Study: Palm Oil and Sorbitan Monostearate
Solid Fat Content Melting Profile
The direct parallel AOCS method (Cd 16b-93) was followed to temper samples containing palm oil and sorbitan monostearate (SMS) in different percentages. After the tempering process, the serial protocol for the measurement was followed. Measurements started at 5°C and finished at 70°C with 5°C intervals. The melting profile was constructed by plotting the SFC as a function of temperature (Fig. 11.12).

The three curves in Fig. 11.12 show that above 25°C the material starts to melt, thus reducing its SFC. A sharp reduction in the SFC takes place above 25°C for the sample with no SMS or with 20% SMS. Pure SMS retains a high value of SFC until 45°C above which it decreases as the temperature increases until it falls to 0% at 60°C.

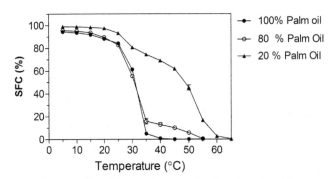

Figure 11.12 Melting profile of three mixtures of palm oil and sorbitan monostearate.

The Avrami Model

A kinetic theory of phase changes was developed by Avrami (1939, 1940, 1941), which describes changes in the volumes of crystals. This theory has been modified for fats and the volume converted to SFC. The Avrami equation is written as (Marangoni, 2005)

$$\frac{SFC(t)}{SFC_{max}} = 1 - e^{-kt^n},\tag{11.18}$$

where SFC (t) is the percentage of solid fat content (SFC/100) at time, t, SFC_{max} is the value of SFC/100 as time approaches infinity. k is the Avrami constant (units of 1/t) and represents the crystallization rate constant, and n is the Avrami exponent that defines the crystal growth mechanism (Marangoni, 2005). Table 11.4 explains the possible values of n and their significance.

Eq. (11.18) was used to fit data collected under the isothermal crystallization protocol. This protocol was followed for mixes of palm oil with SMS at a crystallization temperature of 25°C. The data were then plotted as % of SFC versus time and a nonlinear regression algorithm was used. GraphPad was used for this purpose, but other software such as SigmaPlot or Origin can be used. Fig. 11.13 displays the results for the first 15 min of crystallization.

Table 11.4 Explanation of the Avrami exponent *n*

Avrami exponent (*n*)	Various types of growth and nucleation
1 + 0 = 1	Rodlike growth from instantaneous nuclei
1 + 1 = 2	Rodlike growth from sporadic nuclei
2 + 0 = 2	Disclike growth from instantaneous nuclei
2 + 1 = 3	Disclike growth from sporadic nuclei
3 + 0 = 3	Spherulitic growth from instantaneous nuclei
3 + 1 = 4	Spherulitic growth from sporadic nuclei

Adapted from Sharples, A., 1966. Introduction to Polymer Crystallization. Edward Arnold, Ltd., London, pp. 44—59).

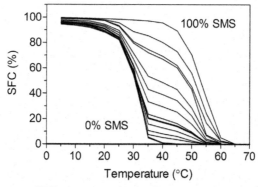

Figure 11.13 Solid fat content (SFC) versus time for the first 15 min of isothermal crystallization at 25°C.

The Avrami model takes into account that crystallization occurs by both nucleation and crystal growth and is based on the assumptions of isothermal transformation conditions, spatially random nucleation, and linear growth kinetics in which the growth rate of the new phase depends only on temperature and not on time. It is also assumed that the density of the growing bodies is constant.

The Avrami parameters obtained for the data in Fig. 11.13 where: n = 1.1 in the 0% and 60% mixes but n = 3.9 for 100% palm oil. This indicates two different nucleation and growth mechanisms. The presence of SMS seems to change the mechanism of growth for the samples: the values of n indicates a rodlike growth from instantaneous nuclei for samples with SMS and a spherulitic growth from sporadic nuclei for pure palm oil. The values obtained for the crystallization constant k were 1.9×10^{-2} min^{-1} for the samples with 0% and 60% palm oil and $3.3 \ 10^{-12}$ min^{-1} for 100% palm oil. These results show that the nucleation rate for the mixes containing SMS is faster than for pure palm oil.

Melting Profiles for Blends

The construction and analysis of an isosolid phase diagram (temperature vs. composition) helps to elucidate whether the components of blends of two different fats are soluble in each other or not. These diagrams are useful to notice incompatibility between two fats (eutectic or peritectic point) or compatible (monotectic). It is known that monotectic systems are those in which the components have similar melting points, molecular volumes, and polymorphic forms.

The same experiment as the one presented in the SFC melting profile section was followed; in this case, 14 mixtures of palm oil and SMS were prepared containing 0, 5, 10, 15, 20, 25, 30, 40, 50, 60, 70, 80, 90, 100% SMS. Results of the SFC versus temperature are shown in Fig. 11.13. The points were then fitted to a cubic spline curve with the assistance of the curve fitting software GraphPad. A cubic spline fitting interpolates points between the data. From the interpolated points, a new graph is constructed: temperature versus composition. To do this, the temperatures corresponding to SFC were extracted in increments of 5% for the SFC. This is a tedious work that needs to be done by hand. Fig. 11.14 shows the temperature as a function of SMS%. Notice that each line in Fig. 11.14 represents a particular SFC%. The original figure would have displayed only scattered points, but Fig. 11.14 shows the data already manipulated with another cubic spline curve. The results in Fig. 11.14 show a monotectic partial solid solution behavior between palm oil and SMS.

The use of ^{1}H-pNMR to measure SFC in the fats and oils industry is a well known and widely used technique due to its ease of use and the repeatability of the measurements.

Even though the relaxation technique is known and used in many areas, it has only recently been used as a tool to investigate the phase composition of lipids

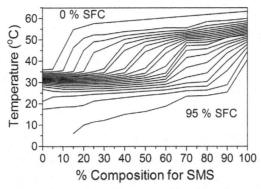

Figure 11.14 Isosolid diagrams in which temperature is plotted as a function of blend composition. Each line is a constant solid fat content (SFC).

(Trezza et al., 2006). These authors measured the T_2 values for four different polymorphic forms in 11 lipid systems. They found a characteristic value of $12.1 \pm 0.8 \, \mu s$ for the α form, a $17 \pm 8 \, \mu s$ for the β form, and a 18 ± 3 for the β' form.

CONTROLLED-STRESS RHEOMETER

Technique principles

The dynamic method: an oscillatory stress is applied and the strain response analyzed. The transient method: a static stress is applied and the strain analyzed. The applied torque is small, with its range between $0.1 \, \mu Nm$ and $200 \, mNm$

Parameters to control

Shear stress
Frequency
Time
Temperature

Relevant information obtained for fats

• For the dynamic method:
  Storage moduli (G')
  Loss moduli (G'')
  Complex moduli (G)
  Phase angle (δ)
• For the transient method:
  Viscosity

Typical output for fats

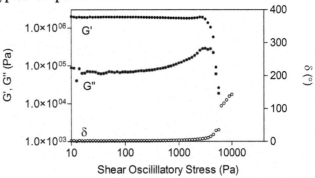

Principles of the Technique

Rheology studies the flow and deformation of matter. Theoretical aspects of rheology focus on the relation of the flow and deformation behavior of material with its internal structure. By studying the relationship between stress (force) and strain (deformation) or strain rate (flow) in a material, it is possible to learn about the internal structure of the material. Elastic, viscous, or plastic behavior can be detected and measured by performing rheological measurements (Steffe, 1996).

Fats have been shown to form a crystalline network that entraps liquid oil (Marangoni, 2005), which makes them behave like rigid solids until a deforming stress exceeds a yield value, at which point the material starts to flow like a viscous liquid (de Man and Beers, 1987). Some of the textural properties in butter, margarine, and spreads, such as spreadability and hardness are largely rheological in nature (Rohm, 1993; Prentice, 1984).

Rheology experiments are separated into large deformation and small deformation processes. This section is dedicated to small deformation rheology, in which small stresses are applied to study the microstructure of the material.

A solid Hookean material will display elastic behavior under rheological experiments, whereas Newtonian fluids show a plastic one. But some materials are viscoelastic, they will show both behaviors. Because fats fall in this category, we will focus the attention on rheological measurements to study viscoelastic materials.

Rheometers can carry out both transient and dynamic experiments. An example of a transient experiment is a creep one, where the applied force is constant and the strain is measured as a function of time. In dynamic experiments, an oscillating stress (or strain) is applied, and the corresponding oscillating strain (or stress) is observed. Dynamic methods are used to learn about the viscoelastic properties of the material, by measuring the amplitude and phase shift, δ, of the strain after a stress was applied. The amplitude of the applied stress in the region of interest is usually very small. The region of interest is where the proportionality between the stress and the strain is constant, which is called the "linear viscoelastic region (LVR)".

To understand rheometer measurements, it is necessary to introduce some basic concepts used to describe the deformation of solids, liquids, and in particular soft matter, such as viscoelastic materials.

A Hookean solid is the one for which the applied force or stress, σ is directly proportional to the deformation or strain, γ via a proportionality factor, G known as the shear modulus, as indicated by Eq. (11.19)

$$\sigma = G\gamma. \tag{11.19}$$

The stress is the force acting per unit area of sample and is responsible for the deformation. It is composed of a normal (tensile or compressive) and a tangential (shear) component with units of pascals (Pa) or newton per square meter (N/m). The strain is the displacement, defined as a relative change in length. Only elastic solids display Hookean behavior. Some solid materials in food behave as ideal solids only for small deformations. When Hookean behavior is no longer observed, the stress is no longer proportional to the strain, although a stressed solid might still return to its original shape once the force is removed. However, there is a point at which the solid cannot return to its original shape, and it will either break or flow (Prentice, 1984). This point is referred to as yield point.

Fluids are different from solids in that they do not possess a rigid structure. Liquids are isotropic so that, on the application of a stress, the material moves without modification of the structure, for as long as the stress remains on it. The characteristic property in this case is that the rate at which the material deforms is proportional to the applied force.

The relation that holds for ideal Newtonian liquids is that the applied stress is proportional, via viscosity η, to the strain rate $\dot{\gamma}$ (Eq. 11.20)

$$\sigma = \eta\dot{\gamma}. \tag{11.20}$$

The rheological properties of Newtonian liquids are independent of the shear rate and the previous shear histories and are dependent only on the temperature and composition (Prentice, 1984). On the other hand, the viscosity of non–Newtonian liquids is not constant and can depend on the rate and/or the time over which the shear force is applied.

The rheological behavior of an "ideal plastic" (or Bingham plastic) is a function of a critical yield stress, σ_c. For values of stress less than σ_c, the application of a deforming force leads to elastic or solid-like behavior of the sample. For values of stress greater than σ_c, the bonds in the sample start to break, leading to its plastic deformation or flow, described as a liquid behavior of the sample.

The rheological behavior of this ideal plastic is described by two equations: Eqs. (11.21) and Eq. (11.22):

$$\sigma = G\gamma \quad (for\ \sigma < \sigma_c) \tag{11.21}$$

$$\sigma - \sigma_0 = \eta\dot{\gamma}(for\sigma > \sigma_c), \tag{11.22}$$

where ideal plastics show a clear division between elastic and viscous behavior, nonideal plastics exhibit these two behaviors simultaneously. These are said to be viscoelastic materials (Prentice, 1984).

Fats exhibit viscoelastic behavior that results from the presence of certain liquid fraction in a solid matrix (Marangoni, 2005). This crystal network consists of a three-dimensional array of polycrystalline particles arranged into clusters that interlink to form separated crystals or bigger aggregates. A deformation of this crystal network will stretch the bonds among the clusters. If the strain is within the elastic region, the material will respond as a solid, thus the stretched bonds will return to their equilibrium. On the other hand, if the stress exceeds the elasticity point, bonds will break and the material will flow: Fats behave as fluids at strains beyond their limit of elasticity. Having in mind that broken bonds can reform once the stress is removed, the material will display viscoelastic behavior (Marangoni, 2005, page 145).

It is the dynamic method that enables the study of the viscoelastic behavior of fats. Small deformation studies on fats are carried out in the region where the material has an elastic behavior. It is in the LVR where the proportionality factor between the stress and strain is a constant. The LVR region can be found by plotting the two variables from Eq. (11.19). The region where a straight line is obtained is the LVR.

In the dynamic method, a sinusoidal perturbation is applied to the sample, which results in an oscillatory strain:

$$\gamma = \gamma_0 \sin(\omega t), \tag{11.23}$$

where γ_0 is the maximum amplitude of the strain, ω is the oscillation frequency expressed in rad/s (in units of hertz, it is $\omega/(2\pi)$ hertz), and t represents time. If nonlinear processes are ignored, so that the system is represented by a differential equation involving a simple harmonic oscillator (energy storage term) and a friction (energy lost) term that is linear in velocity then, the oscillatory strain produces a shear stress, σ, acting on a surface in contact with the material, has been described by Marangoni (2005, Eq. (27), page 168) and is given by

$$\sigma = \sigma_0 \sin(\omega t + \delta). \tag{11.24}$$

Here σ_0 is the maximum amplitude of the shear stress and δ is the phase shift of the stress with respect to the strain. This stress is what will be measured by a detector. Fig. 11.15 shows the curve of an oscillating strain and the resulting oscillating stress as functions of time. When the sample is an ideal (or purely elastic) solid, the maximum strain occurs when the maximum stress is applied. The stress and strain are said to be in phase when $\delta = 0$ degree (Fig. 11.15). If the material is purely viscous, the stress and strain are out of phase by 90 degrees (Fig. 11.15). Viscoelastic materials exhibit a behavior that lays somewhere in between these two extremes, where the phase shift is between 0 and 90 degrees (Fig. 11.15). (Steffe, 1996).

Eq. (11.24) can be manipulated using trigonometric identities to obtain

$$\sigma = \sigma_0 \sin(\omega t)\cos(\delta) + \sigma_0 \sin(\delta)\cos(\omega t), \tag{11.25}$$

which can also be rewritten as

$$\sigma = \gamma_0 \left[\left(\frac{\sigma_0}{\gamma_0} \cos(\delta) \right) \sin(\omega t) + \left(\frac{\sigma_0}{\gamma_0} \sin(\delta) \right) \cos(\omega t) \right], \tag{11.26}$$

which allows us to define

$$G' = \frac{\sigma_0}{\gamma_0} \cos(\delta) \tag{11.27}$$

(A)

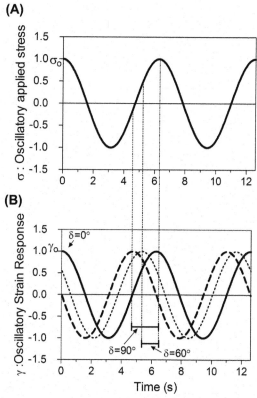

(B)

Figure 11.15 Oscillating strain (γ) curve and the resulting oscillating stress (σ) as functions of time. Three different phase lags are shown.

$$G'' = \frac{\sigma_0}{\gamma_0}\sin(\delta), \tag{11.28}$$

where G' is the shear storage modulus, which provides information about the elastic nature of the material and G'' is the shear loss modulus, which provides information about the dissipation of energy by the material.

One can rewrite Eq. (11.26) to obtain

$$\sigma = \gamma_0 G' \sin(\omega t) + \gamma_0 G'' \cos(\omega t). \tag{11.29}$$

G' and G'' are the variables measured in fat experiments. A purely elastic material has $G'' = 0$ for which $\delta = 0°$, whereas a purely viscous material has $G' = 0$ for which $\delta = 90$ degrees.

A relation between the square of G' and G'' can be expressed as

$$\sqrt{G'^2 + G''^2} = \sqrt{\left(\frac{\sigma_0}{\gamma_0}\cos\delta\right)^2 + \left(\frac{\sigma_0}{\gamma_0}\sin\delta\right)^2} = \frac{\sigma_0}{\gamma_0}. \tag{11.30}$$

The ratio of these two moduli is used to describe viscoelastic behavior, characterized by the ratio of G' to G'':

$$\tan(\delta) = \frac{G''}{G'}. \tag{11.31}$$

$\tan(\delta)$ is directly related to the energy lost divided by the energy stored.

Many food products containing fats, such as chocolates, spreads, or even gels are studied using the dynamic methods that allow the characterization of G', G'', and tan (δ) (Whorlow, 1980). Fig. 11.16 shows the behavior of G' and G'' for three different cases: (1) a stabilized dispersion, (2) a weakly flocculated dispersion, and (3) a strongly flocculated dispersion or gel.

The first case, Fig. 11.16A shows a stabilized dispersion, the simplest case where interparticle forces are negligible between the particles (black dots) that are immersed in a medium (white spaces) so that the system is stabilized. A log-log graph of the values for G' and G'' as functions of angular frequency shows two curves with $G'' > G'$. The values of the slopes can be used to compare which of the two moduli will dominate at high or low frequencies. The second case, Fig. 11.16B shows the case where there is the formation of flocs (black and gray dots sticking together). The interparticle forces are now stronger compared with the first case. A log-log plot of G' and G'' as functions of frequency helps to understand what is happening. $G' > G''$ at high frequencies but remains smaller for low frequencies. The behaviors of both G' and G'' show the viscoelastic nature of the material. The third case, Fig. 11.16C, is for very strong interparticle forces that make the particles flocculate to create a three-dimensional network. The results show a frequency-independent elastic modulus G' that largely exceeds the loss modulus G''. At high frequencies, the value of G'' increases. Fig. 11.16C does not show it, but

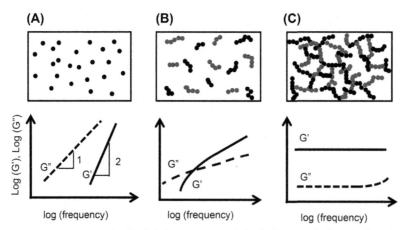

Figure 11.16 Microstructure of colloidal dispersions and the behavior of G' and G'' as functions of angular frequency for (A) a stabilized dispersion, (B) a weakly flocculated dispersion, and (C) a strongly flocculated dispersion or gel. *(Adapted from Khan et al. (1997).)*

when the yield point is reached, G' decreases abruptly and G'' increases, so that the two moduli cross-marking the breaking point of the material (see example in Data Interpretation section).

Instrument Description

Rheometers can be of two types: controlled-stress and controlled-strain rheometers. A controlled-stress rheometer applies a torque either to control the stress at a desired level or to drive the strain to a desired amount. In this kind of rheometer, the material is placed between two surfaces: a bottom surface, which is fixed and a top surface, called the geometry, which rotates. In the controlled-stress rheometer, the torque or stress is the independent variable that is applied to the geometry. In the controlled-strain rate rheometer, the material is placed between two plates. The bottom plate moves at a fixed speed and the torsional force produced on the top plate is measured. Hence, for this case, the strain rate is the independent variable and the stress the dependent one. Here it is worth noting that a controlled-stress rheometer is a better approach for determining yield stress. With these instruments, the stress can be increased in a gradual controlled way until the yield point is observed. With the controlled-strain instruments, the yield has to occur before the measurement can be performed (Semanicik). Fig. 11.17 shows a schematic representation of the geometry and the moving parts in a controlled-stress instrument.

The measurement of the angular displacement is done by an optical encoder device. This device can detect very small movement, as small as 40 nRad. The encoder consists of

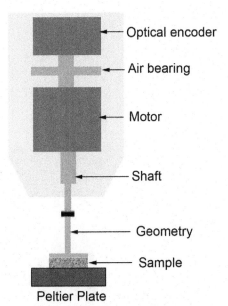

Figure 11.17 Schematic diagram of a stress controlled rheometer main unit.

a noncontacting light source and photocell arranged on either side of a transparent disc attached to the drive shaft. There is also a stationary segment of a similar disc between the light source and encoder disc. The interaction of these two discs results in light patterns that are detected by the photocell. As the encoder disc moves when the sample is strained by the applied stress, these patterns change (AR 2000 Rheometer, 2005). The associated circuitry interpolates and digitizes the resulting signal to produce digital data. These data are directly related to the angular deflection of the disc and therefore the strain of the sample.

Fig. 11.18 shows our own TA AR2000 (TA Instruments, New Castle, DE, USA) rotational controlled-stress rheometer. This rheometer has three main components: (1) the main unit mounted on a cast metal stand that supports the geometry, (2) an electronic control circuitry contained in a separated box (electronic box), and (3) the sample holder (Peltier plate).

The main unit uses an arm that can be moved up and down, either directly with the buttons on the front panel of the main unit or using the software. This arm contains the optical encoder and motor shown in Fig. 11.17. The electronic box is the link between the rheometer and the computer. The sample holder consists of a Peltier plate that uses the Peltier principle to control the temperature. A water bath helps in cooling this device. The Peltier plate can be removed and replaced by concentric cylinders (Fig. 11.19). Concentric cylinders are used with fluid materials for which the temperature must be controlled because the system consists of a water jacket and an inner cup where the sample is placed. The geometry then slides into this cup.

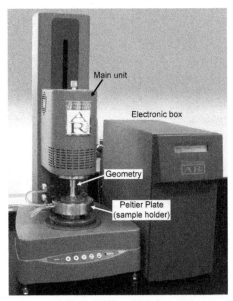

Figure 11.18 TA Rheometer 2000 controlled-stress dynamic shear rheometer.

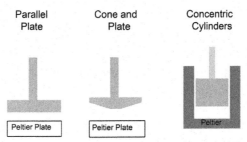

Figure 11.19 Three different geometries and their corresponding base plates, commonly used in rotational rheometers.

During testing, the sample is in contact with two surfaces: a static bottom plate and a moving geometry. Because the bottom plate does not rotate, the key to a useful measurement is the selection of the geometry. The geometry is the piece that is attached to the driving motor spindle and rotates under specific conditions. Geometries are constructed from stainless steel, aluminum, or acrylic. Stainless steel is relatively heavy but has a low coefficient of thermal expansion. Aluminum is less useful because it has a higher thermal coefficient of expansion and is chemically incompatible with many substances. Acrylic and polycarbonate material offer less resistance but can react chemically with the sample under study.

Typical geometries are parallel plate, cone and plate, and concentric cylinders (Fig. 11.19). Concentric cylinders are usually used for low-viscosity liquids; cone-and-plate are used for liquids and dispersions with particles size less than 5 μm; and parallel plates are used for gels, pastes, soft solids, and polymer melts. The rheometer uses the geometry specifications (i.e., diameter, angle of cone, gap) to give a result that is independent of the geometry used. Geometries come in different diameters: larger diameter geometries (60 mm, 40 mm) are used for low-viscosity materials, whereas smaller diameters (20 mm, 15 mm, 10 mm) are used for high-viscosity materials. Because the dimensions of the geometry are taken into account before reporting the results, an important aspect of the measurement is to have a properly loaded sample that shows no excess or shortage of material. The material must cover the whole geometry area. To prevent slippage when working with fat disks (explanation below), 60-grit aluminum oxide sand paper is typically attached to both surfaces using an epoxy glue. TA Instruments also sells geometries with roughness. If sand paper is to be used, it is recommended to cut two circles, one slightly larger than the diameter of the geometry for the Peltier plate and the other one the same diameter as the geometry. Instant Krazy Glue can be used to attach these two circles to both surfaces. To remove the sand paper later on, let it soak in acetone for at least 5 min.

Using the Instrument

In the dynamic mode, the material is subjected to a sinusoidal stress, and the resulting sinusoidal strain or shear rate is measured. Products such as margarine and spreads are required to hold their structure at room temperature, yet they need to be spreadable. These are materials required to exhibit a "yield" value that marks the difference between a solid behavior and a plastic one, but that does not reach the point where the material breaks completely. The material can be characterized by the determination of the yield value as well as comparing the values of G' and G''.

Calibration

Rheometer calibration is carried out by the instrument manufacturer. The user can perform periodic checks to verify the working condition of the rheometer and to set parameters to be used during testing. This is done after turning the equipment on and opening the software. A daily check requires the following: (1) instrument inertia calibration, (2) geometry inertia calibration, (3) mapping or rotational torque calibration, (4) gap zeroing, (5) bearing friction check, and (6) a temperature calibration. All these "checks" are performed without the sample in place.

The "instrument inertia check" must be performed to compensate for any acceleration and deceleration of the motor shaft. The torque output of the motor in a rheometer is composed of the torque required to overcome the instrument inertia and the torque deforming the sample. The instrument inertia takes care of only the torque associated with the instrument, not with the sample.

The "geometry inertia check" is carried out with the geometry attached to the spindle. This step checks the inertia of the geometry by rotating the shaft with the geometry attached to it. In the testing of crystallized fat, it is recommended that the surface of the geometry and the sample holder be covered with sand paper to avoid slippage. The geometry inertia check should be done with the sand paper already in place.

When performing the "mapping check," the instrument checks if there are any variations in behavior during one revolution of the shaft. To create the map, the software rotates the drive shaft at a fixed speed, monitoring the torque required to maintain this speed through a full 360 degrees of rotation. The "bearing friction check" is done to compensate for any friction that might appear when performing a temperature ramp.

The "gap," the distance between the bottom of the geometry and the top of the Peltier plate, might change due to changes in the temperature. When using temperature ramps, it is important to compensate for any thermal expansion or contraction by performing a temperature calibration together with a bearing friction correction. The so-called "zeroing the gap" is not a calibration, but rather a positioning of the gap to tell the instrument what to consider as "0." Every time that a geometry is removed or the instrument turned on, the gap needs to be set to "0." When this option is selected, the geometry is brought into contact with the Peltier plate, which should have the sand paper already attached, if performing studies on disks of fats. Two modes can be selected to zeroing the gap: normal force mode or deceleration. We have our machine set for deceleration.

Experimental Procedure

Creating the Test Program

For the dynamic mode, our TA AR2000 allows the user to choose between frequency sweep, strain sweep, stress sweep, temperature ramp, temperature sweep, time sweep, or manual. The software prompts the user to enter information regarding the experiment to be performed: (1) conditioning, (2) test, and (3) postexperiment.

The information for the conditioning of the sample is done before the real measurement and requires two parameters: the temperature and the Normal force. The Normal force is the force that the geometry will keep on the sample while performing the measurement. The only purpose is to maintain contact without slippage between the geometry and the material and causes no deformation. Alternative, the user may choose to keep the gap constant, which means that the geometry maintains a constant distance from the Peltier plate. Working under a constant gap is set from the instrument window. If Normal force is used, we recommend a value of 4N for fat disks, which is usually 3.2 mm in thickness as described in the sample preparation section (below).

The "test step" is either a frequency sweep or a stress sweep when working with crystallized fats.

In a "frequency sweep" test, the frequency is ramped between a minimum and a maximum value. A sweep between 1 and 1000 Hz is a good first option for fats. In a frequency sweep, the user needs to select one controlled variable from the torque (μN.m), the oscillation stress (Pa), the displacement (rad), the %strain, and the strain. Our laboratory usually uses the oscillation stress.

In a "stress sweep" test, the stress is set for a range of values chosen by the user. 1 and 1000 Pa are typically used for fats. A quick test on a sample will allow the user to determine if there is need to work with higher values. The stress sweep is carried out at one particular frequency that the user must specify, typically 1 Hz for fats.

The "postexperiment" step is performed to prepare the sample holder for the next experiment. The only variable to control is the temperature.

Once the three steps are set, the user must load the sample and perform the experiment. It is important that all dynamic tests are performed within the LVR region of the fat material for which deformation angles should not exceed 1 degree, and the strain should be below 0.01% (Marangoni, 2005, p. 173).

Sample Preparation

This section explains how to make disks of fats from the melt. If the material does not hold its shape, then there is no sample preparation involved because the material can be used as it is. Special molds were designed and built in our laboratory to make solid fat samples starting with melted fat. These molds allow the user to make fat disks of uniform diameters and thicknesses. The molds have three parts: a flat bottom plate, an intermediate template with holes of uniform dimensions, and a flat cover plate (Fig. 11.20). Molds in our laboratory are either 20 mm diameter by 1 mm or 3.2 mm thickness or 10 mm diameter by 3.2 mm thickness. The three plates are screwed in place with butterfly nuts.

The procedure to prepare a solid disk of material using one of the molds requires six simple steps. The first step is to melt the material in the oven to erase all crystal memory. In general, 60°C for 30 min or 80°C for 15 min is used. The second step requires warming up all materials that will be used to assemble the molds: tools to handle the melted material, strips of either parafilm or aluminum paper and the components of the molds. This will avoid crystallization of the material during the procedure. The third step is the preparation of the base of the mold for which a strip of aluminum paper or parafilm, is placed on top of the surface of bottom part of the mold to prevent the fat from sticking to the mold's surface. For fats requiring melting at higher temperatures (e.g., high–melting-point fractions) aluminum is recommended as parafilm melts when

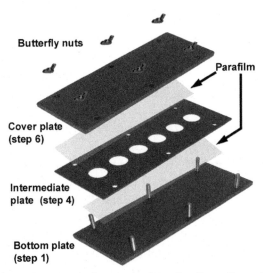

Butterfly nuts

Parafilm

Cover plate
(step 6)

Intermediate
plate (step 4)

Bottom plate
(step 1)

Figure 11.20 PVC molds used to prepare fat sample disks of uniform dimensions used in rheological studies.

in contact with a fat sample above 50°C. The fourth step is the insertion of the middle plate of the mold. Because the base plate of the mold contains screws, the middle plate gets aligned by these screws and is easily slid down. The fifth step consists of pouring the molten sample into the mold holes. A strip of parafilm or aluminum paper is placed on top of the molten fat. The intent is to prevent the sample from sticking to the mold. Trial and error will show the user if it is necessary or not. It is important to avoid air bubbles forming on the surface of the melted fat, something not easy to achieve. The sixth step is to close the mold. Again, the screws attached to the bottom plate are the guides to align the top plate, which is held in place by butterfly nuts. If a particular crystallization temperature is desired, the plastic mold can be placed in an incubator immediately after pouring the melted fat.

One problem with these molds is that the closing of the mold usually creates air bubbles on the surface of the disk. In addition, as the fat crystallizes, it contracts and forms air bubbles. These observations prompted us to not use the top plate of the mold. The fat then crystallizes with no bubbles, but it creates uneven surfaces that require trimming. The trimming is done with a razor and usually only the perimeter requires touching up. The amount of stress or strain that might be introduced because of this trimming is unknown.

Loading the Sample

Loading of samples is different for soft pastelike fats than for solidified disks. When working with soft fats, with the geometry far from the sample, loading of the sample is done by using a spatula and placing enough material between the geometry and the Peltier plate. As the geometry is brought down toward the Peltier plate, it reaches contact with the sample. At this point, it will start to squeeze the material until the geometry reaches the predetermined gap or Normal force. The material that extends beyond the geometry is then trimmed. The geometry used in this case could either be a cone or a parallel plate, although we prefer to use parallel plates.

When working with fats that are crystallized into disk, the geometry is placed far away from the Peltier plate and the disk position on top of the sand paper on it. The geometry is then lowered to bring it into contact with the sample. The closure is performed by either maintaining a constant gap or by applying a constant normal force.

Case Study

We present here results obtained under a dynamic method for fully hydrogenated soybean oil (FHSO) (Bunge, Toronto, ON, Canada) crystallized in 2 cm diameter and 3.2 mm thickness molds. Two measurements were carried out: a frequency sweep and a shear stress sweep. FHSO was tested by making disks of 100% FHSO as well as disks

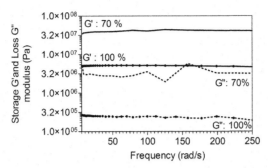

Figure 11.21 Results of a frequency sweep using a controlled-stress rheometer, for two samples made with 100% fully hydrogenated soybean oil (FHSO) and 70% FHSO combined with 30% canola oil.

containing 70% FHSO and 30% canola oil. A Normal force of 4N was selected to close the gap in the rheometer. The temperature was kept constant at 20°C.

A frequency sweep from 1 to 250 rad/s was performed to determine a suitable frequency to carry out the stress sweep. The behavior of G' and G'' for both samples is shown in Fig. 11.21.

The range of frequencies analyzed in these experiments shows that the behaviors of G' and G'' are not frequency-dependent. The constant range is the LVR region. Both G' and G'' have higher values for the sample containing oil than for the 100% FHSO sample. The presence of oil introduces more elasticity into the material as evidenced by the data. Any of the frequencies shown in Fig. 11.21 could be chosen to perform the stress sweep. We choose 1 Hz or 6.28 rad/s, as it has become the typical starting value for our laboratory, yet the user can work in any frequency within the LVR. The stress sweep measurement was then carried out in the region from 1 to 10,000 Pa at a frequency of 1 Hz. Results of the stress sweep are displayed in Fig. 11.22.

The values of interest lay in the LVR region that can be identified with the yield point: the point at which the slope starts to change for G', G'', and δ ("Yield" in Fig. 11.22). The breaking point is the point that causes a permanent deformation in the sample and usually happens before the three curves (G', G'', and δ) intercept and cross each other ("Break" in Fig. 11.22A). In both samples, G' is higher than G'' in the LVR region (region before the "Yield" point), which indicates that the material is more elastic than viscous. Values of both, G' and G'' are higher for the sample that contains a dilution than the pure ones. This is consistent with the notion that the oil introduces "elasticity" to a solid. The sample with the oil is able to withstand higher oscillations before both the yield point and the breaking point appear.

(A)

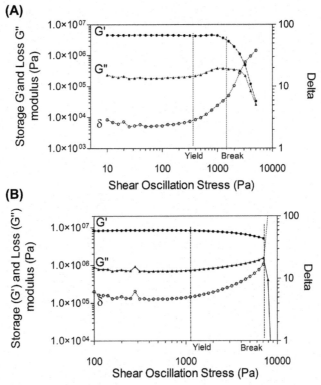

(B)

Figure 11.22 Behavior of G', G'', and δ in the linear viscoelastic region (LVR) for a stress sweep for two samples of fully hydrogenated soybean oil (FHSO). (A) 100% FHSO and (B) 70% FHSO and 30% canola oil.

Table 11.5 shows the results for a system formed from interesterified and noninteresterified tristearin (SSS)–triolein blends crystallized and stored at 20 and 30°C (Rogers et al., 2008a,b). Fat disks were prepared after the sample preparation described earlier and storing the sample at 20 and 30°C for 24 h before the measurement was carried out. A stress sweep was performed at 1 Hz. Results show that G' is higher for the interesterified at 20°C but much lower at 30°C. G' increases as the percentage of solids of tristearin increases and also with the interestification.

Rheological Characteristics of the Fractal Model

Van den Tempel (1961) introduced the first empirical model for fat systems that related viscoelastic parameters with the amount of solids. Over the following 25 years, other

Table 11.5 Storage moduli (G', Pa) for interesterified and noninteresterified tristearin-triolein blends crystallized and stored at 20°C, 30°C for 24 h

%SSS	20°C		30°C	
	Noninteresterified	Interesterified	Noninteresterified	Interesterified
30	$9.4 \times 10^5 \pm$ 2.9×10^4	$1.6 \times 10^6 \pm$ 7.2×10^4	$8.8 \times 10^5 \pm$ 2.2×10^4	NA
40	$1.5 \times 10^6 \pm$ 3.6×10^4	$4.9 \times 10^6 \pm$ 3.0×10^5	$1.7 \times 10^6 \pm$ 1.7×10^5	$2.1 \times 10^6 \pm$ 9.2×10^4
50	$5.6 \times 10^6 \pm$ 3.5×10^5	$7.0 \times 10^6 \pm$ 3.9×10^9	$2.1 \times 10^6 \pm$ 1.4×10^5	$4.1 \times 10^6 \pm$ 3.2×10^5

researchers improved this model (Van den Tempel, 1961, 1979; Shih et al., 1990; Brown and Ball, 1985; Wu and Morbidelli, 2001). In 1992, light scattering revealed that fats were fractal in nature (Vreeker et al., 1992). Marangoni and his group improved even farther the empirical model by introducing fractality (Marangoni and Rousseau, 1999; Narine and Marangoni, 1999a, 1999b; Marangoni, 2000, 2002; Marangoni and Rogers, 2003). The fractal model currently used in our laboratory describes the relationship between the shear storage modulus of networks of particle clusters, the solid fat fraction, and the rheological fractal value for the weak–link regime. For spherical clusters of fat (which are themselves formed from fractal aggregates of smaller units, the primary particles) interacting exclusively via van der Waals forces, the model proposes that the storage modulus, G', is related to the Hamaker coefficient, A_H, by

$$G' = \lambda \phi^{1/(3-D_r)} \tag{11.32}$$

$$\lambda = \frac{A_H}{6\pi a \gamma d_0^2}. \tag{11.33}$$

Here, λ is the slope of $\ln[G']$ with respect to $\ln[\phi]$ where ϕ is the solid fraction (equivalent to fat content divided by 100), D_r is the fractal dimension of the structure of which the floc is composed as obtained by the rheometer, a diameter of the primary particle, γ is the extensional strain at the limit of linearity, and d_0 is the equilibrium distance between the floc surfaces.

Dynamic rheological measurements have become an integral part of fat analyses and give an insight into the mechanical properties of fats. The use of models, such as the fractal model mentioned above, makes the rheometer an important instrument to help in the characterization of fat crystal networks.

DIFFERENTIAL SCANNING CALORIMETER

About the instrument

- Measurement of temperature differences to or from the sample in relation to a reference to compute the heat flowing into the sample.
- The scanning mode of operation changes the temperature linearly.
- Design of the cell allows precise temperature control.
- Requires purge gas to keep the measuring cell clean.

Information obtained for fats

- Heat flow as a function of time or temperature
- Enthalpy of melting or crystallization
- Melting onset, endset and peak
- Crystallization onset, endset, and peak

Parameters to control

- Initial and final temperature.
- Heating or cooling rate.
- Amount of sample.
- Thermal history of sample.

Typical output: melting behavior of a fat

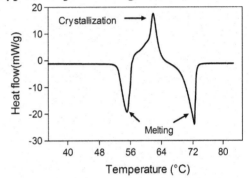

Principles of the Technique

Differential scanning calorimeters (DSCs) are separated into two categories: heat flux DSC and power compensation DSC. These calorimeters differ in the design and measuring principles, but they have in common that the measured signal is proportional to the heat flow rate ϕcp. In a heat flux DSC, the primary measured signal is a temperature difference between a sample S and a reference R. This difference in temperature is converted to a difference in heat flow rate to S and to R, for which it is necessary to use calibration factors previously determined. On the other hand, a power compensation DSC is one in which the power supplied to S and R is monitored. The control loop automatically adjusts the power supplied to S or R so that the temperature difference between S and R is maintained as close to zero as possible. In the power compensation DSC, the recorded output signal is proportional to the differential head flow rate between S and R.

Some of the physical phenomena studied with a DSC are glass transitions, melting profiles, heats of fusion, percent crystallinity, oxidative stability, curing kinetics, crystallization kinetics, and other phase transitions. The focus of this chapter is on the use of a heat flux DSC to determine the thermal behavior of fat samples. Typically, researches look at the temperature at which the fat melts and crystallizes, as well as the enthalpy of the transition. DSC data can also be used to study the reaction kinetics of phase transitions, but this is beyond the scope of this chapter.

When the DSC is used in the scanning mode, the temperature is changed linearly. The heat flow rate ϕ_{cp} (Eq. 11.34) is proportional to the heating rate $\frac{dT}{dt}$

$$\phi_{cp} = c_p \frac{dT}{dt} \tag{11.34}$$

with c_p, the heat capacity as the proportionality factor. In a DSC, the differential heat flow rate depends on the differential heat capacity and heating rate. The measured heat flow in scanning mode is never zero and is made up of three parts (Eq. 11.35):

$$\phi(T, t) = \phi_o(T) + \phi_{cp}(T) + \phi_r(T, t). \tag{11.35}$$

ϕ_o is due to the difference in temperature between the values read at the S and the R positions, ϕ_{cp} is caused by the difference in heat capacity of sample and reference sample, and ϕ_r is the heat flow contribution from the heat of the transition occurring in the sample. The first and second terms define the baseline, and the third term defines the "peak" of the measured curve.

Instrument Description

This section, which discusses the equipment, is based entirely on our own calorimeter, the TA Q2000 DSC (TA Instruments, New Castle, DE, USA), but the principles apply to all DSCs. The readers should refer to their own manuals for instrument/mode/maker specifications. The TA Q2000 DSC consists of four components (Fig. 11.23): a sample cell, a cell controller, an external cooling system (RCS), and a purge system.

The Cell

A DSC is a twin instrument, comprising of individual sample and reference calorimeters within a common thermal enclosure, the cell (Fig. 11.24). Each calorimeter will measure heat flow as a function of temperature or time. The two calorimeters are assumed to be identical. The term "calorimeters" refers to each raised platform and its respective sensor. The output of the DSC is the difference between the heat flows, after converting the measured difference in temperature. Fig. 11.24 shows a schematic diagram of the inside of the cell as well as the location of both pans on the raised platforms where the S and the R are placed. The sensors are attached to these two platforms.

The two calorimeters are part of one bigger unit, the "sensor" (Danley and Caulfield, 1972). The sensor body is made out of constantan (a copper–nickel alloy usually consisting of 55% copper and 45% nickel), consisting of a thick flat base and the two raised platforms (Fig. 11.24 insert). The thin side (wall) of the sensor creates the thermal resistance. This constantan material provides good thermal conduction to S and R. As the

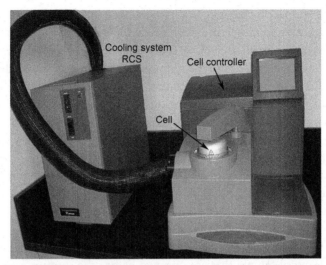

Figure 11.23 TA Q2000 differential scanning calorimeter showing the cell, the cell controller and the cooling system. The purge system enters the cell controller from the back and is directed to the cell and the refrigerated cooling system (RCS) system.

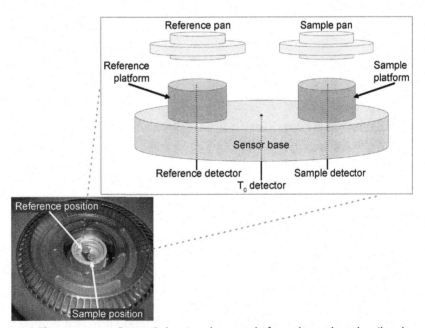

Figure 11.24 The constantan "sensor" showing the two platforms located on the silver base at the bottom of the cell enclosure, holding both the reference and the sample pans. The three detector wires used in the calorimeter are also shown.

temperature of the furnace is changed, heat is transferred from the silver base of the enclosure to the sensor body and thus to the pans. The temperature of the furnace is controlled by the refrigerated cooling system (RCS). Thermocouples, or sensors, on the underside of each platform are used to measure the temperature of the sample and the reference. A third sensor is used to measure the temperature of the sensor base.

Pans could be made from gold, copper, aluminum, graphite, platinum, or Alodine aluminum (Guide for choosing DSC pans, TA Instrument). A pan consists of a bottom and an optional lid. Either the lids can provide a complete enclosure or pin hole can be punch on it to release the pressure that builds up inside as the temperature is increased. Most samples can be run in pans that are closed using a crimping press. If there is a need of atmospheric interaction with the sample, then a pan with no lid is necessary. Crimped pans improve the thermal contact between the sample, pan, and disc, reduce thermal gradients in the sample, minimize spillage, and enable retention of the sample for further study. In our laboratory, fat samples are placed in Alodine aluminum pans containing no more than 10 mg of sample that are crimped before the study is carried out. These pans are coated with an inert fluorophosphate coating, which gives the pans a slightly yellow or gold color rather than the typical silver-aluminum color. This coating renders the pans inert to many chemicals.

The Cell Controller

The cell controller houses all the electronics and is the heart of the DSC. It consists of a touch display window and an automatic cell that can be opened or closed from the commands on the touch screen window. The controller connects to the computer from which all commands are given to the instrument.

The Cooling System and Purge Gas

The RCS provides cooling or heating to the cell to precisely control the temperature of the sample and the reference pans, allowing an operating temperature range from −90 to 550°C. The RCS must be turn on/off directly from its switch and also from the touch screen. This system is refrigerated by two different materials; the first stage uses propylene, whereas the second stage uses a blend of ethylene and propane.

A purge gas is used to continuously purge the cell. This process helps with the heat transfer efficiency toward the pan by maintaining a clean cell, supplying a smooth thermal blanket and by eliminating hot spots. Because the purge gas is continuously circulating throughout the cell, it also helps to remove any moisture or other gases that may accumulated as a result of the pans being heated. Nitrogen is most commonly used as a purge gas as it is inexpensive, inert, and easily available. In addition, it does not interfere with heat measurements due to its low thermal conductivity. This purge gas gets heated before entering the cell to equilibrate its temperature with that in the cell.

The RCS also requires a purge gas, to automatically purge the interior of the RCS cooling head when the cell is open during loading/unloading. At the back of the DSC, it should be seen that the purge gas line coming from the nitrogen cylinder splits into two lines: one going to the cell and the other to the RSC cooling head.

Using the Instrument
Parameters Obtained for Fats

Thermograms associated with phase changes of fat samples are created as the temperature is increased or decreased at a controlled rate. The cooling or heating rates are set by the operator. Thermograms can also be created as heat flows as a function of time at a constant temperature. Typical thermograms display the difference in heat flow on the Y axis, whereas the temperature or time is plotted along the X axis. Parameters commonly analyzed are the temperatures of the summits of (1) the peak of crystallization, T_C, and (2) the peak of melting, T_M, the onset temperatures of crystallization, T_{OC} and of melting, T_{OM} and the enthalpy of crystallization, ΔH_C, and fusion or melting ΔH_M (Fig. 11.25).

The temperature at which the phase change peak (be it melting or crystallization) displays the highest heat flow value is the summit of the peak and is associated with the temperature at which at least half of the lipid species have gone through the phase transition. Associated with this peak is the onset temperature (T_{OM} or T_{OC}), defined as the temperature at which the first crystallites either melt or form, and is observed as a deviation from the baseline. The area under the peak corresponds to the enthalpy of the phase transition ΔH. The enthalpy of a system is equal to the energy added through heat, only if the system is under constant pressure and when the only work done on the system is due to an expansion caused by heating alone. Under these circumstances, the enthalpy is equal to the heat supplied by the system (Q), so that $\Delta H = Q$. Units are typically joules (J) or J/g when specific enthalpy is reported.

TA Instruments' Universal Analysis software for DSC is used to calculate the parameters obtained from the thermograms. This TA software offers four different profiles

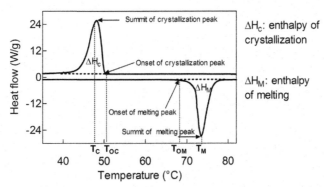

Figure 11.25 Schematic diagram of a thermogram, plotting heat flow as a function of temperature, showing the parameters typically measured for fats.

(fittings) to define the area of the peak. These four options are found under peak analysis > fitting and are named: lineal, sigmoidal horizontal, sigmoidal tangential, and extrapolated. The choice of fitting requires that the user select the region of interest (start and finish temperature). When no difference in the flow rate is observed before and after the melting/crystallizing peak, then the linear fitting is advised. In this case, a straight line will be traced between the two selected points. The sigmoidal horizontal fitting automatically creates a curved line with one inflection between the two points, whereas the sigmoidal tangential allows the user to manually chose the inflection point. Any of these two fittings is advised when there is a marked difference between the baseline heat flow before and after the peak. The extrapolated fitting uses a horizontal line regardless of the shape of the peak between the two temperature points of interest.

Crystallization is an exothermic event, where energy from the sample is released during the process. Our instrument is set to display a positive peak, such as the crystallization peak in Fig. 11.25, for an exothermic event, but the opposite can also be set. The melting of a fat is an endothermic event, where the sample absorbs energy from the environment. To keep consistency with what was said above, the melting peak is a negative one, as shown on Fig. 11.25.

Calibration

To obtain proper heat flow values and temperatures, it is necessary to have a calibrated instrument for the specific experimental conditions. Full calibration of the instrument comprises three runs that can be carried out by following the calibration wizard in the instrument's operating software. Details of what each calibration run are described below. Typically, a calibration is performed once in the course of all experiments, as calibration results are saved and reused each time an experiment is run. The calibration needs to be redone when key experimental conditions are changed (e.g., a different purge gas is used and/or heating/cooling rates are altered) or when the sample cell is burnt. The cell is burnt when there are too many contaminants inside the cell. "Burning" the cell means to clean the cell by raising the temperature as high as possible (350°C for this RCS system) to burn any residues left on the cell.

The TA Q2000 calorimeter has been designed so that the heat flowing into one of the platforms does not affect the temperature of the other. The reference platform is then independent of the sample platform. A constantan and a chromel wire are welded to the center of the sensor base structure and constitute the cell detector that reports T_o. The instrument reports the temperatures, T_s and T_r, of the chromel disks attached to the sample platform and the reference platform, respectively, with respect to the body detector (Danley, 2003); therefore, the differential temperatures are defined as

$$\Delta T_o = T_o - T_s \tag{11.36}$$

$$\Delta T = T_S - T_R, \tag{11.37}$$

where $T_o = T_{chromel\ wire} - T_{constantan}$ is the temperature at the base of the sensor body (see Fig. 11.24) and subscripts R and S refer to reference and sample, respectively.

The algorithm followed by TA Instruments is called the "lumped heat capacity" method (Danley, 2003), where the signal is converted to a heat flow rate (ϕ) using a temperature-dependent proportionality factor so that

$$\phi \propto \Delta T. \tag{11.38}$$

Each calorimeter comprises a thermal resistance, R, and a heat capacity, C. The heat flow rate for the sample and reference is written as

$$\phi_s = \frac{T_s - T_0}{R_s} - C_s \frac{dT_s}{dt} \tag{11.39}$$

$$\phi_r = \frac{T_r - T_0}{R_r} - C_r \frac{dT_r}{dt}. \tag{11.40}$$

Calibration for the TA Q2000 instrument involves the determination of the values for R_s, R_r, C_s, and C_r.

Physically, the operator needs to perform three different experiments: (1) empty cell, no pans; (2) two discs of sapphire are placed on the sample and reference platforms in the cell; and (3) a crimped pan containing a high-purity material, usually indium. Experiments 1 and 2 constitute the "baseline" calibration, whereas experiment 3 is the temperature and heat of fusion calibration.

The baseline calibration compensates for subtle differences between the reference and sample thermocouples (which are located just underneath the constantan of the cylinder that holds the sample pan and the reference pan). The output of this calibration is the calculation of a baseline slope and offset values, which are used to flatten the baseline and zero the heat flow signal. Most instruments perform only one run for this calibration, which consists of heating the empty cell through the instrument's temperature range. The TA Q2000 DSC also includes a second run with two sapphire disks. The disks are not sealed inside a pan, but they are rather used without the pans, because this material gives equal heating rate as it has no discontinuous (first order) phase transitions in this temperature range. Our lab usually performs these two calibrations for the maximum temperature range of interest for fats (−70°C to 350°C). The manual recommends a heating ramp of 20°C/min. The temperature and heat of fusion calibrations ensure that the sample thermocouple reading is correct under the experimental conditions chosen. It involves the use of standard materials, such as high-purity metals, indium, and gallium, which are melted at the same heating rate used in the analysis of future samples. High-purity materials are used because they give narrow peaks.

The calibration performed on the TA Instruments gives three different outputs: the cell constant, the onset slope, and the calibration temperature (Thermal Advantage Manual, 2000). The cell constant is the ratio between the theoretical heat of fusion and the measured experimental heat of fusion of the standard. The onset slope or thermal resistance is a measure of the temperature drop that occurs in a melting sample in relation to the thermocouple. Theoretically, a standard high-purity sample should melt at a unique temperature. However, as it melts and draws more heat, a temperature difference develops between the sample and the sample thermocouple. The thermal resistance between these two points is calculated as the onset slope of the heat flow versus temperature curve on the low-temperature side of the melting peak. The temperature calibration makes a correction based on the difference between the observed and the theoretical melting temperature of the high-purity metal used. If only one standard is used, then the calibration shifts the sample temperature by a constant value. A two-point calibration—that is, a calibration based on corrections not at only one temperature as described above, but at two temperatures—shifts the temperature with a linear correction (straight line) and projects this correction to temperatures above and below the two calibration points.

Experimental Procedure

Calorimetric measurements require that the instrument is set to its experimental mode.

The user can use the instrument's operation software wizard to define and edit an experimental protocol. The software from TA is self-explanatory and uses one-line predetermined instructions or segments that aid in the edition of the protocol.

Sample Preparation

The sample preparation involves the encapsulation of the fat in pans designed for calorimetric analysis. The pan requires between 1 and 10 mg of sample. The amount of sample depends on the material used. Care should be taken not to overfill the pan to the point that the sample spills when the pan is being closed using the lid. Overfilled pan will leak material, which will lead to erroneous readings and most critically, contamination of the cell. The software uses both the weight of the empty pan as well as the weight of the sample. It is important to accurately measure these values so that the transition enthalpy is correctly reported in units of J/g. Once the sample is deposited in the indent of the bottom pan, the pan is crimped using a press specially designed for this purpose. It is recommended that gloves or tweezers be used all the time when encapsulating the material to prevent any contamination from oily residues on the surfaces of the hands. The manipulation and transportation of the pans from the weighing station or encapsulation station to the cell of the DSC should be done without introducing any contamination.

When the temperature of crystallization or storage is different than room temperature, all tools and pans must be precool or preheat. Depending on the experiment, it may also be advisable to prepare (and seal) pans in a temperature-controlled environment. The DSC cell should also be pretempered. The user must keep in mind that any temperature variations will affect the thermal behavior of the sample (be it crystallization or melting) and hence affect results.

Creating the Test Program

The two phenomena/phase changes most studied in the area of fats are crystallization and melting.

The melting properties of a crystallized fat are investigated by heating the sample at a controlled rate until the fat is totally melted. The thermal history of the sample is crucial in the final results. The history includes the conditions of crystallization before the melt, the storage temperature, and the storage time. All of these variables, together with the heating rate will determine the outcome of the melting process.

In the crystallization process, care is taken to erase all crystal memory from the sample (at the recommended 80°C for 15 min), thus results will not be affected by thermal history. If the sample was not completely melted, then the solids present will act as nucleation sites and when lowering the temperature, they could induce a different crystallization as if the material was completely melted.

An example of a method for studying fully hydrogenated soybean oil (FHSO) (Bungee, Toronto, ON, Canada), which has a melting point of approximately 65°C is given in Table 11.6. The objective of this experiment was to find the temperatures at which the fat undergoes phase changes from the liquid to solid and from solid to liquid state. In our laboratory fats are commonly studied using 5°C/min for both melting and cooling. This rate seems to be good compromise between the run time and the precision of the results.

What follows is a step-by-step explanation of the method outlined in Table 11.6:

- Step 1 is performed to ensure that the cell is at the correct temperature.
- Step 2 brings the cell temperature from 24°C to 80°C at a rate of 5°C/min.
- Step 3 holds the cell isothermally at 80°C for 15 min to erase all memory of the crystal structure.

Table 11.6 Method Log to study the melting and crystallization behavior of fully hydrogenated soybean oil

Step	Action
1	Initial temperature 24°C
2	Ramp 5°C/min to 80°C
3	Isothermal for 15 min
4	Ramp 5°C/min to 24°C
5	Isothermal for 15 min
6	Ramp 5°C/min to 80°C

- Step 4 reduces the temperature from 80°C to 24°C at a speed of 5°C/min. The sample will crystallize during this step, but it is unknown at which temperature this will happen.
- Step 5 maintains the temperature at 24°C for 15 min. Step 5 is to ensure complete crystallization of the sample as well to equilibrate the material at this final temperature. If the user wants to study storage time, then the pan is removed after this step and moved to an incubator at 24degrees to be melted at a later point in time. For the purpose of this example, it was enough to keep it 15 min; this allows for a uniform equilibration of the temperature.
- Step 6 completes the thermal study of the sample. During this ramp, the material will be melt again and the onset or peak-melting values can be obtained.

Loading the Sample

The sample can be loaded any time that the measuring cell is not in use. Care should be taken not to introduce any thermal changes in the sample when it is deposited inside the cell. It is recommended to preheat or precool the cell to the desired starting temperature before opening the cell to deposit the pan. Both reference and sample pans should be well centered into each corresponding platform (Fig. 11.24). The reference pan is placed in the rear position of the cell. Tweezers and/or gloves might be necessary when handling the pans so as not to introduce any contamination.

Data Interpretation

Three different studies on FHSO (Bungee, Toronto, ON, Canada) describing the effect of melting/cooling rate on melting, crystallization, and recrystallization are presented below. The aim is to illustrate the challenges that can be encountered when using a DSC for the study of fats and to highlight the importance of thermal history as well as experimental conditions on the results.

Different Melting Rates

Specimens of FHSO, as received from the supplier and after 1 year of storage were crimped into Alodine aluminum pans. Three samples were analyzed after a melting ramp that started at 24°C and finished at 80°C (Steps 1 and 2 in Table 11.6). Three different melting rates were used: 1, 5, and 10°C/min. Each thermogram datum was analyzed using a linear fit. The results of the melt are shown on Fig. 11.26.

Fig. 11.26 shows the results of the melting behavior of FHSO after it was melted at three different rates. The faster the melting rate, the higher the temperature that corresponds to the summit or the melting peak. The obtained summit values were 68.2, 69.5, and 71.6 for 1, 5, and 10°C/min, respectively. The onset value for the three melting profiles was ~66.3°C. These are the values reported directly by the software when using the linear fit. It can be observed that the parameter that stays constant with the different

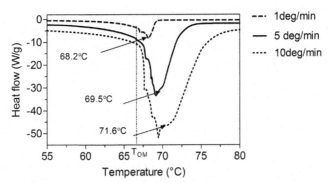

Figure 11.26 Melting profiles for fully hydrogenated soybean oil, as it was received from the supplier, performed at three different melting rates.

cooling rates is the onset. Simulations of melting behavior (Höhne et al., 2003) had shown that the extrapolated peak onset temperature is relatively independent of experimental parameters.

Crystallization Under Different Cooling Rates

Steps 3 and 4 were followed after steps 1 and 2 (Table 11.6). Step 3 allows the equilibration of the sample at this temperature for 15 min, and step 4 cools down the sample, hence allowing for crystallization to occur. Crystallization was carried out at four difference cooling rates: 0.5, 1.0, 5.0, and 10.0°C/min. Each cooling rate was carried out using its corresponding calibration file. As before, each thermogram was analyzed using a linear fit. The results are shown in Fig. 11.27.

The parameter values obtained from Fig. 11.27 are reported in Table 11.7. The results show that cooling rates affect the position of the crystallization peak maximum (summit), while the onset crystallization temperature is similar for the four cases.

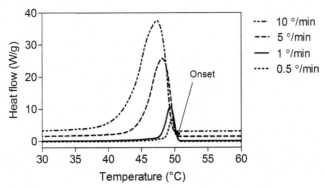

Figure 11.27 Crystallization profile of fully hydrogenated soybean oil following four different cooling rates, 0.5, 1.0, 5.0, and 10.0°C/min.

Table 11.7 Parameters obtained for fully hydrogenated soybean oil when crystallized at different cooling rates

Cooling rate (°/min)	Onset temperature (°C)	Summit value (°C)	Enthalpy (J/g)
0.5	50.38	49.35	124.3
1.0	49.99	48.20	140.1
5.0	49.48	47.38	141.7
10.0	50.49	49.56	147.3

The crystallization summit values for crystallization are not the same as the summit melting values observed under the same cooling rates (Fig. 11.26). Each run in this experiment was calibrated using temperature data for *heating* experiments. TA Instruments reports (Cassel, 1972) that unless a complex multiple rate calibration is carried out, there will be an error of about $1-2°C$ for the cooling data (Menczel and Leslie, 1990; Menczel, 1997).

Recrystallization

This example shows the reader one of the problems associated with working with fats: recrystallization. Steps 5 and 6 were followed (Table 11.6) on the same FHSO sample studied above. The goal was to melt the sample, after it was crystallized under known controlled conditions. The results of the melting behavior are shown in Fig. 11.28.

Fig. 11.28 shows a melting peak with a summit at $52.1°C$ and a second one at $62.3°C$. Between these two peaks, an exothermic peak is observed. This can be explained by assuming that a change in the polymorphic form of the sample took place. The first observed melting peak might be due to the melting of one of the metastable polymorphic forms of a fat, either α or β'. The subsequent increase in temperature seems to trigger a reordering of the sample, with a peak that indicates crystallization. This might cause the

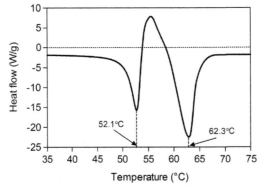

Figure 11.28 Melting behavior of a sample of fully hydrogenated soybean oil using the method log of Table 11.6.

rearrangement of the molecules to form a crystalline form that melts at a higher temperature, as shown by the second melting peak. This is a good example of a sample that should be further studied using other techniques, such as XRD, to discover the true polymorphic form present in the original sample.

Storage Time

This example emphasizes the changes that can come about due to storage time. Storing a sample for a long time might allow it to equilibrate and undergo a transition to the stable polymorphic form. In this example, we show what happened to two samples that had the same thermal history but were stored at the same temperature for different periods of time.

Alodine aluminum pan were used to encapsulate the raw material. Steps 1 to 5 were followed (Table 11.6) using a cooling rate of 5°C/min. Two different crystallization/storage times were evaluated. One sample was kept isothermally at 24°C in the DSC cell for 15 min before the melt (solid line in Fig. 11.29). The other sample was transferred to a temperature-controlled incubator set at 24°C for 72 h and subsequently melted in the DSC (dashed line in Fig. 11.29). A rate of 5°C/min was used to melt both samples.

The result for the sample stored for 15 min shows two peaks that were identified as α and β' (Small, 1966). The result for the sample that was stored for 72 h also showed these two peaks together with a third one that was identified as the β polymorphic form (Small, 1966). It appears that having kept the sample at a constant temperature for a 72 h induced a solid–state polymorphic transformation. These two experiments show the importance of knowing the history of the sample and understanding if the polymorphic forms are already present in the sample or if they are created through polymorphic transformations during the melting progresses.

One has to have in mind the importance of calibrating the DSC, as this is crucial to obtain accurate values. Whenever possible, the operator should use a calibration routine

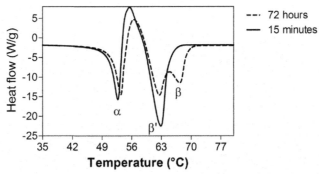

Figure 11.29 Two melting profiles of 5 degrees/min for a sample of fully hydrogenated soybean oil that has the same crystallization history but different storage time.

that calibrates simultaneously for heating and cooling. The onset of crystallization is a parameter that can be used to report both the starting of melting as well as the starting of crystallization, as it is not heating rate/cooling rate—dependent. Results obtained for the enthalpy do not show the same value for the different cooling/heating rates. The enthalpy is the energy that a system absorbs or releases during a phase transition. Small variations in the instrumentation or in the handling of the sample, the pans, and the cell might lead to discrepancies in the obtained values of the enthalpy for different heating/cooling rates.

MICROSCOPY

About the instrument
Microscopes use either light or electrons to reveal an image of the sample. Each kind of microscope has its own magnification and resolution.

Parameters to control
- Intensity of light or voltage of the electron beam
- Sample preparation

Information obtained for fats
Crystal sizes
Crystal morphologies

Principles of the Technique

There are a number of different microscopic techniques that can help understand structure in soft materials such as edible fats. Microscopy is used to determine particle's sizes and shapes. By particles, one can think of a single unit or of the aggregation of many units, depending on the length scale of interest. Length scale refers to the size, e.g., nanometers, micrometers, millimeters. The sizes measured can be used to compute aspect ratios for different populations found on one sample or, to find size or shape distribution or, to compute a 2D fractal dimension.

Microscopy can be divided into three categories:
1. Optical microscopy
2. Electron microscopy (EM)
3. Mechanical "touch"

Optical microscopy involves diffraction, reflection, or refraction of visible light. Visible light corresponds to wavelengths from 390 to 700 nm. The wavelength can be fine-tuned by using particular light sources or filters along the beam path. Optical microscopy requires the collection of the scattered electromagnetic radiation originated at the sample after the beam of light coming from the source either passes by the sample (transmission microscopy) or is reflected by the sample (reflection or fluorescent microscopy). Optical lenses are glass lens specially designed to collimate the beam of light. A 2D image of the object of interest is obtained. The most common microscope used for routine analysis is the bright field (BF) compound light microscope. Transmission BF microscopes allow the light source to go through the sample before the image is analyzed. Compound means that there are ocular glass lenses in the binocular eyepieces (or the camera) and in the objective or nosepiece mounted on a rotating turret closer to the sample. Optical microscopes also cover confocal microscopes. This is a special case of a light microscope in which the light source is a laser. By using a laser, parallelism between the beams of light is achieved. The technique captures 2D images at different depths, which enables the reconstruction of a 3D structure. A confocal microscope uses two pin holes, one in front of the laser source and the second one before the detector.

EM uses a beam of electrons to illuminate the sample. It involves the detection of those electrons that are scattered by the sample to create the image. Both types of EM have an electron gun, which contains an electron source (a filament that produces a cloud of electrons), a Wehnelt cylinder (to form the beam, the analogy of an objective), and an anode (to accelerate the beam). EM uses coils and apertures instead of glass lenses to control the beam of electrons. Scanning electron microscopy (SEM) looks at the surface of a sample by focusing the beam of electron into a small spot that is moved across the surface of the sample doing what is named a *rast scan*. An SEM image is formed from signals that are scattered from the sample as a result of the specimen—beam interaction. Detectors must be strategically positioned to capture the scattered radiation without being in the path of the electron beam. In contrast to SEM, transmission electron microscopy (TEM) works by forcing the beam of electrons to pass through a thin film of the sample before the image is capture onto a screen or a camera/detector. It is then imperative to have a thin layer (~ 100 nm) or the material will absorb too much of the electron beam. Objects with different internal structures can be differentiated because they give different projections.

The mechanical touch microscopy involves the touching of the surface with a mechanical probe. In this kind of microscope, the detector records movement of the probe.

Fig. 11.30 shows the analogy between a compound light microscope, and SEM and TEM.

Magnification

Magnification refers to a number. It is a physical term defined as the ratio of the size of the image to the size of the object. This can be expressed as image size = magnification × object real size. The magnification is denoted by a number followed by the letter "×.".

In a compound light microscope, the magnification is obtained by multiplying the power of the objective lens by the power of the eyepiece, which is typically 10 times. When a camera is used, then the power of the eyepiece is replaced by the power of the camera. In electron microscopes, the magnification is given by the accelerated voltage used (more information below). Light microscopes could have up to 1500× magnification, while electron microscopes could be magnified up to 500,000×.

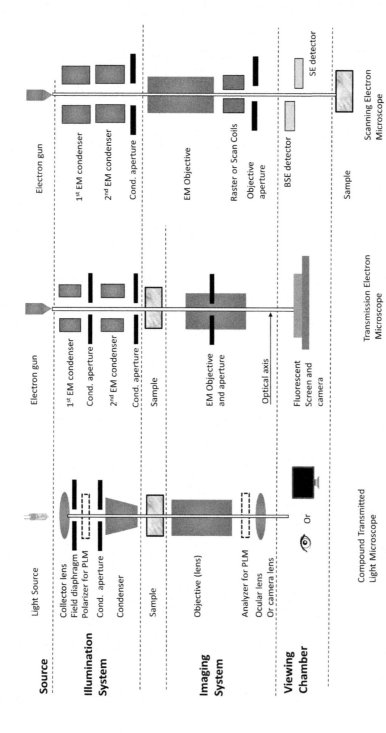

Figure 11.30 Simple cartoon representation of the illumination, imaging, and viewing sections in a compound transmitted light microscope, a transmission electron microscopy and a scanning electron microscopy. *EM*, electron microscopy; *PLM*, polarized light microscopy.

Magnification in a Light Microscope

An objective in a light microscope is basically a tube with two or more lenses, depending on the manufacture and the application. The lens in the objective closes to the sample gives a magnified image of the sample (second focal plane) at a distant L from the objective. The eyepiece or camera farther magnifies this image.

Objectives are mounted on a rotating turret, which allows choose among the different magnifications. Typically, one starts with a low magnification and once the feature is identified, an objective with a higher magnified is used.

The distance L from the first image plane to the second focus plane is known as the tube length. This image focus plane gets formed between the top of the objective and the beginning of the eye piece lens. This value L is marked on the barrel of the objective along with the numerical aperture and the magnification. Nowadays, many objectives are made with an infinite tube length, which means that the second focal plane is focused at infinity. This allows for easy interchange of objectives among different brands.

Another parameter of importance is the *working distance* value, which is also reported on the objective. The smaller the working distance gets, the higher the magnification used. Most objectives have a colored band around the bottom circumference (Fig. 11.31) for easy identification. The color code magnification follows both the Deutsches Industrie Norm (DIN) and the Japanese Industrial Standard (JIS).

Manufacturers also place other kinds of information on the objective, like if it is to be used in oil. The user must get familiar with the particularities according to the manufacture.

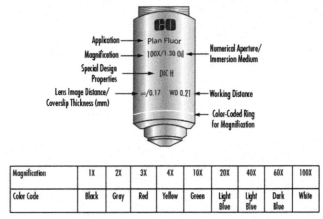

Magnification	1X	2X	3X	4X	10X	20X	40X	60X	100X
Color Code	Black	Gray	Red	Yellow	Green	Light Blue	Light Blue	Dark Blue	White

Figure 11.31 Typical transmissive microscope objective. *(Image courtesy of Edmund Optics)*

On the other hand, the magnification of an EM microscope is given by the accelerating power used. The higher the voltage, the more the magnification. EM microscopes are high-vacuum instruments. The vacuum is needed to allow electrons to travel within the instrument unimpeded, something necessary to guarantee the desire magnification.

Illumination

BF on a compound microscopy requires alignment of the field and condenser apertures as well as finding the focal plane that the condenser lens creates. Efforts to improve light microscopes are continuous regarding optical aberrations and diffraction of light in areas not desired; but the resolution, which is due to the wavelength of the light source, cannot be improved. A common problem that users encounter is related to the lack of clarity of an image due to the out-of-focus light coming from points not in the focal plane. This is why it is important to align the beam of light every time that a microscope is turned on.

Koehler (or Köhler) illumination and *critical illumination* are two techniques to align the beam of light. *Koehler illumination* creates a superior even illumination of the sample compared with the critical illumination and many manufactures suggest this way of alignment.

Steps to Follow to Achieve Köhler Illumination

1. Place a microscope slide containing a specimen on the stage.
2. Use the desire objective. Start with a low magnification.
3. Bring the condenser as close to the slide as possible.
4. Open the field diaphragm.
5. Move the stage up or down until the image is on focus.
6. Close the field diaphragm to reveal a hexagon.
7. Move the condenser up or down until the edges of the hexagon are sharp and well defined.
8. Center the hexagon.
9. Open the field diaphragm enough to cover the field of view.
10. Remove an eye piece.
11. Find the hexagon and open or close the condenser diaphragm until the hexagon covers three-fourths of the field of view.
12. If the objectives to use are not parfocal (same distance from the sample to the objective), then steps 2 to 11 need to be performed with every objective, and switching from one objective to another is not possible without loosing the focus. Computerized microscope tends to have all this information stored, so when one switches the objective, the height of the stage gets adjusted accordingly.

Illumination on an EM is achieved by controlling the EM lenses. TEven, although EM microscopes do not use glass lenses, one refers to the electromagnetic coils as "lenses." In EM, the same as with light microscopes, the condenser lenses (2–4 depending on the microscope) are responsible for the amount of illumination that reaches the sample and control the beam intensity or brightness. The objective lens focuses the beam of electrons onto the sample and applies a small amount of magnification. The intermediate and projector lenses magnify the beam and project it onto the camera (CCD or film) or screen to form an image. With the computing power available today, it is possible to adjust the lenses simultaneously to find the optimum combination that illuminates the sample correctly and gives superresolution.

Resolution

Resolution is the ability to separate (resolve) two closely spaced points (particles) as two separate entities.

There are mathematical formulas that use the wavelength and the numerical aperture—the microscope's ability to gather light—to calculate resolution.

When light from the various points of a specimen passes through the objective and is reconstituted as an image, the various points of the specimen appear in the image as small patterns (not points) known as Airy patterns. This phenomenon is caused by diffraction or scattering of the light, as it passes through the minute parts and spaces in the specimen and the circular rear aperture of the objective. The limit up to which two small objects are still seen as separate entities is used as a measure of the resolving power of a microscope.

Earnest Abbe in the late 19th century introduced a formula that is used for the resolution of a microscope. Here we mention resolution as the distance on the xy plane, not the z-axis, which is the perpendicular to the specimen plane. The resolution or minimum detectable separation (d) in the xy plane is given by

$$d = 1.22\lambda/(NA_{Obj} + NA_{Con}),\qquad(11.41)$$

where λ is the wavelength of the light used in the microscope and NA is the numerical aperture for the objective (*Obj*) and condenser (*Con*). 1.22 comes from using the diffraction rings in the Airy disks generated by those two points to separate (see more http://zeiss-campus.magnet.fsu.edu/articles/basics/resolution.html).

The numerical aperture NA is given by the product of the refractive index of the imaging medium and $\sin(\alpha)$ where α is one-half of the objective's or condenser's opening angle.

A light microscope resolution is not less than 0.6 µm. On the other hand, EM can have a resolution between 0.2 and 0.05 nm. The resolution on an EM microscope is given by the accelerating voltage. The accelerating voltage on a scanning electron microscope can be varied between 500–30,000 V. An electron accelerated by a potential of 30 kV has a shorter wavelength than one accelerated by a 5 kV potential. Thus, the 30 kV electron beam will present a better resolution than a 5 kV electron beam. TEM can handle voltages between 40 and 100 kV, hence, offering a higher resolution than SEM.

Microcopy for Edible Fats

Edible fat crystals are birefringent, which means that the crystals appear shinny under polarized light. A bright field microscope is easily adapted to work under polarized light. Two extra lenses are needed, the polarizer, which is placed between the field diaphragm and the condenser, and the analyzer, which is placed after the objective and before the ocular or the detector in the camera. Polarized light microscopy (PLM) can easily distinguish between edible fat crystals that appeared bright compared with the liquid phase, which remains dark (Chawla and deMan, 1990; Rousseau et al., 1996, Shi et al., 2005). The lowest limit of resolution for this technique is in the order of 1 μm.

As mentioned before, SEM and TEM had a greater resolution compared with PLM. In TEM, the crystals appear darker than the background. The SEM image is inverted compared with the TEM. Bright areas of the image are the result of more electrons being scattered either from topography or heavy element staining. In this case, a good understanding of the sample is needed before doing SEM. More than once, cryo-SEM/TEM is used, where all components in the edible fat matrix are frozen before imagining. Fig. 11.32 highlights the different structures observed by each technique. The reader can also appreciate the differences in magnification and resolution.

Sample Preparation for Edible Fats

Observing edible fats under a microscope is not easy due to the lack of contrast between the solid and the liquid fat. The tendency is to remove the liquid oil and to observe the solid structure or to use low temperatures after sublimating the water.

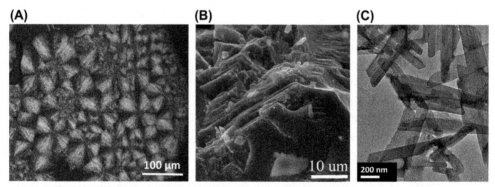

Figure 11.32 Three images of cocoa butter. (A) Crystallized untampered cocoa butter. Image obtained using polarized light microscopy in a Leica DMRX A2 microscope at the Food, Health and Aging Laboratory at the University of Guelph, Canada. (B) After crystallization under laminar shear using scanning electron microscopy (Marangoni et al., 2012). (C) After crystallization under laminar shear using transmission electron microscopy (Marangoni et al., 2012).

Preparing a sample for microscopic observation can be as simple as adding a dye or as complicated as processing the sample to remove the oil, followed by either sublimation or vitrification at low temperature, coating, and fracturing.

Depending on the particular system to study, one might need to resource to chemical fixation, cryofixation, dehydration, embedding, sectioning, staining, and freeze fracturing.

Sample Mounting for Light Microscopy

A bright field microscope will make use of a flat or concave microscope slide made of soda lime glass, borosilicate glass, or plastic, with ground edges. The user needs to pay attention to the sample that they are trying to image not to create artificial interaction between the sample and the subtract.

Cover slips, made of borosilicate or silicate glass are usually place on top of the sample to create a semitranslucent thin sample and to prevent contamination. Concave microscope slides contain one or more surface depressions that are ideal for liquid solutions and larger specimens. The user must have in mind that most microscopes are designed to take into account that the right cover slip thickness is used. Manufactures matched the cover slips to the objective to be used. Expect no clear image if a cover slip of different thickness is used. The standard thickness is 170 μm. In general, the rule of thumb for objectives having a numerical aperture of 0.7 or higher is that they can tolerate a variation of 10 μm, whereas lower numerical aperture objectives (0.3–0.7) can tolerate a higher level of deviation, usually up to 30 μm.

Dry mount: This method of mounting a sample consists in simply positioning a thinly sliced section on the center of the slide, and place a cover slip over the sample. The cover slip is then gently press down making sure not to leave any fingerprint marks.

Wet mount: In this method, the specimen is suspended in fluids such as water, brine, glycerin, or immersion oil. A wet mount usually requires the specimen to be pipetted in place. One should avoid placing a large amount of the specimen, as it can spill over the edges of the slide and contaminate the microscope.

Smear slides: This method works well with semisolid materials. A thin layer of the sample is created on the slide by using either a second slide or a spatula. It is important to make sure that the material is even and does not drip on the slide sides. A cover slip is placed on top of the sample creating a semitranslucent sample. One needs to be careful not to trap air bubbles.

Staining: This technique is used to achieve contrast through color. The color helps to reveal structural details undetected by other ways of preparing the sample. Staining solutions such as iodine, methylene blue, and crystal violet can be added to wet or dry mounts.

Sample Mounting for Scanning Electron Microscopy

Specimen going into an SEM is mounted on special "stubs." A stub is a small sample holder that can sustain the sample preparation and vacuum at once inside the SEM chamber. Special tapes, carbon or cooper compatibles and special glues are needed to maintain the sample in position. Samples that are not conductive, such as edible fats, need to be coated with some conductive material. The coating thickness is also important as too little might not give enough information and too much might interfere with the morphology that one is trying to see. If working at cryotemperatures, after mounting the sample, the stub is frozen by submerging it into a slurry of liquid nitrogen or another liquid. After complete freezing of the sample is achieved, the samples if fractured to expose its internal morphology. Finally, the fracture surface is coated before taking it into the SEM chamber. As with light microscopy, samples might need to be stained to see the desire features.

Sample Mounting for Transmission Electron Microscopy

TEM requires that the specimen is thin enough for the electron beam to go through it. This is why TEM grids are used to support the specimen. Grids come in different sizes, mesh, and shapes and could be made with different materials, such as stainless steel, cooper, titanium, carbon, etc. High-atomic-number stains are used to enhance contrast. The stain can either absorb or scatter electrons. The idea is to help image the desire feature. Uranyl acetate and lead citrate are two common stains.

The work carried out by manufactures and users over the last 30 years on cryo-TEM has focused on obtaining electron microscopic images of macromolecules in their native hydrated state. When a macromolecule is embedding in vitrified water, a frozen-hydrated sample can be directly introduced and visualized in the TEM microscope. Vitrification, achieved by flash freezing (10,000°C/sec), freezes the sample without water crystallization and ensures that molecular rearrangements during freezing are insignificant. Computerized image processing is used to recreate the macromolecule's 3D structure. Edible fat systems containing no water had been visualized following a strict sample preparation protocol in which the oil is removed as explained in Chapter 1.

CONCLUSIONS

As we move forward in the electronic era, more and more innovative and new equipment appears in the market. Developing faster and more reliable equipment is the ultimate goal of many companies. But, if the operator neglects to understand the basic mechanism and theory behind the instrument's capabilities, all the manufacturer's efforts will be lost. It is imperative that every operator understands not only how to run a sample but also why the equipment was designed the way it was. This understanding will help the operator make educated guesses regarding the variables that need to be kept fixed and the ones that can be changed.

It was shown in this chapter that the different techniques can successfully help to elucidate the microstructure of fats and oils. Each technique contributes to one piece of the puzzle in the quest of understanding these materials from the nano- to the meso- to the micro- to the macroscale.

Every operator is encouraged to spend some time familiarizing themselves with the equipment by thoroughly reading the manuals, before performing any experiment. Investing time to understand the basic functioning of the equipment, as well as the theory behind the technique, will guarantee the success of the measurements. Avoiding this step may lead to erroneous results that might put the validity of the results in jeopardy. Training sessions with experts in the area, specifically with the ones that built the equipment, might be a reasonable way to answer some of the inevitable questions. A good result is not equipment-dependent, and its reproducibility should be universal, assuming that the correct technique has been used.

ACKNOWLEDGMENT

Dr. Peyronel would like to thank Dr. Maria del Rosario Morales Armenta for her help with the photographs and the editing of some of the figures.

REFERENCES

AOCS Official Method Cj 2-95.X-Ray Diffraction Analysis of fats. Official Methods and Recommended Practices of the AOCS. AOCS, Urbana, Illinois. Reapproved 2009.

AOCS Official Method Cd 16-81.Solid Fat Content (SFC) by Low-Resolution Nuclear Magnetic Resonance: The Indirect Method. Revision 2000.

AOCS Official Method Cd 16b-93. Solid Fat Content (SFC) by Low-Resolution Nuclear Magnetic Resonance: The Direct Method. Revision 2000.

Assifau, A., Champion, D., Chiotti, E., Verel, A., 2006. Characterization of water mobility in biscuits dough using a low-field 1H NMR technique. Carbohydr. Polym. 64, 197−204.

AR2000 Rheometer, 2005. Operator's Manual. TA Instruments.

Avrami, M., 1939. Kinetics of phase change I. General Theory. J. Chem. Phys. 7, 1103−1112.

Avrami, M., 1940. Kinetics of phase change I. I. Transformation-time relations for random distribution of nuclei. J. Chem. Phys. 8, 212−224.

Avrami, M., 1941. Kinetics of phase change I. I. I. Granulation, phase change, and microstructure. J. Chem. Phys. 9, 177−184.

Blumich, B., 2005. Essential NMR for Scientists and Engineers. Springer Verlag, Berlin.

Brown, W.D., Ball, R.C., 1985. Computer simulation of chemically limited aggregation. J. Phys. 517−521.

Capozzi, F., Cremonini, M.A., 2009. Nuclear magnetic resonance spectroscopy in food analysis. In: Otles, S. (Ed.), Handbook of Food Analysis Instruments. Taylor and Francis, Boca Raton, pp. 281−318.

Carr, H.Y., Purcell, E.M., 1954. Effects of diffusion on free precession in nuclear magnetic resonance experiments. Phys. Rev. 94, 630−638.

Cassel, R.B., 1972. Communication from TA: How Tzero™ Technology Improves DSC Performance Part VI: Simplifying Temperature Calibration for Cooling Experiments. TA Instruments, Inc., New Castle, DE.

Chawla, P., deMan, J.M., 1990. Measurement of the size distribution of fat crystals using a laser particle counter. J. Am. Oil Chem. Soc. 67, 329−332.

Coupland, J., 2001. Determination of Solid Fat Content by Nuclear Magnetic Resonance, Current Protocols in Food Analytical Chemistry. John Wiley & Sons, Inc.

Cullity, B.D., Stock, S.R., 2001. Elements of X-Ray Diffraction, third ed. Prentice Hall Inc.

Danley, R., 2003. New heat flux DSC measurement technique. Thermochim. Acta 295, 201—208.

Danley, R.L., Caulfield, P.A., 1972. DSC Baseline Improvements Obtained by a New Heat Flow Measurement Technique. TA Instruments, New Castle DE.

Deleanu, C., Pare, J.R.J., 1997. Nuclear magnetic resonance spectroscopy (NMR): principles and applications. In: Pare, J.R.J., Belanger, J.M.R. (Eds.), Instrumental Methods in Food Analysis. Elsevier Science B.V, Amsterdam, pp. 179—238.

deMan, J.M., deMan, L., 2001. In: Marangoni, A., Narine, S. (Eds.), Texture of Fats in: Physical Properties of Lipids. Marcel Dekker, pp. 191—217.

deMan, J.M., Beers, A.M., 1987. Fat crystal networks: structure and rheological properties. J. Texture Stud. 18, 303—318.

Forshult, S.E., 2001. Quantitative Analysis with Pulsed NMR and the CONTIN Computer Program Karlstad. Karlstad Universitet, Sweden.

Guide for choosing DSC pans. TA Instruments Thermal Applications Note. TN-12. TA Instruments, New Castle DE.

Goldstein, A., Seetharaman, S., 2011. Effect of a novel monoglyceride stabilized oil in water emulsion shortening on cookie properties. Food Res. Int. 44, 1476—1481.

Hahn, E.L., 1950. Spin echoes. Phys. Rev. 80, 580—594.

Höhne, G.W.H., Hemmiger, W.F., Flammersheim, H.J., 2003. Differential Scanning Calorimetry. Springer-Verlag Berlin, Heidelberg.

IUPAC. International Union of Pure and Applied Chemistry Standard Methods for the Analysis of Oils, Fats & Derivatives, International Union of Pure and Applied chemistry. Blackwell Scientific Publications, Oxford, UK. Solid content determination in fats by NMR, standard method 2.150.

Khan, S.A., Royer, J.R., Raghana, S.R., 1997. Rheology: Tools and Methods. Aviation Fuels with Improved Fire Safety: A Proceeding. The National Academy of Sciences.

Klug, H.P., Alexander, L.E., 1977. X-Ray. Diffraction Procedures for Polycrystalline and Amorphous Material, second ed. John Wiley.

Laredo, T., Barbut, S., Marangoni, A.G., 2011. Molecular interactions of polymer oleogelation. Soft Matter 7, 2734.

Larsson, K., 1966. Classification of glyceride crystal forms. Acta Chem. Scand. 20, 2555—2560.

Marangoni, A.G., Rousseau, D., 1999. Plastic fat rheology is governed by the fractal nature of the fat crystal network and by crystal habit. In: Wildak, N. (Ed.), Physical Properties of Fats, Oils and Emulsifiers. AOCS Press, Champaign, IL, pp. 96—111.

Marangoni, A.G., 2000. Elasticity of high volume-fraction fractal aggregates networks: a thermodynamic approach. Phys. Rev. B 62, 13951—13955.

Marangoni, A.G., 2005. Fat Crystal Networks. Marcel Dekker, New York.

Marangoni, A.G., 2002. The nature of fractality in fat crystal networks. Trends Food Sci. Technol. 13, 37—47.

Marangoni, A.G., Rogers, M., 2003. Structural basis for the yield stress in plastic disperse systems. Appl. Phys. Lett. 82, 3239—3241.

Marangoni, A.G., Acevedo, N., Maleky, F., Co, E., Peyronel, F., Mazzanti, G., Quinn, B., Pink, D., 2012. Structure and functionality of edible fats. Soft Matter 8, 1275.

Menczel, J.D., Leslie, T.M., 1990. Temperature calibration of a power compensation DSC on cooling. Thermochim. Acta 166, 309—317.

Menczel, J.D., 1997. Temperature calibration of heat flux DSC's on cooling. J. Therm. Anal. 49, 193—199.

Minispec User's Manual. Bruker, 1989. The Applications Group Bruker Spectrospin. Ltd. Milton, Canada.

Narine, S.S., Marangoni, A.G., 1999a. Fractal nature of fat crystal networks. Phys. Rev. E 59, 1908—1920.

Narine, S.S., Marangoni, A.G., 1999b. Mechanical and structural model of fractal networks of fat crystals at low deformations. Phys. Rev. E 60, 6991—7000.

Pecharsky, V.K., Zavalij, P.Y., 2009. Fundamentals of Powder Diffraction and Structural Characterization of Materials, second ed. Springer Science +Business Media, New York, USA.

Prentice, J.H., 1984. Measurements in the Rheology of Food Stuffs. Elsevier Applied Science, London.

Provencher, S.W., 1982. A general purpose constrained regularization program for inverting noisy linear algebraic and integral equations. Comput. Phys. Commun. 27, 213−229.

Rogers, M.A., Wright, A.J., Marangoni, A.G., 2008a. Food Res. Int. 41, 1026−1034.

Rogers, M.A., Tang, D., Ahmadi, L., Marangoni, A.G., 2008b. Fat crystal networks. In: Aguilera, J.M., Lillford, P.J. (Eds.), Food Materials Science.

Rohm, H., Weidinger, K.H., 1993. Rheological behavior at small deformations. J. Texture Stud. 24, 157−172.

Rousseau, D., Hill, A.R., Marangoni, A.G., 1996. Restructuring butterfat through blending and chemical interestification. 2. Morphology and polymorphism. J. Am. Oil Chem. Soc. 73, 973−981.

Scherrer, P., 1918. Bestimmung der Grösse und der inneren Struktur von Kolloidteilchen mittels Röntgensrahlen ([Determination of the size and internal structure of colloidal particles using X-rays]). Nachr Ges Wiss Goettingen, Math-Phys Kl 98−100 (In German).

Sharples, A., 1966. Introduction to Polymer Crystallization. Edward Arnold, Ltd., London, pp. 44−59.

Small, D.M., 1966. Handbook of Lipid Research. Plenum Press, New York and London.

Semanicik, J.R. Yield Stress Measurements Using a Controlled Stress Rheometer. TA Instruments. Publication RH-058.

Shih, W.H., Shih, W.Y., Kim, S.I., Liu, J., Aksay, I.A., 1990. Scaling behavior of the elastic properties of colloidal gels. Phys Rev. A 42, 4772−4779.

Shi, Y., Liang, B., Hartel, R.W., 2005. Crystal morphology, microstructure, and textural properties of model lipid system. J. Am. Oil Chem. Soc. 82, 399−408.

Steffe, J.F., 1996. Rheological Methods in Food Process Engineering, second ed. Freeman Press, MI.

Thermal Advantage. DSC, 2000. User Reference Guide. TA Instruments, New Castle, DE.

Timms, R., 2005. SFI or SFC? Why the difference? Inform 16, 2.

Thomson, J.J., 1906. Conduction of Electricity through Gases, second ed. Cambridge University press, London, p. 325.

Traficante, D., 2005. Relaxation: An Introduction. Encyclopedia of Magnetic Resonance. Wiley & Sons Ltd.

Trezza, E., Haiduc, A.M.W., Goudappel, G.J., van Duynhoven, J.P.M., 2006. Rapid phase-compositional assessment of lipid-based food products by time domain NMR Magnetic Resonance in chemistry. Magn. Reson. Chem. 44, 1023−1030.

Van Putte, K., van den Enden, J., 1974. Fully automated determination of solid fat content by pulsed NMR. J. Am. Oil Chem. Soc. 51, 316−321.

Van den Tempel, M., 1961. Mechanical properties of plastic disperse systems at very small deformations. J. Colloid Sci. 16, 284−296.

Van den Tempel, M., 1979. Rheology of concentrated suspensions. J Colloid Interface Sci. 71, 18−20.

Vreeker, R., Hoekstra, L.L., den Boer, D.C., Agterof, W.G.M., 1992. The fractal nature of fat crystal networks. Colloids Surf., A 65, 185−189.

West, A.R., 1984. In: West, A.R. (Ed.), Solid State Chemistry and its Applications. John Wiley & Sons, Chichester, West Sussex, England.

Whorlow, R.W., 1980. Rheological Techniques. Ellis Horwood, Chichester.

Wilson, A.J.C., Prince, E. (Eds.), 1999. International Tables for Crystallography, vol. C, second ed. Publish jointly with the International Union of Crystallography by Springer.

Wu, H., Morbidelli, M., 2001. A model relating structure of colloidal gels to their elastic properties. Langmuir 17, 1030−1036.

INDEX

'Note: Page numbers followed by "f" indicate figures, "t" indicate tables.'